Cancer Biomarkers

Cancer Biomarkers

Editor
Alexandros G. Georgakilas
Biology Department
East Carolina University
Greenville, NC
USA

CRC Press
Taylor & Francis Group
Boca Raton London New York

CRC Press is an imprint of the
Taylor & Francis Group, an **informa** business
A SCIENCE PUBLISHERS BOOK

CRC Press
Taylor & Francis Group
6000 Broken Sound Parkway NW, Suite 300
Boca Raton, FL 33487-2742

First issued in hardback 2019

CRC Press is an imprint of Taylor & Francis Group, an Informa business

ISBN-13: 978-1-57808-761-7 (Hardback)

Library of Congress Cataloging-in-Publication Data

Cancer biomarkers / editor, Alexandros Georgakilas.
p. ; cm.
Includes bibliographical references and index.
ISBN 978-1-57808-761-7 (hardcover : alk. paper)
I. Georgakilas, Alexandros.
[DNLM: 1. Tumor Markers, Biological. 2. Early Detection of Cancer. 3. Neoplasms--diagnosis. 4. Neoplasms--metabolism. 5. Neoplastic Processes. QZ 241]
616.99'4--dc23

2012036496

Visit the Taylor & Francis Web site at
http://www.taylorandfrancis.com

and the CRC Press Web site at
http://www.crcpress.com

Preface

Human beings are exposed on a daily basis to a variety of genotoxic agents such as chemicals and ionizing radiations. The additional deleterious effects of mutations in key genes controlling proper DNA replication, oxidative stress defense mechanisms, DNA repair and apoptosis create a synergistic lethal weapon against organism's homeostasis and well-being. Changes in gene sequences and expression levels as well as protein structure and/or function have been associated with cancer predisposition and prognosis. These 'molecular markers' or else biomarkers can be useful in early detection of cancer, determining prognosis and monitoring disease progression or therapeutic response. Because of the limitations of conventional epidemiological studies in accurately measuring exposure to many pollutants, scientists have developed molecular epidemiological approaches where they have incorporated molecular and cellular markers for improving disease incidence and thus mortality. It is therefore of outmost importance to present the current state of dependable bio-marker development and their use to human cancer predisposition, prognosis and treatment. Eminent experts from various scientific fields cover biomarker-related topics such as application to: cancer risk, cancer predisposition genes, early diagnosis and personalized treatment as well as current advances from the clinic.

Our book comprises a synthesis of concise and constructive review articles. *Dr. Zhu* and colleagues critically discuss the need for improving research on biomarkers for early detection and screening of cancers while *Dr. Franco and colleagues* review several biomarkers of oxidative stress in carcinogenesis. *Dr. Bishayee's* chapter focuses on the exciting potential of β-catenin as a novel biomarker and therapeutic target in liver cancer. On the same direction, *Dr. Lu and colleagues* report on new innovative and promising urine non-invasive biomarkers for prostate cancer detection while *Dr. Ruiz-Echevarria and colleagues* focus on ectodomain shedding as a mechanism to generate cancer biomarkers (Ectomarkers). On a more mechanistic pathway direction, *Drs. Mastellos and Nounesis'* comprehensive review article discusses the breast cancer 1 carboxyl- terminal (BRCT) signaling 'landscape' and associated emerging concepts and implications in tumorigenesis and cancer diagnostics. *Dr. Kletsas'* extensive work targets cellular senescence and cancer development and specifically some 'antagonistic and synergistic relations' while *Dr. Balatsos and colleagues* report on the new and emerging era of microRNAs (miRNAs) and specifically miRNAs from regulators of gene expression to potential use as cancer biomarkers. Finally, on a more clinical direction and relating to personalized therapy, *Dr. Oliver and colleagues*, shed light on the potential application of targeting DNA repair protein PARP-1 in breast cancer and the paradigm of a DNA repair-based therapeutic strategy and *Dr. Tran and Aziz's* study targets the drugs statins in prostate cancer from the aspect of prevention and treatment. Last but not least, *Dr. Panayiotidis and colleagues*, discuss realistically mechanistic aspects of oxidative stress involvement in human environmental and occupational carcinogenesis.

May 2012

Alexandros G. Georgakilas, PhD
Biology Department
1000 E 5th Str.
Howell Science Complex
East Carolina University
Greenville, NC 27858
USA

Contents

List of Contributors

Aziz, Khaled

Department of Radiation Oncology and Molecular Radiation Sciences, Sidney Kimmel Comprehensive Cancer Center, Johns Hopkins School of Medicine, 1550 Orleans Street, CRB2, Rm 406, Baltimore, MD, 21231, USA; E-mail: *Aziz.khaled@mayo.edu*

Balatsos, Nikolaos A.A.

Department of Biochemistry and Biotechnology, University of Thessaly, Ploutonos 26, 412 21 Larissa, Greece; E-mail: *balatsos@bio.uth.gr*

Berg, Christine D.

Division of Cancer Prevention, National Cancer Institute, 6130 Executive Boulevard, Rockville, MD 20852, USA; E-mail: *bergc@mail.nih.gov*

Bishayee, Anupam

Department of Pharmaceutical Sciences, School of Pharmacy, American University of Health Sciences, 1600 East Hill Street, Signal Hill, CA 90755, USA; E-mail: *abishayee@auhs.edu*

Chen, Xiaofei

Department of Biochemistry and Molecular Biology, Brody School of Medicine, East Carolina University, 600 Moye Blvd., Greenville, NC 27834, USA; E-mail: *CHENX07@students.ecu.edu*

Chen, Yan-Hua

Department of Anatomy and Cell Biology, 7N100 Brody Medical Sciences Building; and Leo Jenkins Cancer Center, The Brody School of Medicine, 600 Moye Boulevard, East Carolina University, Greenville, North Carolina 27834, USA; E-mail: *cheny@ecu.edu*

Franco, Rodrigo

Redox Biology Center and School of Veterinary Medicine and Biomedical Sciences, 114 VBS 0905, University of Nebraska-Lincoln, Lincoln, NE 68583, USA; E-mail: *rfrancocruz2@unl.edu*

Garcia-Garcia, Aracely

Redox Biology Center and School of Veterinary Medicine and Biomedical Sciences, University of Nebraska-Lincoln, Lincoln, NE 68583, USA; E-mail: *agarcia-garcia2@unl.edu*

Giannouli, Stamatina

Department of Biochemistry and Biotechnology, University of Thessaly, Ploutonos 26, 412 21 Larissa, Greece; E-mail: *stgianno@bio.uth.gr*

González-Flores, Ariannys

Instituto López Neyra de Parasitología y Biomedicina, Avda. Conocimiento, 18100 Armilla, Granada, Spain; E-mail: *ariannys@ipb.csic.es*

Gourgoulianis, Konstantinos I.

Respiratory Medicine Department, University of Thessaly Medical School, University Hospital of Larissa, Biopolis, 411 10 Larissa, Greece; E-mail: *kgourg@med.uth.gr*

Kletsas, Dimitris

Laboratory of Cell Proliferation & Ageing, Institute of Biology, National Centre for Scientific Research "Demokritos", Athens, Greece; E-mail: *dkletsas@bio.demokritos.gr*

López, Laura

Instituto López Neyra de Parasitología y Biomedicina, Avda. Conocimiento, 18100 Armilla, Granada, Spain; E-mail: *laural@ipb.csic.es*

Lu, Qun

Department of Anatomy and Cell Biology, 7N100 Brody Medical Sciences Building; and Leo Jenkins Cancer Center, The Brody School of Medicine, 600 Moye Boulevard, East Carolina University, Greenville, North Carolina 27834, USA; E-mail: *luq@ecu.edu*

Majuelos-Melguizo, Jara

Instituto López Neyra de Parasitología y Biomedicina, Avda. Conocimiento, 18100 Armilla, Granada, Spain; E-mail: *jara_mm@ipb.csic.es*

Maragozidis, Panagiotis

Department of Biochemistry and Biotechnology, University of Thessaly, Ploutonos 26, 412 21 Larissa, Greece; and Respiratory Medicine Department, University of Thessaly Medical School, University Hospital of Larissa, Biopolis, 411 10 Larissa, Greece; E-mail: *pmarago@med.uth.gr*

Mastellos, Dimitris

Protein Chemistry Laboratory, IRRP, National Centre for Scientific Research Demokritos, 153 10 Aghia Paraskevi, Greece; E-mail: *mastellos@rrp.demokritos.gr*

Nounesis, George

Biomolecular Physics Lab, IRRP, National Centre for Scientific Research Demokritos, 153 10 Aghia Paraskevi, Greece; E-mail: *nounesis@rrp.demokritos.gr*

Oliver, Francisco Javier

Instituto López Neyra de Parasitología y Biomedicina, Avda. Conocimiento, 18100 Armilla, Granada, Spain

Current address: Dept. de Bioquímica y Biología Molecular I, Facultad de Ciencias, Universidad de Granada, Avda. Fuentenueva s/n, 18071 Granada, Spain; E-mail: *joliver@ipb.csic.es*

Overcash, Ryan F.

Department of Biochemistry and Molecular Biology, Brody School of Medicine, East Carolina University, 600 Moye Blvd., Greenville, NC 27834, USA; E-mail: *OVERCASHR03@students.ecu.edu*

Panayiotidis, Mihalis I.

Department of Pathology, Medical School, University of Ioannina, University Campus, 45110, Ioannina, Greece; E-mail: *mpanagiot@cc.uoi.gr*

Pappa, Aglaia

Department of Molecular Biology & Genetics, Democritus University of Thrace, University Campus, 68100, Alexandroupolis, Greece; E-mail: *apappa@mbg.duth.gr*

Peralta-Leal, A.

Instituto López Neyra de Parasitología y Biomedicina, Avda. Conocimiento, 18100 Armilla, Granada, Spain; E-mail: *andreperalta@ipb.csic.es*

Pinsky, Paul F.

Division of Cancer Prevention, National Cancer Institute, 6130 Executive Boulevard, Rockville, MD 20852, USA; E-mail: *pinskyp@mail.nih.gov*

Rodriguez, Art A.

Department of Chemistry, 300 Science and Technology Building, Thomas Harriot College of Arts and Sciences, East Carolina University, Greenville, North Carolina 27858, USA; E-mail: *rodriguezar@ecu.edu*

Rodríguez, María Isabel

Instituto López Neyra de Parasitología y Biomedicina, Avda. Conocimiento, 18100 Armilla, Granada, Spain

Current address: Dept. de Bioquímica y Biología Molecular I, Facultad de Ciencias, Universidad de Granada, Avda. Fuentenueva s/n, 18071 Granada, Spain; E-mail: *mirlara@ipb.csic.es*

Rodriguez-Rocha, Humberto

Redox Biology Center and School of Veterinary Medicine and Biomedical Sciences, University of Nebraska-Lincoln, Lincoln, NE 68583, USA; E-mail: *hrodriguezrocha2@unl.edu*

Rodriguez-Vargas, José Manuel

Instituto López Neyra de Parasitología y Biomedicina, Avda. Conocimiento, 18100 Armilla, Granada, Spain; E-mail: *jmrodriguez@ipb.csic.es*

Ruiz-Echevarria, Maria J.

Department of Oncology; and Department of Anatomy and Cell Biology, Brody School of Medicine, East Carolina University, 600 Moye Blvd., Greenville, NC 27834; and Lineberger Comprehensive Cancer Center, University of North Carolina at Chapel Hill, Chapel Hill, NC 27599, USA; E-mail: *ruizechevarriam@ecu.edu*

Tran, Phuoc T.

Department of Radiation Oncology and Molecular Radiation Sciences, Sidney Kimmel Comprehensive Cancer Center, Johns Hopkins School of Medicine, 1550 Orleans Street, CRB2, Rm 406, Baltimore, MD, 21231, USA; E-mail: *tranp@jhmi.edu*

Zhu, Claire S.

Division of Cancer Prevention, National Cancer Institute, 6130 Executive Boulevard, Rockville, MD 20852, USA; E-mail: *zhucla@mail.nih.gov*

Ziech, Dominique

Faculty of Medicine, School of Medicine, Universidad Diego Portales, Santiago, Chile; E-mail: *dziech@gmail.com*

CHAPTER 1

Improving Research on Biomarkers for Early Detection and Screening of Cancers

Claire S. Zhu*, Paul F. Pinsky and Christine D. Berg
Division of Cancer Prevention, National Cancer Institute
6130 Executive Boulevard, Rockville, MD 20852, USA

ABSTRACT

For early detection to work, a biomarker needs to be able to identify cancer at an early stage, before manifestation of clinical symptoms. Therefore, a biomarker for early detection of cancer must show a measurable signal in persons without symptoms. An effective, reliable, non-invasive early detection biomarker is the key to the success of cancer screening programs. Despite decades of research, there is currently no FDA-approved blood-based test for cancer screening. A critical aspect of biomarker research involves the quality of the biospecimens utilized for marker analysis.

This review is focused on biomarkers that have potential for early detection and screening, particularly those that can be measured in body fluid such as blood or urine. We conducted an extensive literature review to 1) summarize the state of current research in early detection biomarkers and 2) assess to what degree information about the biospecimens are reported in published biomarker research, using the Biospecimen Reporting for Improved Study Quality (BRISQ) guidelines as a standard. The goal is to identify ways to improve future research. To that end, we highlight the unique challenges in identifying early detection biomarkers, and some of the critically important, but often over-looked, aspects of biomarker research that are uniquely relevant to early detection and screening. We identify some emerging trends in early detection biomarker research that may lead the field in the right direction.

**Corresponding author*: E-mail: zhucla@mail.nih.gov

In conclusion, developing effective early detection biomarkers for cancer screening is a multi-disciplinary undertaking that spans the continuum of basic and clinical sciences.

INTRODUCTION

Cancer biomarkers can be broadly categorized into groups by their application: biomarkers for risk, biomarkers for early detection, biomarkers for diagnosis, biomarkers for prognosis, biomarkers for treatment response, and biomarkers for sub-typing. There are many different molecular types of cancer biomarkers, including protein, DNA, RNA, gene-expression signature, proteomic signature, methylation pattern, and more. Molecular probes used by various *in vivo* imaging techniques (e.g. MRI, PET, Zn sensing) can also be considered biomarkers.

Considerable progress has been made in cancer biomarker research over the last decade, as high through-put molecular techniques made it possible to analyze the whole genome at once (Institute of Medicine 2006; Kulasingam and Diamandis 2008). Hundreds of tumor-associated biomarkers have been reported, and some of these have successfully made their way to clinical use (Sturgeon et al. 2008). Table 1 shows some of the FDA-approved biomarker-based tests for cancer diagnosis, prognosis, disease monitoring, or therapy response. Examples include PSA for prostate cancer diagnosis and monitoring, CA125 and HE4 for ovarian cancer diagnosis and monitoring, AFP for nonseminomatous testicular cancer diagnosis and monitoring, and CA15-3 for breast cancer prognosis.

However, a biomarker test for cancer screening, one that can readily detect cancer in an early and treatable stage and that is relatively non-invasive – two basic criteria for a cancer screening test – is still beyond reach. There is currently no FDA-approved blood-based test for cancer screening.

This review summarizes the state of current research in early detection biomarkers, highlighting emerging trends, and most importantly, drawing attention to some of the critically important but often overlooked aspects of biomarker research that are uniquely relevant to early detection and screening. The primary focus is thus on biomarkers that have potential for early detection and screening, particularly those that can be measured in body fluids such as blood or urine. These biomarkers generally belong to one of the following categories: secreted proteins, shed membrane-associated proteins, circulating tumor cells, circulating DNA/RNA molecules, autoantibodies and

Table 1 Some FDA-approved cancer biomarker tests.

Biomarker	*Type*	*Condition*	*Indicated use*	*Sample*
AFP (immunoassay)	Tumor-derived plasma protein	Nonseminomatous testicular cancer, fetal neural tube defects	Patient management, monitoring of disease progression	Serum
CA125 (immunoassay)	Tumor-derived plasma protein	Ovarian cancer	Patient management, monitoring of disease progression	Serum
CA15-3 (immunoassay)	Tumor-derived plasma protein	Breast cancer	Patient management, monitoring of disease progression	Serum
CA19-9 (immunoassay)	Tumor-derived plasma protein	Pancreatic cancer	Aid in patient management	Serum/plasma
CEA (immunoassay)	Tumor-derived plasma protein	Multiple Cancers	Patient management, monitoring of disease progression	Serum/plasma/urine
HE4 (immunoassay)	Tumor-derived plasma protein	Ovarian cancer	Patient management, monitoring of disease progression	Serum
OVA1 (5-marker panel: APOA1, B2M, CA125, TF, TTR, immunoassay)	Tumor-derived plasma protein	Ovarian cancer	Aid in diagnosis	Serum
PSA (immunoassay)	Tumor-derived plasma protein	Prostate cancer	Diagnosis, patient management, monitoring of disease progression	Serum
ROMA† (2-Marker panel: CA125, HE4, immunoassay)	Tumor-derived plasma protein	Ovarian cancer	Diagnosis	Serum
BRAF mutation (Real-time PCR)	Tumor DNA mutation	Melanoma	Treatment response	FFPE Tissue

Table 1 Contd. ...

Table 1 Contd. ...

C-KIT (IHC)	Tumor-derived plasma protein	Gastrointestinal cancers	Diagnosis, treatment response	FFPE Tissue
EGFR (IHC)	Tumor-derived plasma protein	Colorectal cancer	Treatment response	FFPE Tissue
Mammaprint (DNA microarray)	Gene expression signature	Breast cancer	Prognosis	Fresh Tissue
Breast Lymph Node Assay (DNA microarray)	Gene expression signature	Breast cancer	Metastatic status	Fresh Tissue
HER2/neu (FISH, IHC)	Chromosomal aberration, tumor associated protein	Breast cancer	Treatment response, prognosis	FFPE Tissue
TOP2A (FISH)	Chromosomal aberration (copy number changes)	Breast cancer	Prognosis	FFPE Tissue
Tissue of Origin (DNA microarray)	Gene expression signature	Multiple cancers	Diagnosis	FFPE Tissue
Fecal Occult blood test	Blood	Gastrointestinal disorders	Diagnosis	Feces
Urovysion bladder cancer kit (FISH)	Chromosomal aberration (aneuploidy for chromosomes 3, 7, 17 and loss of the 9p21 locus)	Bladder cancer	Diagnosis, monitoring for recurrence	Urine
NMP 22 (immunoassay)	Tumor-derived urine protein	Bladder cancer	Diagnosis	Urine

Source: http://www.fda.gov/MedicalDevices/DeviceRegulationandGuidance/Databases
†**ROMA:** Risk of Ovarian Malignancy Algorithm
IHC: Immunohistochemistry test
FISH: Fluorescent *in situ* Hybridization test

metabolites. Biomarkers for prognosis, drug response, disease sub-typing or other tumor-based biomarkers are out of the scope of this article.

OVERVIEW OF EARLY DETECTION BIOMARKER RESEARCH

Biomarker discovery generally follows two approaches. One is the "candidate gene" approach, driven by knowledge about the functions of the genes; the other is the so called "omics", or hypothesis-free approach, in which the entire composite of biomolecules in a system is compared between cases and controls. The latter is often conducted using high through-put technologies, such as gene-expression microarrays, mass spectrometry-based proteomics, high density protein microarrays, and more recently, metabolomics and glycomics to identify differentially expressed molecules in blood, urine, saliva, and tumor tissues.

Candidate biomarkers are usually selected based on biological functions related to cancer biology (Hanahan and Weinberg 2011); some are selected based on previously published preliminary results associating the biomarkers to the cancer. Classical tumor markers are often selected as potential early detection biomarkers. These include CA125 (ovarian cancer), CA19-9 (pancreatic cancer), CEA (gastrointestinal tract cancers), AFP (HCC), PSA (prostate cancer), and have been around for decades. Many of them have been shown to be effective for disease monitoring and for assessing treatment response. As mentioned earlier, none of these and other well-known tumor markers have been proven definitively to be effective for early detection in a screening setting.

Tumor suppressor genes (*P53, APC, RB*), oncogenes (*KRAS, MYC, RAF, EGFR, PTEN*), as well as genes involved in cell proliferation (*EGFR*), apoptosis (*IGF-1, IGF-1R, TNF-α, TP53*), cell senescence (telomerase), angiogenesis (*VEGF*), and cell adhesion (integrins, cadherins, matrix metalloproteinases), are often chosen as potential early markers. Testing mutations or epigenetic changes in these genes in circulating cell-free DNA in blood or urine is a promising approach to early detection of cancer (Kohler et al. 2011).

Deficient DNA repair, along with inactivation of tumor suppressor genes and activation of oncogenes, are thought to be early events in carcinogenesis. Epidemiological studies have consistently associated DNA repair capacity with cancer. Thus, measurement of global DNA damage (e.g. 8-oxodG), microsatellite instability, and various aspects of DNA repair activities in blood or urine may be potential early detection biomarkers (Kryston et al. 2011).

Recently, there is growing literature suggesting the involvement of the immune system in cancer initiation through growth-promoting cytokines and tumor angiogenesis; inflammation also produces reactive oxygen species (ROS) and reactive nitrogen intermediates (RNI) that can cause DNA damage (Grivennikov et al. 2010; Kryston et al. 2011). Higher levels of pro-inflammation markers in serum have been associated with several cancers including lung (IL-6, IL-8, CRP), colorectal (CRP), and endometrial (CRP) cancers (Pine et al. 2011; Tsilidis et al. 2008; Wang et al. 2011).

The advent of high through-put technologies produced hundreds of new potential cancer biomarkers, most of which are either unknown genes, or known genes previously not linked to cancer. Serum/plasma proteomics and gene expression microarray analysis have been applied to virtually all cancers (Bakry et al. 2011). Many studies have reported that signatures or profiles of proteins or mRNAs can distinguish cancers from controls. MammaPrint, the first FDA-approved test of its kind, is a gene-expression signature test, based on 70 genes, for breast cancer prognosis (van 't Veer et al. 2002).

In recent years, several new types of cancer biomarkers have emerged. These include microRNAs, metabolomic biomarkers, and glycoproteomic biomarkers. Disregulation of microRNAs (miRNA) has been associated with various cancers, including ovarian, pancreatic, colorectal, breast and esophageal cancers. MicroRNAs have been detected in serum, plasma, and stool samples; changes in their levels have been shown to be predictive of cancer (Huang et al. 2009; Link et al. 2010; Taylor and Gercel-Taylor 2008; Wang et al. 2009; Zhang et al. 2010).

Metabolomic profiling has been conducted in serum, urine and tissue samples for biomarker discovery. One of the first metabolomic biomarkers in cancer, sarcosine, an N-methyl derivative of the amino acid glycine, was shown to be detectable in urine, and is predictive of prostate cancer in patients with moderately elevated PSA (Sreekumar et al. 2009). Other metabolic biomarkers have been reported for colorectal, bladder and pancreatic cancers (Issaq et al. 2008; Ritchie et al. 2010).

Perhaps the newest focus in cancer biomarker discovery is glycomics, the comprehensive analysis the entire glycan structures in serum or plasma. In 2006, the NCI established The Alliance of Glycobiologists for Detection of Cancer and Cancer Risk (http://glycomics.cancer.gov/home) to facilitate research in this area. To date, serum glycomic profiling has been conducted on esophageal, breast, liver, and prostate cancers (Goetz et al. 2009; Goldman et al. 2009; Isailovic et al. 2008; Mann et al. 2010).

To provide a more quantitative analysis of the early detection marker research landscape, we conducted a literature search. Specifically, we searched PubMed using the keywords "cancer", "biomarker" and "early detection" or "early diagnosis", limiting to only primary research articles on humans published in English in the last 10 years. This search strategy yielded 800+ primary research articles (after excluding reviews, commentaries, perspectives, etc). Only articles explicitly mentioning "early detection" or "early diagnosis" as the potential use of the biomarkers were included. We acknowledge that this search strategy is by no means all encompassing, and may have missed some relevant articles; nevertheless, we believe that it does provide a reasonably complete and accurate summary of recent published work in the field.

Figure 1 shows a steady upward trend in the number of articles published in each year over the last decade. Figure 2 shows the proportion of publications by cancer site. Ovarian, lung, breast, prostate and colorectal cancers are on the top, together representing about 50% of all publications. We categorized the biomarkers by sample sources, i.e. whether the biomarker was measured in serum/plasma, tumor tissue, urine, proximal fluid (e.g. pancreatic juice, ascites), exhaled breath, or other type of samples, regardless of the molecular types (Fig. 3).

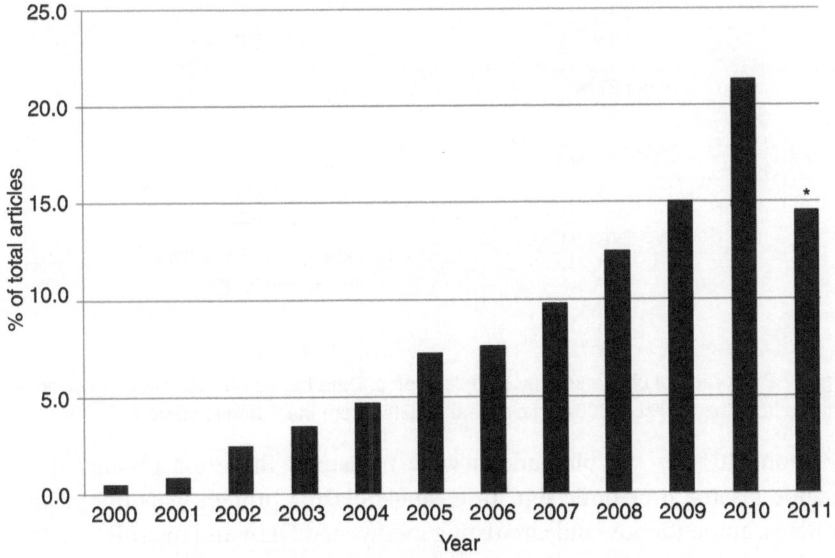

Fig. 1 Proportion of publications by year. Data based on 858 articles that resulted from our PubMed search strategy. * indicates number of articles as of July 2011.

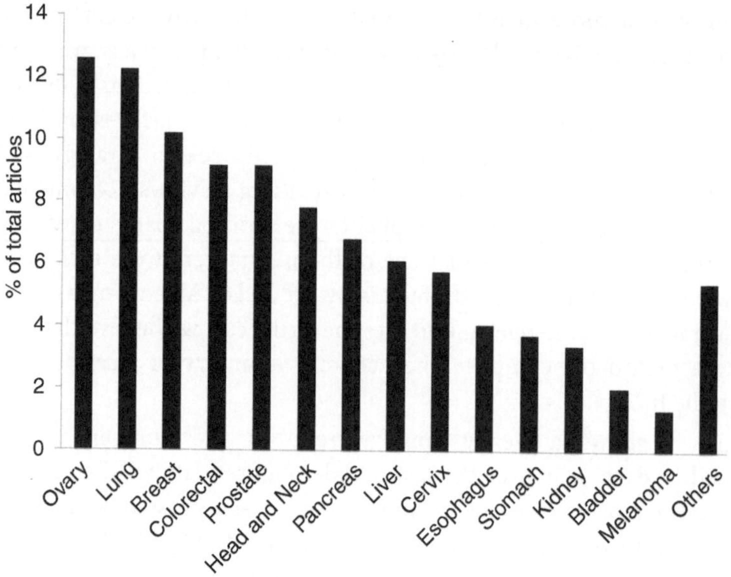

Fig. 2 Proportion of publications by cancer site. Data is based on 300 articles randomly selected from the full list of 800+ articles that resulted from the PubMed search.

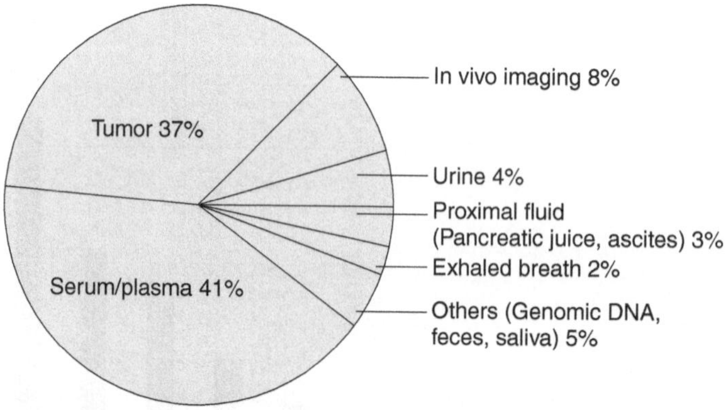

Fig. 3 Proportion of biomarkers by sample source. Data based on 150 randomly selected articles from the full list of 800+ articles that resulted from the PubMed search.

About 40% of the biomarkers were measured in serum/plasma; these include plasma proteomic signature, single plasma protein, panel of plasma proteins, autoantibody, and circulating methylated DNA and microRNA. And another 40% of the biomarkers were measured in tumor tissues, these include tissue proteomic signature, protein, mRNA, genome-wide gene expression signature, DNA mutation, methylated DNA, and microRNA. Various *in vivo* molecular imaging studies comprise 8% of the total publications. The rest of the biomarkers were measured in a variety of other body fluids such as urine, pancreatic juice, ascites, feces and saliva (5%), and exhaled breath (2%). Figure 4 shows the proportion of biomarkers by molecular type. Proteomic signatures are the most predominant category, comprising 36% of the total biomarkers (19% for plasma proteomics and 17% for tissue proteomics), reflecting the rapid advancement in proteomic technologies in the last decade. Other types such as metabolomics and microRNA comprise a small portion of the total biomarkers (3% each), reflecting the emerging nature of the research fields.

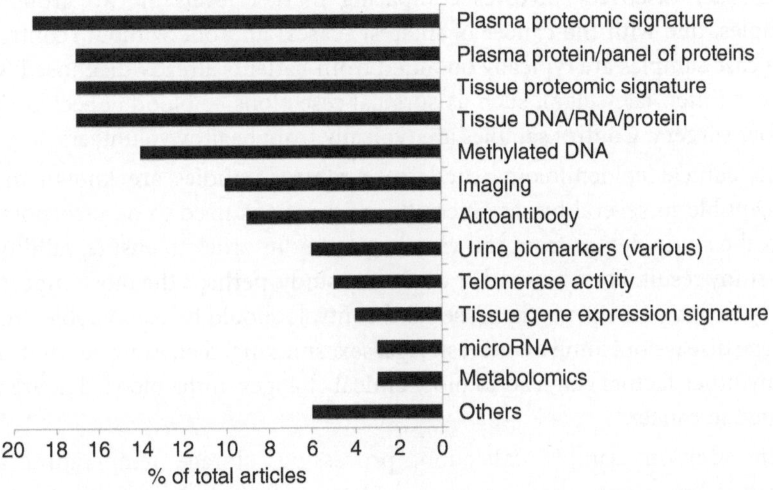

Fig. 4 Relative abundance of each biomarker category in the literature published in the last 10 years. Data based on 150 randomly selected articles from the full list of 800+ articles that resulted from the PubMed search.

SOME PROBLEMS IN RESEARCH QUALITY

Despite the large number of potential early detection biomarkers reported in the literature, very few have undergone definitive validation – a necessary step toward clinical application. A few attempts to validate some of the "proteomic signatures" in independent studies have resulted in disappointment (Baggerly et al. 2004; McLerran et al. 2008). The complexity of the "omics" data presents an additional challenge: data analysis is so complex and requires such specialized skills that mistakes often escape notice by peer-review, or even by the researchers themselves (Goldberg 2011).

A recent multi-institutional validation study evaluated 28 top candidate markers and various combinations of them for their performance in detecting pre-clinical ovarian cancers, using pre-diagnostic serum samples from the Prostate, Lung, Colorectal and Ovarian Cancer (PLCO) Screening Trial. None of the markers or marker combinations performed better than CA125 alone (Cramer et al. 2011; Zhu et al. 2011).

One reason for the lack of validated early detection biomarkers may lie in the way most cancer biomarkers were identified. A standard method for biomarker discovery involves comparing marker levels in two groups of samples, one with the cancer of interest (cases) and one without (controls). The case samples are typically obtained from patients already diagnosed with cancer (often late stage), such as surgical resections or blood collected right before surgery. Control samples are typically from healthy volunteers.

In clinical epidemiology, such "case-control" studies are known to be susceptible to several biases. Special considerations need to be incorporated into the study design from the very beginning in order to ensure validity of the study result. For a biomarker discovery study, perhaps the most important study design principle is that cases and controls should be comparable except for the disease outcome of interest. Age, sex, smoking, diet, exercise habit, and many other factors can lead to biochemical changes in the blood that are not related to cancer.

In addition, sample collection, processing, storage temperature and duration, and number of freeze/thaw cycles can all affect the validity of a study (Ransohoff and Gourlay 2010; Thorpe et al. 2007). For example, if samples from cases are collected and processed differently than the samples from controls, there may be systematic differences between the sample groups that are not related to cancer. However, seldom is this information reported in the literature in sufficient detail.

In an effort to improve biomarker research quality in general, a reporting standard has recently been proposed for all research that involves the use of human biospecimens (Moore et al. 2011). The Biospecimen Reporting for Improved Study Quality (BRISQ) outlines the data elements that are considered important when evaluating the validity of a study. The data elements are organized along the life cycle of biospecimens: Pre-acquisition, Acquisition, Stabilization/Preservation, Storage/Transport, Pre-analytic quality assessment, as well as level of importance (Tier I – Tier III).

We conducted a literature review to assess how much information regarding biospecimens is currently reported in the literature (manuscript in preparation). We used the same PubMed search strategy mentioned in the previous section, but limited to the articles that used serum or plasma samples, and have publically available full text. We evaluated 100+ articles against 20 BRISQ Tier I and Tier II items that are relevant to blood samples, including sample processing protocol, storage temperature, storage duration, ascertainment of disease status, timing of the sample relative to diagnosis, type of blood collection tubes, patient recruitment scheme, source of the samples.

We estimate that very few articles, if any, reported all of the BRISQ recommended items. A majority (> 70%) of the articles do not mention one or more of the following: how long the samples have been stored before laboratory analysis, what type of tubes were used for blood draw, time interval between blood draw and diagnosis, time interval between blood draw and processing, or freeze/thaw cycles. Although less prevalent, a significant number of articles (>10%) failed to provide any information about clinical or demographic characteristics of the study subjects (such as age and sex), or diagnostic information (whether the diagnosis was confirmed by histological examination by a pathologist).

We also found that about half of the articles do not provide enough information on *how* the patients and the samples were selected for the study. Usually there are no inclusion or exclusion criteria cited, nor eligibility criteria. The context in which the samples were obtained is rarely mentioned. Typically, only the names of the institutions that provided the samples are listed, without information on the original purpose of the collection. Often, the samples come from multiple institutions, including commercial vendors, without any detail about whether or not these samples are comparable in terms of sample processing protocol, storage condition and duration, or underlying patient population.

In most cases, the reporting deficiency highlighted above is a reflection of poor study design. We believe that in many cases, the investigators simply did not think of them as important or necessary for their study. Reporting guidelines not only can help readers, reviewers, and journal editors to better evaluate and interpret the study results, they can also feedback into study design and help improve research quality as they become adopted by the research community. Similar reporting guidelines exist for randomized trials (CONSORT), diagnostic markers (STARD), prognostic markers (REMARK), and observational studies (STROBE).

UNIQUE CHALLENGES IN IDENTIFYING EARLY DETECTION BIOMARKERS

For a biomarker to work for early detection, it must show a measurable signal in apparently healthy persons without symptoms. This requirement poses unique challenges to biomarker discovery and validation. For biomarker discovery, it is common to use clinical samples, i.e. samples collected after clinical diagnosis of cancer, usually due to symptoms. Such an approach may yield biomarkers that are abundant in clinical samples and absent in control samples. However, this approach may not be optimal or efficient for identifying early detection biomarkers, because biomarkers present in high quantity in clinical samples may not be present in measurable amounts in samples from early disease; conversely, biomarkers of early occult disease may not be present at high levels in clinical samples collected at diagnosis when the disease became clinically apparent.

Even when a study is conducted with the utmost care and best study design, a biomarker discovered in clinical samples must be validated in pre-diagnostic samples, i.e. samples collected from healthy subjects who later developed the cancer. Our experience suggests that biomarkers identified in clinical samples tend not to hold up in pre-diagnostic samples. There are several possible reasons for this. First, when using samples at the extreme end of a spectrum, more abundant molecules can easily overwhelm much less abundant ones. The abundant molecules may be the result of exponential tumor growth, but are not present at high enough level during early stages of carcinogenesis. Meanwhile, the less abundant, "true signals" of early events may be missed.

Second, the temporal expression pattern of most cancer biomarkers is unknown. Although most tumor-associated markers are up-regulated in tumors, some genes or proteins have been reported to be down-regulated in tumors as compared to normal tissues (Ali et al. 2006; Zhao et al. 2010). It is possible that some biomarkers, such as those associated with tumor tissue microenvironment, show differential signals in the early events of carcinogenesis, but the signals attenuate as the tumor progresses.

Early stage tumors are likely to be very small, in the millimeter range. Using mathematical modeling, it was estimated that occult early stage serous ovarian cancers have a median diameter of less than 3 mm (Brown and Palmer 2009). The ability to detect a biomarker secreted by a tumor this small depends not only on assay sensitivity (limit of detection), but also on physiological factors such as, how much of the biomarker is secreted by the tumor, how stable the protein is in the circulation, and many other factors. According to another study that relates tumor size and biomarker level in the blood, based also on mathematical modeling, detecting tumors in the millimeter range (diameter) is theoretically possible with current technologies, but only under unlikely physiological scenario of zero background noise, a high rate of biomarker secretion by the tumor, and a high proportion of the biomarker reaching the blood (Lutz et al. 2008) – an unlikely situation for most biomarkers in the blood.

In addition, tumor heterogeneity poses another challenge. For example, molecular genetic analyses showed that different ovarian cancer histological subtypes contain different mutations spectra (Lalwani et al. 2011). Furthermore, a recent Cancer Genome Atlas project on high grade serous ovarian cancer showed significant molecular heterogeneity within the subtype (Cancer Genome Atlas Research Network 2010). For these reasons, it is commonly considered that no single marker can detect a cancer with high enough sensitivity, and that a panel of biomarkers, combined in a model, is necessary in order to achieve requisite performance characteristics.

EMERGING TRENDS IN EARLY MARKER DISCOVERY

Using Pre-diagnostic Samples

To identify blood biomarkers of preclinical cancer, it seems a logical approach to use serum/plasma samples collected before clinical diagnosis (pre-diagnostic samples) for discovery. One argument against this approach is that such biomarkers are likely in very low concentrations so that current discovery technologies will not be able to detect them. A recent article by Pitteri et al., however, reported that changes in the level of EGFR can be detected in pre-diagnostic plasma collected from future breast cancer patients using current proteomic technology (Pitteri et al. 2010).

Limited availability of pre-diagnostic samples is another major obstacle. Prospective sample collection is a daunting task, involving a large number of people and many years of follow-up in order to accumulate enough cancers for a study. For example, in the PLCO intervention arm, about 39,000 women were followed for an average of 13 years; only about 200 ovarian cancer cases were identified. For rarer cancers, the number is much smaller – 55 mesothelioma cases in about 77,000 men and women in the PLCO intervention arm.

Several large, population-based prospective collections currently exist that can provide pre-diagnostic samples. Table 2 lists several large cancer screening or prevention trials that can provide pre-diagnostic biospecimens. Pooling

Table 2 Cancer screening and prevention trials that have prospectively collected biospecimens.

Trial Name	*Trial Endpoint*	*Biospecimens*	*Website*
ASCUS/LSIL Triage Study (ALTS)	Cervical cancer	Serum, tissue	http://dcp.cancer.gov/programs-resources/groups/bgcrg/alts
Prostate, Lung, Colorectal, and Ovarian Cancer Screening Trial (PLCO)	Prostate, Lung, Colorectal, Ovarian cancer	Serum, plasma, DNA, tumor tissue	http://www.plcostars.com
National Lung Screening Trial (NLST)	Lung cancer	Plasma DNA, urine, sputum	http://www.acrin.org/ACRIN-NLSTBIOREPOSITORY.aspx
UK Collaborative Trial of Ovarian Cancer Screening (UKCTOCS)	Ovarian Cancer	Serum	http://www.instituteforwomenshealth.ucl.ac.uk/academic_research/gynaecologicalcancer/gcrc/ukctocs
Breast Cancer Prevention Trial (BCPT)	Breast cancer	Plasma, DNA	http://www.nsabp.pitt.edu/
Study of Tamoxifen and Raloxifene (STAR)	Breast cancer	Plasma, DNA, tumor tissue	http://www.nsabp.pitt.edu/
Prostate Cancer Prevention Trial (PCPT)	Prostate cancer	Serum, DNA, tumor tissue	http://www.swog.org/Visitors/PCPT.asp#BRUG
Selenium and Vitamin E Cancer Prevention Trial (SELECT)	Prostate cancer	Plasma, toenails, tumor tissue	http://www.swog.org/Visitors/select/
Carotene and Retinol Efficacy Trial (CARET)	Lung cancer	Serum, plasma, DNA, tumor tissue	http://www.compass.fhcrc.org/caretweb/
Alpha-Tocopherol, Beta-Carotene Cancer Prevention study (ATBC)	Lung cancer	Serum, plasma, whole blood	http://atbcstudy.cancer.gov/
Women's Health Initiatives (WHI)	Breast cancer	Serum, plasma	http://www.whiscience.org/

of these resources may be necessary for less common cancers. A coordinated approach, such as a virtual national biorepository with centralized access and review procedures, should be implemented to maximize the scientific utilization of these precious resources (Goodman et al. 2006).

Using Proximal Fluids

Tumor-derived biomarkers may be present in the blood at a very low concentration when the tumor is sub-clinical, thus very small. Proximal fluids, such as pancreatic juice and ascites, contain higher concentration of tumor-secreted markers, and with less complexity, compared to circulating blood. Pancreatic juice can be collected during endoscopic procedures for treatment of pancreatic cancer or benign pancreatic diseases. Several proteomic analyses of pancreatic juice have been conducted (Chen et al. 2010; Gronborg et al. 2004; Shirai et al. 2008; Tian et al. 2008). Ascitic fluids, which can be collected during surgery or by paracentesis, have been used for ovarian cancer biomarker discovery (Elschenbroich et al. 2011; Gortzak-Uzan et al. 2008; Kuk et al. 2009).

Tumor interstitial fluid (TIF) was recently proposed as a suitable proximal fluid for biomarker discovery (Celis et al. 2004). TIF is collected by soaking the tissue, cut into small pieces, in PBS buffer at 37°C for 1 hour. It is shown to contain complex protein mixtures secreted by different cell types within the tumor microenvironment. Proteomic analyses of TIF have been conducted for breast cancer (Gromov et al. 2009) and kidney cancer (Teng et al. 2010).

In an interesting study to identify lung cancer biomarkers derived from the tumor microenvironment, the investigators used paired pulmonary venous and radial arterial blood from the same lung cancer patients collected during surgery (Yee et al. 2009). The pulmonary venous bloods received drainage directly from the tumor containing lung segment, and should thus contain the highest concentration of candidate biomarkers; while the radial artery blood represents the systemic circulation, serving as a within person control.

Targeting Pre-malignant Lesions

The most effective early detection/screening programs are the ones that identify pre-malignant lesions. Pap smear (or HPV tests) for cervical cancer screening and colonoscopy for colorectal cancer screening are examples. Both have been shown to be effective to reduce disease-specific mortality. Both tests detect pre-malignant lesions: cervical intraepithelial neoplasia (CIN) for cervical cancer and colorectal adenomas for colorectal cancer. Removal of these pre-malignant lesions has been shown to prevent cancer.

For cancers with well-defined precursors, more research is being focused on identifying biomarkers that can detect the presence of pre-malignant lesions. For examples, several pancreatic cancer studies examined biomarkers in pancreatic intraepithelial neoplasia (PanIN) and intraductal papillary mucinous neoplasm (IPMN), and found several biomarkers that are up-regulated in these lesions (Cheung et al. 2008; du Rieu et al. 2010; Gold et al. 2010). A glycomic analysis revealed specific glycans that can distinguish between patients with high grade dysplasia (HGD) or Barrett's esophagus, and patients with esophageal adenocarcinoma (Mechref et al. 2009). For many cancers, however, there are no known pre-malignant lesions. For these cancers, understanding the origin and etiology of the cancer is a necessary step towards developing effective early detection biomarkers.

CONCLUDING REMARKS

Early detection has been proven to save lives from colorectal, cervical and breast cancers (National Institutes of Health Consensus Development Panel 1996; U.S. Preventive Services Task Force 2008; U.S. Preventive Services Task Force 2009). However, for most other cancers, there is currently no effective tool for early detection or screening. An effective, reliable, non-invasive early detection biomarker is the key to the success of a cancer screening program. However, developing such a biomarker has proven to be more challenging than many laboratory scientists had expected.

Even if a biomarker has excellent performance in terms of sensitivity and specificity, its effectiveness as a screening tool depends on many other factors such as disease prevalence, effectiveness of treatments, natural history of the disease, etc. The gold standard for evaluating a screening test is a randomized controlled trial with cancer-specific mortality as the endpoint. Such cancer screening trials may take years and cost hundreds of millions of dollars, thus are not feasible under most circumstances. Cost-benefit evaluation also influences whether a test should be used as a screening tool even if it is effective.

Targeted screening, in which a sub-group of the population at elevated risk is identified and screened, may be necessary for rare cancers like ovarian cancer. Genetic predispositions, such as BRCA1/2 mutations, family history, smoking, and obesity are well-known risk factors for many cancers, and may be useful for stratifying the population. In recent years, genome-wide associate studies (GWAS) have revealed many common genetic variations associated with various cancers (http://www.ncbi.nlm.nih.gov/gap). Developing risk models that incorporate genetic, epidemiological and biochemical biomarkers may lead to better risk assessment and stratification.

In conclusion, developing effective early detection markers for cancer screening is an undertaking requiring a truly multi-disciplinary mindset, and an integrated team with expertise in molecular biology, clinical epidemiology, statistics, pathology, clinical oncology, bioinformatics and more. Knowledge in population sciences and cancer screening is particularly important not only for conducting a trial, but also for guiding the biomarker development and validation process to the right track from the very beginning.

REFERENCES

Ali, I.U., Z. Xiao, W. Malone, M. Smith, T.P. Conrads, T.D. Veenstra, P. Greenwald, B.T. Luke, and J.W. McLarty. 2006. Plasma proteomic profiling: search for lung cancer diagnostic and early detection markers. Oncol Rep, 15: 1367-1372.

Baggerly, K.A., J.S. Morris, and K.R. Coombes. 2004. Reproducibility of SELDI-TOF protein patterns in serum: comparing datasets from different experiments. Bioinformatics, 20: 777-785.

Bakry, R., M. Rainer, C.W. Huck, and G.K. Bonn. 2011. Protein profiling for cancer biomarker discovery using matrix-assisted laser desorption/ionization time-of-flight mass spectrometry and infrared imaging: a review. Anal Chim Acta, 690: 26-34.

Brown, P.O., and C. Palmer. 2009. The preclinical natural history of serous ovarian cancer: defining the target for early detection. PLoS Med, 6: e1000114.

Cancer Genome Atlas Research Network. 2010. Integrated genomic analyses of ovarian carcinoma. Nature, 474: 609-615.

Celis, J.E., P. Gromov, T. Cabezon, J.M. Moreira, N. Ambartsumian, K. Sandelin, F. Rank, and I. Gromova. 2004. Proteomic characterization of the interstitial fluid perfusing the breast tumor microenvironment: a novel resource for biomarker and therapeutic target discovery. Mol Cell Proteomics, 3: 327-344.

Chen, R., S. Pan, X. Duan, B.H. Nelson, R.A. Sahota, S. de Rham, R.A. Kozarek, M. McIntosh, and T.A. Brentnall. 2010. Elevated level of anterior gradient-2 in pancreatic juice from patients with pre-malignant pancreatic neoplasia. Mol Cancer, 9: 149.

Cheung, W., M.M. Darfler, H. Alvarez, B.L. Hood, T.P. Conrads, N. Habbe, D.B. Krizman, J. Mollenhauer, G. Feldmann, and A. Maitra. 2008. Application of a global proteomic approach to archival precursor lesions: deleted in malignant brain tumors 1 and tissue transglutaminase 2 are upregulated in pancreatic cancer precursors. Pancreatology, 8: 608-616.

Cramer, D.W., R.C. Bast, Jr., C.D. Berg, E.P. Diamandis, A.K. Godwin, P. Hartge, A.E. Lokshin, K.H. Lu, M.W. McIntosh, G. Mor, C. Patriotis, P.F. Pinsky, M.D. Thornquist, N. Scholler, S.J. Skates, P.M. Sluss, S. Srivastava, D.C. Ward, Z. Zhang, C.S. Zhu, and N. Urban. 2011. Ovarian cancer biomarker performance in prostate, lung, colorectal, and ovarian cancer screening trial specimens. Cancer Prev Res (Phila), 4: 365-374.

du Rieu, M.C., J. Torrisani, J. Selves, T. Al Saati, A. Souque, M. Dufresne, G.J. Tsongalis, A.A. Suriawinata, N. Carrere, L. Buscail, and P. Cordelier. 2010. MicroRNA-21 is induced early in pancreatic ductal adenocarcinoma precursor lesions. Clin Chem, 56: 603-612.

Elschenbroich, S., V. Ignatchenko, B. Clarke, S.E. Kalloger, P.C. Boutros, A.O. Gramolini, P. Shaw, I. Jurisica, and T. Kislinger. 2011. In-depth proteomics of ovarian cancer ascites: combining shotgun proteomics and selected reaction monitoring mass spectrometry. J Proteome Res, 10: 2286-2299.

Goetz, J.A., Y. Mechref, P. Kang, M.H. Jeng, and M.V. Novotny. 2009. Glycomic profiling of invasive and non-invasive breast cancer cells. Glycoconj J, 26: 117-131.

Gold, D.V., M. Goggins, D.E. Modrak, G. Newsome, M. Liu, C. Shi, R.H. Hruban, and D.M. Goldenberg. 2010. Detection of early-stage pancreatic adenocarcinoma. Cancer Epidemiol Biomarkers Prev, 19: 2786-2794.

Goldberg, P. 2011. IOM Committee Will Probe Duke Scandal Together With Other "Omics" Case Studies. Cancer Letter, 37.

Goldman, R., H.W. Ressom, R.S. Varghese, L. Goldman, G. Bascug, C.A. Loffredo, M. Abdel-Hamid, I. Gouda, S. Ezzat, Z. Kyselova, Y. Mechref, and M.V. Novotny. 2009. Detection of hepatocellular carcinoma using glycomic analysis. Clin Cancer Res, 15: 1808-1813.

Goodman, G.E., M.D. Thornquist, C. Edelstein, and G.S. Omenn. 2006. Biorepositories: let's not lose what we have so carefully gathered! Cancer Epidemiol Biomarkers Prev, 15: 599-601.

Gortzak-Uzan, L., A. Ignatchenko, A.I. Evangelou, M. Agochiya, K.A. Brown, P. St Onge, I. Kireeva, G. Schmitt-Ulms, T.J. Brown, J. Murphy, B. Rosen, P. Shaw, I. Jurisica, and T. Kislinger. 2008. A proteome resource of ovarian cancer ascites: integrated proteomic and bioinformatic analyses to identify putative biomarkers. J Proteome Res, 7: 339-351.

Grivennikov, S.I., F.R. Greten, and M. Karin. 2010. Immunity, inflammation, and cancer. Cell, 140: 883-899.

Gromov, P., I. Gromova, J. Bunkenborg, T. Cabezon, J.M. Moreira, V. Timmermans-Wielenga, P. Roepstorff, F. Rank, and J.E. Celis. 2009. Up-regulated proteins in the fluid bathing the tumour cell microenvironment as potential serological markers for early detection of cancer of the breast. Mol Oncol, 4: 65-89.

Gronborg, M., J. Bunkenborg, T.Z. Kristiansen, O.N. Jensen, C.J. Yeo, R.H. Hruban, A. Maitra, M.G. Goggins, and A. Pandey. 2004. Comprehensive proteomic analysis of human pancreatic juice. J Proteome Res, 3: 1042-1055.

Hanahan, D., and R.A. Weinberg. 2011. Hallmarks of cancer: the next generation. Cell, 144: 646-674.

Huang, Z., D. Huang, S. Ni, Z. Peng, W. Sheng, and X. Du. 2009. Plasma microRNAs are promising novel biomarkers for early detection of colorectal cancer. Int J Cancer, 127: 118-126.

Institute of Medicine. 2006. Developing Biomarker-Based Tools for Cancer Screening, Diagnosis, and Treatment. Washington DC: The National Academies Press.

Isailovic, D., R.T. Kurulugama, M.D. Plasencia, S.T. Stokes, Z. Kyselova, R. Goldman, Y. Mechref, M.V. Novotny, and D.E. Clemmer. 2008. Profiling of human serum glycans associated with liver cancer and cirrhosis by IMS-MS. J Proteome Res, 7: 1109-1117.

Issaq, H.J., O. Nativ, T. Waybright, B. Luke, T.D. Veenstra, E.J. Issaq, A. Kravstov, and M. Mullerad. 2008. Detection of bladder cancer in human urine by metabolomic profiling using high performance liquid chromatography/mass spectrometry. J Urol, 179: 2422-2426.

Kohler, C., Z. Barekati, R. Radpour, and X.Y. Zhong. 2011. Cell-free DNA in the circulation as a potential cancer biomarker. Anticancer Res, 31: 2623-2628.

Kryston, T.B., A.B. Georgiev, P. Pissis, and A.G. Georgakilas. 2011. Role of oxidative stress and DNA damage in human carcinogenesis. Mutat Res, 711: 193-201.

Kuk, C., V. Kulasingam, C.G. Gunawardana, C.R. Smith, I. Batruch, and E.P. Diamandis. 2009. Mining the ovarian cancer ascites proteome for potential ovarian cancer biomarkers. Mol Cell Proteomics, 8: 661-669.

Kulasingam, V., and E.P. Diamandis. 2008. Strategies for discovering novel cancer biomarkers through utilization of emerging technologies. Nat Clin Pract Oncol, 5: 588-599.

Lalwani, N., S.R. Prasad, R. Vikram, A.K. Shanbhogue, P.C. Huettner, and N. Fasih. 2011. Histologic, molecular, and cytogenetic features of ovarian cancers: implications for diagnosis and treatment. Radiographics, 31: 625-646.

Link, A., F. Balaguer, Y. Shen, T. Nagasaka, J.J. Lozano, C.R. Boland, and A. Goel. 2010. Fecal MicroRNAs as novel biomarkers for colon cancer screening. Cancer Epidemiol Biomarkers Prev, 19: 1766-1774.

Lutz, A.M., J.K. Willmann, F.V. Cochran, P. Ray, and S.S. Gambhir. 2008. Cancer screening: a mathematical model relating secreted blood biomarker levels to tumor sizes. PLoS Med, 5: e170.

Mann, B., M. Madera, I. Klouckova, Y. Mechref, L.E. Dobrolecki, R.J. Hickey, Z.T. Hammoud, and M.V. Novotny. 2010. A quantitative investigation of fucosylated serum glycoproteins with application to esophageal adenocarcinoma. Electrophoresis, 31: 1833-1841.

McLerran, D., W.E. Grizzle, Z. Feng, W.L. Bigbee, L.L. Banez, L.H. Cazares, D.W. Chan, J. Diaz, E. Izbicka, J. Kagan, D.E. Malehorn, G. Malik, D. Oelschlager, A. Partin, T. Randolph, N. Rosenzweig, S. Srivastava, I.M. Thompson, M. Thornquist, D. Troyer, Y. Yasui, Z. Zhang, L. Zhu, and O.J. Semmes. 2008. Analytical validation of serum proteomic profiling for diagnosis of prostate cancer: sources of sample bias. Clin Chem, 54: 44-52.

Mechref, Y., A. Hussein, S. Bekesova, V. Pungpapong, M. Zhang, L.E. Dobrolecki, R.J. Hickey, Z.T. Hammoud, and M.V. Novotny. 2009. Quantitative serum glycomics of esophageal adenocarcinoma and other esophageal disease onsets. J Proteome Res, 8: 2656-2666.

Moore, H.M., A. Kelly, S.D. Jewell, L.M. McShane, D.P. Clark, R. Greenspan, P. Hainaut, D.F. Hayes, P. Kim, E. Mansfield, O. Potapova, P. Riegman, Y. Rubinstein, E. Seijo, S. Somiari, P. Watson, H.U. Weier, C. Zhu, and J. Vaught. 2011. Biospecimen Reporting for Improved Study Quality. Biopreserv Biobank, 9: 57-70.

National Institutes of Health Consensus Development Panel. 1996. National Institutes of Health Consensus Development Conference Statement: cervical cancer, April 1-3, 1996. . J Natl Cancer Inst Monogr: vii-xix.

Pine, S.R., L.E. Mechanic, L. Enewold, A.K. Chaturvedi, H.A. Katki, Y.L. Zheng, E.D. Bowman, E.A. Engels, N.E. Caporaso, and C.C. Harris. 2011. Increased levels of circulating interleukin 6, interleukin 8, C-reactive protein, and risk of lung cancer. J Natl Cancer Inst, 103: 1112-1122.

Pitteri, S.J., L.M. Amon, T. Busald Buson, Y. Zhang, M.M. Johnson, A. Chin, J. Kennedy, C.H. Wong, Q. Zhang, H. Wang, P.D. Lampe, R.L. Prentice, M.W. McIntosh, S.M. Hanash, and C.I. Li. 2010. Detection of elevated plasma levels of epidermal growth factor receptor before breast cancer diagnosis among hormone therapy users. Cancer Res, 70: 8598-8606.

Ransohoff, D.F., and M.L. Gourlay. 2010. Sources of bias in specimens for research about molecular markers for cancer. J Clin Oncol, 28: 698-704.

Ritchie, S.A., P.W. Ahiahonu, D. Jayasinghe, D. Heath, J. Liu, Y. Lu, W. Jin, A. Kavianpour, Y. Yamazaki, A.M. Khan, M. Hossain, K.K. Su-Myat, P.L. Wood, K. Krenitsky, I. Takemasa, M. Miyake, M. Sekimoto, M. Monden, H. Matsubara, F. Nomura, and D.B. Goodenowe. 2010. Reduced levels of hydroxylated, polyunsaturated ultra long-chain fatty acids in the serum of colorectal cancer patients: implications for early screening and detection. BMC Med, 8: 13.

Shirai, Y., K. Sogawa, T. Yamaguchi, K. Sudo, A. Nakagawa, Y. Sakai, T. Ishihara, M. Sunaga, M. Nezu, T. Tomonaga, M. Miyazaki, H. Saisho, and F. Nomura. 2008. Protein profiling in pancreatic juice for detection of intraductal papillary mucinous neoplasm of the pancreas. Hepatogastroenterology, 55: 1824-1829.

Sreekumar, A., L.M. Poisson, T.M. Rajendiran, A.P. Khan, Q. Cao, J. Yu, B. Laxman, R. Mehra, R.J. Lonigro, Y. Li, M.K. Nyati, A. Ahsan, S. Kalyana-Sundaram, B. Han, X. Cao, J. Byun, G.S. Omenn, D. Ghosh, S. Pennathur, D.C. Alexander, A. Berger,

J.R. Shuster, J.T. Wei, S. Varambally, C. Beecher, and A.M. Chinnaiyan. 2009. Metabolomic profiles delineate potential role for sarcosine in prostate cancer progression. Nature, 457: 910-914.

Sturgeon, C.M., M.J. Duffy, U.H. Stenman, H. Lilja, N. Brunner, D.W. Chan, R. Babaian, R.C. Bast, Jr., B. Dowell, F.J. Esteva, C. Haglund, N. Harbeck, D.F. Hayes, M. Holten-Andersen, G.G. Klee, R. Lamerz, L.H. Looijenga, R. Molina, H.J. Nielsen, H. Rittenhouse, A. Semjonow, M. Shih Ie, P. Sibley, G. Soletormos, C. Stephan, L. Sokoll, B.R. Hoffman, E.P. Diamandis, and B. National Academy of Clinical. 2008. National Academy of Clinical Biochemistry laboratory medicine practice guidelines for use of tumor markers in testicular, prostate, colorectal, breast, and ovarian cancers. Clin Chem, 54: e11-79.

Taylor, D.D., and C. Gercel-Taylor. 2008. MicroRNA signatures of tumor-derived exosomes as diagnostic biomarkers of ovarian cancer. Gynecol Oncol, 110: 13-21.

Teng, P.N., B.L. Hood, M. Sun, R. Dhir, and T.P. Conrads. 2010. Differential proteomic analysis of renal cell carcinoma tissue interstitial fluid. J Proteome Res, 10: 1333-1342.

Thorpe, J.D., X. Duan, R. Forrest, K. Lowe, L. Brown, E. Segal, B. Nelson, G.L. Anderson, M. McIntosh, and N. Urban. 2007. Effects of blood collection conditions on ovarian cancer serum markers. PLoS One, 2: e1281.

Tian, M., Y.Z. Cui, G.H. Song, M.J. Zong, X.Y. Zhou, Y. Chen, and J.X. Han. 2008. Proteomic analysis identifies MMP-9, DJ-1 and A1BG as overexpressed proteins in pancreatic juice from pancreatic ductal adenocarcinoma patients. BMC Cancer, 8: 241.

Tsilidis, K.K., C. Branchini, E. Guallar, K.J. Helzlsouer, T.P. Erlinger, and E.A. Platz. 2008. C-reactive protein and colorectal cancer risk: a systematic review of prospective studies. Int J Cancer, 123: 1133-1140.

U.S. Preventive Services Task Force. 2008. Screening for colorectal cancer: U.S. Preventive Services Task Force recommendation statement. Ann Intern Med, 149: 627-637.

U.S. Preventive Services Task Force. 2009. Screening for breast cancer: U.S. Preventive Services Task Force recommendation statement. Ann Intern Med, 151: 716-726, W-236.

van 't Veer, L.J., H. Dai, M.J. van de Vijver, Y.D. He, A.A. Hart, M. Mao, H.L. Peterse, K. van der Kooy, M.J. Marton, A.T. Witteveen, G.J. Schreiber, R.M. Kerkhoven, C. Roberts, P.S. Linsley, R. Bernards, and S.H. Friend. 2002. Gene expression profiling predicts clinical outcome of breast cancer. Nature, 415: 530-536.

Wang, J., J. Chen, P. Chang, A. LeBlanc, D. Li, J.L. Abbruzzesse, M.L. Frazier, A.M. Killary, and S. Sen. 2009. MicroRNAs in plasma of pancreatic ductal adenocarcinoma patients as novel blood-based biomarkers of disease. Cancer Prev Res (Phila), 2: 807-813.

Wang, T., T.E. Rohan, M.J. Gunter, X. Xue, J. Wactawski-Wende, S.N. Rajpathak, M. Cushman, H.D. Strickler, R.C. Kaplan, S. Wassertheil-Smoller, P.E. Scherer, and G.Y. Ho. 2011. A prospective study of inflammation markers and endometrial cancer risk in postmenopausal hormone nonusers. Cancer Epidemiol Biomarkers Prev, 20: 971-977.

Yee, J., M.D. Sadar, D.D. Sin, M. Kuzyk, L. Xing, J. Kondra, A. McWilliams, S.F. Man, and S. Lam. 2009. Connective tissue-activating peptide III: a novel blood biomarker for early lung cancer detection. J Clin Oncol, 27: 2787-2792.

Zhang, C., C. Wang, X. Chen, C. Yang, K. Li, J. Wang, J. Dai, Z. Hu, X. Zhou, L. Chen, Y. Zhang, Y. Li, H. Qiu, J. Xing, Z. Liang, B. Ren, K. Zen, and C.Y. Zhang. 2010. Expression profile of microRNAs in serum: a fingerprint for esophageal squamous cell carcinoma. Clin Chem, 56: 1871-1879.

Zhao, H., J. Shen, L. Medico, D. Wang, C.B. Ambrosone, and S. Liu. 2010. A pilot study of circulating miRNAs as potential biomarkers of early stage breast cancer. PLoS One, 5: e13735.

Zhu, C.S., P.F. Pinsky, D.W. Cramer, D.F. Ransohoff, P. Hartge, R.M. Pfeiffer, N. Urban, G. Mor, R.C. Bast, Jr., L.E. Moore, A.E. Lokshin, M.W. McIntosh, S.J. Skates, A. Vitonis, Z. Zhang, D.C. Ward, J.T. Symanowski, A. Lomakin, E.T. Fung, P.M. Sluss, N. Scholler, K.H. Lu, A.M. Marrangoni, C. Patriotis, S. Srivastava, S.S. Buys, C.D. Berg for the PLCO Project Team. 2011. A framework for evaluating biomarkers for early detection: validation of biomarker panels for ovarian cancer. Cancer Prev Res (Phila), 4: 375-383.

CHAPTER 2

Biomarkers of Oxidative Stress in Carcinogenesis

Humberto Rodriguez-Rocha, Aracely Garcia-Garcia and Rodrigo Franco*
Redox Biology Center and School of Veterinary Medicine and Biomedical Sciences,
University of Nebraska-Lincoln, Lincoln, NE 68583, USA

ABSTRACT

Oxidative stress is caused by an imbalance between the production of reactive species of oxygen and nitrogen and the ability to either detoxify these reactive intermediates, or repair the resulting damage. Ultimately, the reaction of reactive species with biomolecules including lipids, nucleic acids and proteins leads to alterations in cellular function, which might participate in the progression of a variety of human diseases including cancer. Thus, recent research efforts have been directed to identify the role of specific oxidative modifications in the signal transduction events mediating the etiology of cancer progression. Furthermore, the identification of theses oxidative modifications has raised the possibility of using this knowledge to develop new methods to diagnose cancer. In this chapter, we summarize the mechanisms by which reactive species generate distinct oxidative modifications in biomolecules, and the potential of these oxidative modifications to be used as biomarkers during the carcinogenic process.

REDOX ENVIRONMENT AND CANCER

Carcinogenesis leads to a self-replicating transformed cell, which conveys the generation of a larger number of cells with an increasing accumulation of abnormalities. Numerous factors contribute to the initiation, development and progression of cancer including those of genetic, environmental and

**Corresponding author*: E-mail: rfrancocruz2@unl.edu

dietary influences. Oxidative stress and reactive species (RS) participate at all stages of the malignancy/carcinogenic process aroused from both inherited and environmental factors (Klaunig et al. 2010). Oxidative stress is caused by the imbalance between the production of reactive species of oxygen (ROS) and nitrogen (RNS) and the ability to: 1) detoxify the reactive intermediates produced, or 2) repair the resulting damage. Ultimately, oxidative stress conveys the alteration in cellular function caused by the reaction of RS with cellular constituents or biomolecules.

Cancer cells often exhibit increased generation of RS compared with normal cells which provides favorable conditions for cancer cell growth, genetic instability, and survival (Fig. 1). This increased generation of RS is given particularly by alterations in the protein levels of RS generating enzymes and/or mutations in the mitochondrial electron transport chain (ETC). Cancer cells have increased levels of RS, which are associated with increased NADPH-oxidase (NOX) activity/expression (Szatrowski and Nathan 1991; Arnold et al. 2001; Laurent et al. 2005). Mutations in nuclear or mitochondrial genes encoding components of the ETC also lead to an increase in ROS generation by inhibition of electron transfer, with subsequent accumulation of electrons that can be captured by O_2. ETC gene defects have been directly linked to some hereditary cancers (Brandon et al. 2006; Slane et al. 2006; Ishii et al. 2007; Szeto et al. 2007; Guzy et al. 2008; Sudarshan et al. 2009). Depending on tumor type and stage, the expression levels and activities of nitric oxide synthase (NOSs) are also increased in tumor cells compared with normal tissues (Qiu et al. 2003; Wang et al. 2005; Fukumura et al. 2006; Ridnour et al. 2006; Olson and Garban 2008; Tanriover et al. 2008; Lahdenranta et al. 2009). Polymorphisms in myeloperoxidase (MPO), an enzyme that produces hypochlorous acid (HOCl) from hydrogen peroxide (H_2O_2) and chloride anion (Cl^-), have been associated with the etiology of lung, larynx, pancreatic and breast cancer, and also found in leukemia cells (Cascorbi et al. 2000; Ahn et al. 2004; Ambrosone et al. 2005). Cyclooxygenases (COX) produce ROS as byproducts of the metabolism of arachidonic acid. COX1 overexpression has been shown to mediate tumorigenic transformation of immortalized ECV endotheial cells (Narko et al. 1997), while COX2 expression/activity is been linked to a variety of cancer types (Costa et al. 2002; Menter et al. 2010).

Redox imbalance in cancer cells is linked to alterations in antioxidant systems. Elevated levels of glutathione (GSH), the major antioxidant defense in cells, are observed in various types of cancerous cells and solid tumors (Estrela et al. 2006; Roshchupkina et al. 2008). However, the ratio between GSH and its oxidized form, glutathione disulfide (GSSG),

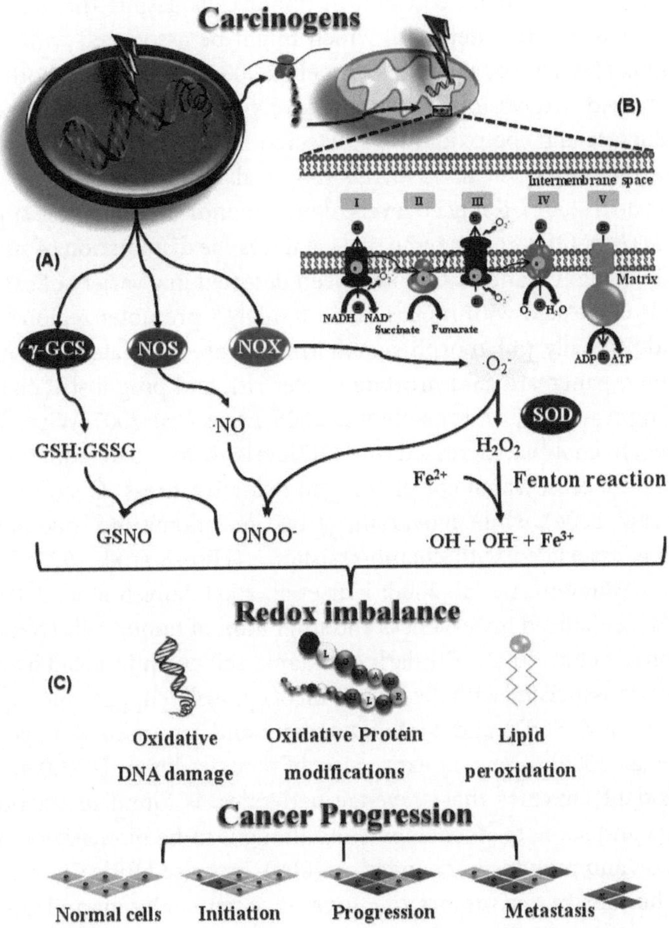

Fig. 1 Cancer cells exhibit increased generation of ROS and RNS compared with normal cells, which provides favorable conditions for cancer initiation, progression and metastasis. Increased generation of RS is given particularly by alterations in the protein levels of RS generating enzymes such as NOX and NOS (A). In the mitochondria, $O_2^{\bullet-}$ is formed by monoelectronic reduction of O_2 by Complexes I and III of the respiratory chain. Mutations in nuclear or mitochondrial genes encoding components of the ETC can also lead to an increase in ROS generation by inhibition of electron transfer (B). In addition, alterations in the levels of antioxidant enzymes such as SODs and γ-GCS (involved in GSH synthesis) also disrupt the redox balance in cancer cells. Although oxidative DNA-damage is a central element in the transformation process towards malignancy, it is also evident that RS can promote/predispose to cancer development by oxidative damage/modification to lipids and proteins/enzymes (C).

Color image of this figure appears in the color plate section at the end of the book.

has been shown to remain constant in tumor cells despite the high levels of RS and oxidative stress detected, which might be associated with increased GSSG efflux (Navarro et al. 1999; Oltra et al. 2001). In contrast, both elevated and decreased expression/enzymatic levels of the antioxidant enzymes GSH-reductase and -peroxidase are detected in transformed cells and tumor tissues (el-Sharabasy et al. 1993; Di Ilio et al. 1995; Esworthy et al. 1995; Hasegawa et al. 2002). Reduced levels of mitochondrial manganese superoxide dismutase (MnSOD), the enzyme that catalyzes the dismutation of superoxide anion ($O_2^{\bullet-}$) into O_2 and H_2O_2, have been detected in a variety of cancer cells and this is correlated with mutations in the DNA promoter region (Xu et al. 2008). Additionally, polymorphisms in MnSOD are associated with increased breast, lung, pancreatic and prostate cancer risk and prognosis (Zhang et al. 1999; Wang et al. 2001; Ambrosone et al. 2005; Ergen et al. 2007; Wheatley-Price et al. 2008). In contrast, increased MnSOD levels are found in brain tumors and mesothelioma cells, which correlates with oxidative stress (Cobbs et al. 1996; Kinnula et al. 2004), while in ovarian, gastric, melanoma and leukemia cancer cells this is associated with chemo-resistance (Hirose et al. 1993; Kuninaka et al. 2000; Mantymaa et al. 2000; Hur et al. 2003; Suresh et al. 2003; Yeung et al. 2008), enhanced invasiveness and migration of tumor cells (Nelson et al. 2003; Connor et al. 2007). Similarly, leukemia cell death induced by estrogen derivatives is associated with the inhibition of cytosolic copper/zinc superoxide dismutase (CuZnSOD) and MnSOD activity and increased ROS generation (Huang et al. 2000). Aberrant expression of peroxiredoxins (PRDXs), a group of antioxidant enzymes that scavenge peroxides, is found in various kinds of cancers and some of its members are thought to be biomarkers of cancer progression and prognosis (Zhang et al. 2009). Increased PRDX1 expression is found in human thyroid tumors, oral squamous cell carcinoma and lung cancer (Yanagawa et al. 1999; Yanagawa et al. 2000; Chang et al. 2001; Kim et al. 2008). Several PRDX isoforms have also been shown to be overexpressed in malignant mesothelioma, brain tumors, breast and cervical cancer (Noh et al. 2001; Kinnula et al. 2002; Karihtala et al. 2003; Chang et al. 2007; Kim et al. 2009; Jarvela et al. 2010).

Tumor growth induces inflammatory processes. Recent studies demonstrate that the presence of a tumor induces an increased oxidative stress status in surrounding tissues, which seems to be mediated primarily by inflammatory cells such as macrophages and mediators such as chemokines, cytokines and prostaglandins (Kryston et al. 2011).

BIOLOGICAL TARGETS OF OXIDATIVE STRESS

Oxidative Post-translational Protein Modifications (OPMs)

Proteins are the main target for oxidative modifications. RS can react either directly with the side chains of amino acid residues, the peptidic backbone of proteins. Lipid and carbohydrate oxidation can generate lipid peroxidation and glycoxidation adducts on proteins, respectively (Berlett and Stadtman 1997; Beckman and Ames 1998). Compared to RNA, DNA and lipids, proteins are the most likely cellular targets of oxidation as a result of their abundance, and it is estimated that nearly 68% of oxidation process is carried out in proteins (Davies 2005). ROS dependent modification of proteins can be classified as cleavage of peptide bonds, and oxidation of amino acid residue side chains.

Peptide Bond Cleavage

Peptide bond cleavage of alkoxyl derivatives of proteins takes place by three different mechanisms: 1) the diamide pathway, where the carboxy-terminal amino acid of the fragment originated from the amino-terminal region of the protein is a diamide derivative, and the amino acid of the fragment formed from the carboxy-terminal region is an isocyanate derivative; 2) the α-amidation pathway, where the carboxy-terminal amino acid of the fragment originated from the amino-terminal region of the protein is an amide derivative, and the amino acid of the fragment formed from the carboxy-terminal region is an α-keto-acyl derivative, (Berlett and Stadtman 1997; Stadtman and Levine 2003); and 3) by ROS-mediated attack of glutamyl, aspartyl, and prolyl side chains, where the amino-terminal amino acid of the carboxy-terminal fragment is a N-pyruvyl derivative from oxidation of glutamyl and aspartyl residues (Garrison 1987), while proline oxidation leads to the formation of 2-pyrrolidone derivative (Uchida et al. 1990).

Oxidation of Amino Acid Residue Side Chains

All the amino acids residues within a protein are vulnerable to oxidative modifications by RS, particularly cysteine and methionine (Berlett and Stadtman 1997).

Cysteine oxidation: Cysteine-based redox reactions are difficult to predict because, so far, a consensus sequence(s) has (have) not been identified. One characteristic of cysteines (Cys-SH) involved in sensing redox changes is the close proximity of polar amino acids, resulting in stabilization of the reactive thiolate anion form (Hirotsu et al. 1999; Schroder et al. 2000). There are reversible and irreversible cysteine modifications. The reversible modifications include disulfide bonds (Cys-SS-Cys), S-nitrosylation (Cys-SNO) (Hess et al. 2005; Hao et al. 2006), S-glutathionylation (Cys-SSG) (Ghezzi 2005) and sulfenic acid formation (Cys-SOH). Irreversible modifications include sulfinic (Cys-SO_2H) or sulfonic acid (Cys-SO_3H) formation (Chae and Rhee 1994; Rabilloud et al. 2002). One electron cysteine oxidation by RS such as $O_2^{\bullet -}$ produces thyil radical (Cys-$S^{\bullet}$), which by its radical nature can amplify oxidative damage. Two electron oxidative attack of Cys-SH by H_2O_2, alkyl hydroperoxides, peroxynitrite ($ONOO^-$), and HOCl mediates Cys-SOH formation (Ellis and Poole 1997; Bryk et al. 2000; Raftery et al. 2001). Cys-SOH can be reduced back to its thiol state by reducing agents or form intra-or intermolecular Cys-SS-Cys with neighboring thiol groups. If oxidative conditions persist, Cys-SOH is hyperoxidized to Cys-SO_2H or Cys-SO_3H acids, which are considered biologically irreversible (Boschi-Muller et al. 2000; Cross and Templeton 2004; Poole et al. 2004). However, in PRDXs (2-Cys Prx), Cys-SO_2H formation is reduced back by sulfiredoxin (Srx) in the presence of ATP (Biteau et al. 2003; Chang et al. 2004; Woo et al. 2005). Cys-SSG is the covalent attachment of GSH to cysteine residues through a mixed disulfide bond, and either GSSG can react with a Cys-SH or an oxidized protein cysteine (Cys-SOH or Cys-$S^{\bullet}$) can react with GSH (Klatt and Lamas 2000; Poole et al. 2004). Cys-SSG linkages are removed by changes in the intracellular GSH/GSSG balance and/or the activity of glutaredoxin enzymes. However, glutaredoxins can mediate both Cys-SSG and their deglutathionylation depending on the GSH/GSSG ratio and the presence of Cys-$S^{\bullet}$ (Beer et al. 2004; Mieyal et al. 2008). Cys-SNO comprises the coupling of a nitrosyl group to a Cys-SH to generate a S-nitrosothiol (SNO). So far, the only mechanism described for Cys-SNO *in vivo* is the reversible trans-nitrosylation of protein thiols by S-nitrosoglutathione (GSNO).

Methionine oxidation: Methionine is one of the most vulnerable amino acid residues to oxidation by RS (Levine et al. 1996). Addition of an extra O_2 atom oxidizes methionine to methionine sulfoxide (MetSO), and a strong oxidant might further oxidize MetSO to methionine sulfone (MetSO_2) (Hoshi and Heinemann 2001). Two different stereoisomers of methionine sulfoxide, L-Met-S-SO and L-Met-R-SO, can be found at physiological conditions (Ripp et al. 1999; Sharov et al. 1999; Dever and Elfarra 2006). Methionine oxidation is reverted by MetSO reductases, which are thiol oxidoreductases.

Tyrosine nitration: Tyrosine nitration or 3-nitrotyrosine (3-NT) formation is an irreversible process (Berlett and Stadtman 1997) that is mediated by $ONOO^-$ (Ischiropoulos et al. 1992), and Fe^{3+}EDTA (Beckman et al. 1992). 3-NT has been shown to be a distinguishable sign of $ONOO^-$ formation (Ischiropoulos et al. 1992). However, different mechanisms for 3-NT formation have been shown. For instance, peroxidases, free heme, or metal ions (ferrous, ferric, and cupric ions) catalyze 3-NT (Thomas et al. 2002). It has also been demonstrated that 3-NT formation by concurrent generation of nitric oxide ($NO^•$) and $O_2^{•-}$, is to a less efficient compared to $ONOO^-$ (Pfeiffer and Mayer 1998). Tyrosyl radicals and nitrogen dioxide ($NO_2^•$) produced by either $ONOO^-$ or hemeperoxidase pathways, are combined to form 3-NT (Radi 2004). Oxidants such as carbonate radicals ($CO_3^{•-}$) or oxo-metal complexes and, to a lesser extent, hydroxyl radical ($OH^•$), also promote tyrosyl radical formation. Reaction of tyrosyl radicals with $NO_2^•$ forms 3-NT. Formation of 3-hydroxytyrosine can also occur through $OH^•$ and oxo-metal complexes (Brennan et al. 2002). Alternatively, tyrosyl radicals might react with $NO^•$, forming 3-nitrosotyrosine with subsequent two-electron oxidation to 3-NT by tyrosine iminoxyl radical, which can take place in the presence of transition metals that can readily oxidize 3-NT before the latter can reversibly generate tyrosyl radicals and $NO^•$ (Gunther et al. 1997).

Chlorination

MPO catalyzes the generation of HOCl, which is an extremely toxic oxidant (Harrison and Schultz 1976). Protein oxidation mediated by HOCl has been shown to induce: 1) side-chain modification in amine (lysine and histidine) and sulfur-containing amino acid residues (cysteine and methionine); 2) protein fragmentation produced by nitrogen-centered radicals derived from lysine residues; and 3) crosslinking/aggregation by oxidation of sulfhydryl groups (Knox et al. 1948; Winterbourn 1985; Hawkins and Davies 1998; Pattison and Davies 2001).

Protein Carbonylation

Protein carbonylation occurs through different reactions. Metal ion-catalyzed oxidation (MCO) of proteins is the most common reaction that mediates carbonylation, and involves $OH^•$ generated when Fe^{3+} or Cu^{2+} react with H_2O_2. $OH^•$ affects contiguous amino acids residues, which may be modified to carbonyl-containing derivatives (Berlett and Stadtman 1997; Stadtman and

Berlett 1998). The amino acid residue side chains of histidine, arginine, lysine, and proline are target of MCO (Amici et al. 1989). Glutamic semialdehyde and aminoadipic semialdehyde are the principal carbonyl derivatives of proline/arginine and lysine residues MCO, respectively (Amici et al. 1989; Requena et al. 2001). Additionally, proline oxidation produces hydroxyproline and glutamic acid, while histidine oxidation produces aspartic acid (Amici et al. 1989). Carbonyl derivatives can also be generated by direct interaction with lipid hydroperoxides. Reaction of proteins with reactive carbonyl derivatives (ketoamines, ketoaldehydes, deoxyosones) generated as a result of the reaction of reducing sugars or other oxidation products with the amino group of lysine residues of proteins during glycation and glycoxidation, can also introduce carbonyl groups (Baynes 1991; Kristal and Yu 1992; Dalle-Donne et al. 2003).

Protein Cross-linking Interactions

The protein cross-linking interactions are generated in different ways, for example, between two carbon-centered radicals produced by $OH^•$ (Garrison 1987), oxidation of sulfhydryl groups, dimerization of tyrosine radicals (Pfeiffer et al. 2000), and cross-linking of carbonyls (Grandhee and Monnier 1991; Wells-Knecht et al. 1995; Verzijl et al. 2000) or aldehyde groups with lysine amino groups (Uchida and Stadtman 1993; Friguet et al. 1994).

DNA Oxidation

Oxidative conditions are of significant importance to the carcinogenic process due to their genotoxic effects. DNA is damaged by oxidative stress generated within cells by stressful conditions as well as during physiological oxidation processes (Totter 1980; Ames 1983). 8-hydroxydeoxyguanosine (8-OHdG) and 2, 6-diamino-4-hydroxy-5-formamidopyrimidine (FapydG) are the major types of lesions generated from the damage of guanine by oxidative conditions (Cadet et al. 1997). Considerable attention is given to DNA lesions derived from 2'-deoxyguanosine, which has a highly favorable oxidation potential (Neeley and Essigmann 2006). 8-OHdG is one of the key DNA lesions generated from deoxyguanosine by oxidative stress (Kasai 1997) and is used as an indicator for oxidative stress-induced DNA damage. 8-OHdG is produced by enzymatic cleavage after 8-hydroxylation of the guanine base (Kasai et al. 1986). FapydG lesions are produced by oxidative stress via imidazole ring opening in purines (Dizdaroglu et al. 2008). There are two types of FapydG lesions (β-C-FapydG and carbocyclic FapydG) described, where the ribose is replaced by a pentane ring, and these lesions are used to detect oxidative DNA damage because of their stability (Ober et al. 2003). However, quantification of DNA lesions is very difficult to perform as different research groups have detected variable levels of 8-OHdG in cells from the same tissue samples (Cadet et al. 2002; Collins et al. 2002; Dizdaroglu et al. 2002). 8-OHdG and FapydG are repaired by the same DNA glycosylases. In humans, this type of damage is repaired by hOgg1 (8-Oxoguanine DNA glycosylase) proteins (van der Kemp et al. 1996; Radicella et al. 1997). FapydG are more abundant in cancerous compared to non-cancerous human tissues (Olinski et al. 1992). An example of this are human leukemia cell lines exposed to γ-radiolysis, where FapydG is found at 3 times higher concentrations than 8-OHdG (Pouget et al. 2000). In colon, stomach, ovary, brain and lung, elevated amounts of FapydG and 8-OHdG were found in cancerous tissues compared with their cancer-free surrounding tissues where the frequency of FapydG is higher than 8-OHdG (Olinski et al. 1992).

Lipid Peroxidation

The abundance of polyunsaturated fatty acids (PUFAs) is very high in cellular membranes and lipoproteins (Dekkers et al. 1996). The mechanism of lipid peroxidation takes place when a free-radical transfers its electrons, and initiates an attack to the cell membrane oxidizing membrane lipids (Halliwell and Gutteridge 1985). ROS mainly target the carbon-carbon double bond of PUFAs where a free radical will take a single electron from the hydrogen. Consequently, the double bond on the carbons loses strength and weakens the carbon-hydrogen bond for easy dissociation of the hydrogen. This reaction provides an unpaired electron to the carbon, which becomes a free radical. The unpaired electron tries to stabilize the bond inducing molecular rearrangements that originate a new molecule called a conjugated diene that easily reacts with O_2 to generate a peroxyl radical. The peroxyl radical takes one electron from a lipid in a process called propagation, and the chain reaction starts oxidizing other lipids (Halliwell and Gutteridge 1985).

Lipid peroxidation generates conjugated dienes and lipid hydroperoxides as initial products, with subsequent decomposition into advanced products of oxidation such as alkanes, aldehydes, and isoprostanes (Moore and Roberts 1998). Lipoxidation of PUFAs in cell membranes generates alkanes including ethane and pentane. Alkanes are metabolized to alkyl alcohols by cytochrome P450 (CYP)-mixed oxidase enzymes (Remmer H et al. 1984).

Aldehydes produced during lipid peroxidation are relatively stable and highly reactive, and are divided into three families: 2-alkenals, 4-hydroxy-2-alkenals, and ketoaldehydes (Uchida 2000; Uchida 2003). 2-Alkenals include 2-hexenal, acrolein and crotonaldehyde (Uchida 2003). The largest family of aldehydes generated from lipid peroxidation is constituted by 4-hydroxy-2-alkenals, from which the most abundant member is 4-hydroxy-2-nonenal (4-HNE) (Esterbauer et al. 1991). 4-HNE can be formed from peroxidation of ω-6 fatty acids such as linoleic acid and arachidonic acid (Esterbauer et al. 1991; Schneider et al. 2001). Ketoaldehydes include malondialdehyde (MDA), glyoxal, and 4-oxo-2-nonenal (ONE) (Uchida 2003). The prostaglandin F2-like compounds, F2-isoprostanes (F2-IsoPs), are stable and specific products of lipid peroxidation which are formed *in vivo* by nonenzymatic free radical-induced peroxidation of arachidonic acid (Morrow et al. 1990; L. Jackson Roberts 1999). First, F2-IsoPs are produced esterified on phospholipids and subsequently released in free form by phospholipases (Morrow et al. 1992). Likewise, MCO of PUFAs promotes protein carbonyl formation by Michael addition reactions of lipid oxidation products with either the amino group of lysine, the imidazole group of histidine, or the sulphydryl group of cysteine residues, and also by interactions with alkoxyl radicals obtained by Fe^{2+}-induced cleavage of lipid hydroperoxides. In addition, saturated aldehydes derived from the PUFAs react with lysine residues to form Schiff base adducts (Refsgaard et al. 2000; Grimsrud et al. 2008).

4-HNE is one of the major byproducts of lipid peroxidation. 4-HNE reacts with thiol and amino groups at physiological pH levels (Schaur 2003). There are three functional groups within 4-HNE: the aldehyde group, the C = C double bond and the hydroxyl group. These three groups can participate alone or in sequence, in chemical reactions (Esterbauer et al. 1991). 4-HNE is an electrophilic molecule, it highly reacts with low molecular weight molecules such as GSH, proteins and DNA (Esterbauer et al. 1991; Uchida 2003). 4-HNE induces etheno (ε) DNA adducts, these adducts are formed after peroxidation of arachidonic acid and microsomal membranes (el Ghissassi et al. 1995). 4-HNE also reacts via epoxidation with nucleic acids to induce the formation of DNA adducts such as 1, N^6-ethenodeoxy-adenosine (εdA) (Jones and Dedon 1999), 3, N^4-ethenodeoxycytidine (εdC) and 1, N^2-ethenodeoxyguanosine (εdG) (Chung et al. 1996).

BIOLOGICAL MARKERS (BIOMARKERS) OF OXIDATIVE STRESS IN CARCINOGENESIS

Biomarkers are defined as cellular, biochemical or molecular alterations in biological media such as human tissues, cells, or fluids (Hulka and Wilcosky 1988), that are objectively measured and evaluated as indicators of normal/pathogenic processes or pharmacological responses to therapies (Naylor 2003). Epidemiologists, physicians, and researchers are currently using biomarkers, in the diagnosis and management of pathologies, such as cancer, infections, immunological and genetic disorders (Hulka and Wilcosky 1988; Perera and Weinstein 2000; Naylor 2003). Blood, cerebrospinal fluid, muscle, nerve, skin, and urine samples are employed to obtain information about healthy and diseased states. Several points must be considered before using biomarkers in clinical studies including potential measurement errors, bias, confounding, cost and acceptability (Mayeux 2004; Ioannidis 2011), and their validity is a complex process. Therefore, the identification of potential biomarkers provides a dynamic and outstanding tool to diagnose asymptomatic disease states and treat them accurately (Schulte 1992).

Biomarkers of oxidative modification of biomolecules have been examined in distinct disease states (Giustarini et al. 2009; Ziech et al. 2010). However, the clinical analyses of such biomarkers depend on the sample type (e.g., tissue, plasma, and urine), requirement of sample pretreatment, performance of analytical technique, and method to normalize concentration among different biological samples, which are relevant factors to ensure reliable quantification (Tamura et al. 2006).

Oxidative stress is a common hallmark of a variety of diseases ranging from neurodegenerative disorders, diabetes and cancer. There is a number of studies demonstrating an association with increased oxidative stress and carcinogenesis. However, the detection and/or use of biomarkers of oxidative stress in cancer treatment/diagnosis will only be relevant if the biomarkers detected can give us relevant information regarding the stage/malignancy of the cancer/tumor or the therapeutic outcome and prognosis to be expected. In this section we will only focus on those biomarkers of oxidative stress that have the potential to provide such information.

Biomarkers of Protein Oxidation in Carcinogenesis

Among oxidative stress biomarkers, the protein-derived biomarkers have been identified in higher number, probably because proteins represent the main constituent and functional biomolecule within cells (Dalle-

Donne et al. 2003; Ptolemy et al. 2007). It is noteworthy that not one but several proteins might be target of oxidative modifications. Among the oxidative modifications of proteins, carbonyls are more commonly used as oxidative stress biomarkers because of their stability (Dalle-Donne et al. 2003). Increased levels of protein carbonyls have been associated with cancer (Pignatelli et al. 2001; Rossner et al. 2007). Therefore, and extensive work on redox proteomics related to protein and amino acids modifications have been undertaken in order to assess biomarkers for oxidative stress (Dalle-Donne et al. 2006).

Considerable effort has been made to identify cancer biomarkers. Increased protein carbonyl content has been associated with bladder and breast cancer, Hodgkin's lymphoma, glioblastoma multiforme and transitional meningioma tissues (Yilmaz et al. 2003; Morabito et al. 2004; Tanriverdi et al. 2007; Mannello et al. 2009). Cys-SSG levels were reported to be increased during oral carcinogenesis (Huang et al. 2007) while increased $ONOO^-$-mediated 3-NT of Src kinase has been reported in human pancreatic ductal adenocarcinoma (MacMillan-Crow et al. 2000). Carcinogens such as hexavalent chromium, cigarette smoke, arsenic and UVB radiation induce protein carbonyl accumulation (Kadirvel et al. 2007; Wang et al. 2007; Mattagajasingh et al. 2008; Lin et al. 2009; Colombo et al. 2010; Perluigi et al. 2010). Recently, chrysotile asbestos were shown to induce carbonylation of glucose 6-phosphate dehydrogenase (Ogasawara and Ishii 2010). Cigarette smoke has also been shown to increase RNS resulting in increased plasma 3-NT levels (Pignatelli et al. 2001). A significant increase in red blood cell and plasma concentrations of Cys-SSG proteins was also found in cigarette smokers (Muscat et al. 2004). 3-NT of fibrinogen, transferrin, plasminogen, and ceruloplasmin, and carbonylation of fibrinogen, were identified in human plasma from lung cancer patients (Pignatelli et al. 2001), while increased plasma protein carbonyl levels were recently associated with breast cancer risk (Rossner et al. 2007). Disruption of Cys-SSG levels and increased protein carbonylation in plasma samples have been described as risk factors for colorectal cancer development (Yeh et al. 2010). Interestingly, salivary samples from oral squamous cell carcinoma patients were reported to present high levels of carbonylated proteins, which were concomitant to high RNS formation and decreased content of antioxidants (Bahar et al. 2007).

Biomarkers of Lipid Peroxidation in Carcinogenesis

A common approach to detect RS formation and oxidative stress is to determine the accumulation of stable byproducts. For example, when oxidized, fatty acids are decomposed in aldehydes, alkanes and/or isoprostanes. Byproducts of aldehydes, such as 4-HNE and MDA, are widely accepted as biomarkers of lipid peroxidation (Pellegrini et al. 2005). Under normal conditions, 4-HNE can be found at low concentrations in human tissues and plasma (Esterbauer et al. 1991). In healthy tissues the concentration of 4-HNE is close to 0.07–2.8 μM and an increase in 4-HNE adducts has been reported in early stage breast cancer development (Albright et al. 2005; Steiner et al. 2007) and it was also associated with an increased risk of lung cancer (Vineis and Husgafvel-Pursiainen 2005). High plasma levels of MDA in bone, breast cancer, Burkitt's and large cell lymphoma (Yazdanpanah et al. 1997) patients were found when compared with control groups (Boyd and McGuire 1991). Patients with chronic myeloid leukemia, oral cancer (Ahmad et al. 2008; Chole et al. 2010), poor prognosis of oropharyngeal cancer (Salzman et al. 2009), as well as with fibroadenoma and adenocarcinoma present high levels of MDA in plasma and erythrocytes (Kumaraguruparan et al. 2002). Similarly, high levels of MDA have been found in relapse patients of acute myeloid leukemia (Zhou et al. 2010), in patients with advanced but not localized prostate cancer (Yossepowitch et al. 2007) and in patients with advanced laryngeal cancer (Taysi et al. 2003).

F2-IsoPs are byproducts of lipid peroxidation suitable to be used as oxidative stress biomarkers *in vivo* because of their stability and detectability (Morrow et al. 1990). F2-IsoPs can be found in all human tissues including plasma, urine, bronchoalveolar lavage fluid, cerebrospinal fluid and bile (Halliwell and Gutteridge 1990; Kadiiska et al. 2005a; Kadiiska et al. 2005b; Milne et al. 2007). Patients with prostate cancer have higher levels of F2-IsoPs in urine when compared to control samples supporting the use of F2-IsoPs as biomarkers of oxidative stress in cancer (Barocas et al. 2011).

Biomarkers of DNA Oxidation in Carcinogenesis

8-OHdG is a well known oxidative stress biomarker in tissue or body fluids (Richter et al. 1988; Shigenaga et al. 1989; Fraga et al. 1990). For instance, an animal model treated with $KBrO_3$, a renal carcinogen, showed high levels of 8-OHdG in cellular DNA in the kidneys, but not in other organs (Kasai et al. 1987). Because 8-OHdG is an unstable molecule, it is hydrolyzed to 8-hydroxyguanine (8-OHG), and both 8-OHdG and 8-OHG are released into the bloodstream and secreted in urine. As a biomarker, 8-OHdG has been used to estimate the DNA damage in humans after exposure to carcinogenic agents, such as tobacco smoke, asbestos fibers, heavy metals, and polycyclic aromatic hydrocarbons. In addition, 8-OHdG has also been widely used as a risk factor for cancer (Valavanidis et al. 2009). Increased 8-OHdG levels are associated with acute myeloid leukemia development and relapse (Zhou et al. 2010), as well as with poor prognosis of epithelial ovarian cancer (Pylvas et al. 2011). Urinary 8-OHdG levels are found increased in metastatic lung cancer and primary lung cancer compared to that of benign disease. Furthermore, the urinary 8-OHdG levels in stages II-IV lung cancer patients were found lower than those of patients at stage I of the disease progression (Yano et al. 2009).

CONCLUSIONS AND PERSPECTIVES

This chapter is focused on analyzing the potential of byproducts from oxidative damage as oxidative stress biomarkers in carcinogenesis. Oxidative stress targets biomolecules such as proteins, lipids and nucleic acids mediating the formation of a variety of reactive byproducts. Oxidatively modified proteins, lipids and DNA have been extensively used as biomarkers of oxidative stress in a number of human diseases including neurodegenerative disorders, cancer and diabetes. Thus, the sole detection of oxidative stress/damage in blood/serum, urine, sputum and/or saliva from patients, by itself, is not informative enough to accurately diagnose cancer. However, recent studies have demonstrated that increased levels of oxidized proteins (carbonyls), lipids (4-HNE and MDA) and DNA (8-OHdG) might provide relevant information regarding stage/malignancy of the cancerous tissue or the therapeutic outcome/prognosis to be expected. From all the biomarkers summarized here, oxidative post-translational protein modifications have the potential to be used as more specific biomarkers to detect cancer disorders, diagnose malignant stage, and also identify the best possible therapeutic treatment. Depending on the functional identity of the protein modified and the effect of the post-translational modification in protein function, protein biomarkers could be correlated more accurately with the type of organ affected and the status of the disorder, and may also be used clinically to assess the therapy efficacy. Future research must be focused then on two distinct aspects. First, a more accurate correlation between biomarkers of oxidative stress and cancer malignancy, stage and prognosis; and second, the molecular identification of oxidatively modified proteins and their functional role in the carcinogenic process.

ACKNOWLEDGEMENTS

This work was supported by the National Institutes of Health Grant P20RR17675 Centers of Biomedical Research Excellence (COBRE), the Research Council

Interdisciplinary Grant, and the Life Sciences Grant Program of the University of Nebraska-Lincoln.

REFERENCES

Ahmad, R., A.K. Tripathi, et al. 2008. Malondialdehyde and protein carbonyl as biomarkers for oxidative stress and disease progression in patients with chronic myeloid leukemia. In vivo 22: 525-528.

Ahn, J., M.D. Gammon, et al. 2004. Myeloperoxidase genotype, fruit and vegetable consumption, and breast cancer risk. Cancer Res 64: 7634-7639.

Albright, C.D., E. Klem, et al. 2005. Breast cancer cell-targeted oxidative stress: enhancement of cancer cell uptake of conjugated linoleic acid, activation of p53, and inhibition of proliferation. Exp Mol Pathol 79: 118-125.

Ambrosone, C.B., J. Ahn, et al. 2005. Polymorphisms in genes related to oxidative stress (MPO, MnSOD, CAT) and survival after treatment for breast cancer. Cancer Res 65: 1105-1111.

Ames, B.N. 1983. Dietary carcinogens and anticarcinogens. Oxygen radicals and degenerative diseases. Science 221: 1256-1264.

Amici, A., R.L. Levine, et al. 1989. Conversion of amino acid residues in proteins and amino acid homopolymers to carbonyl derivatives by metal-catalyzed oxidation reactions. J Biol Chem 264: 3341-3346.

Arnold, R.S., J. Shi, et al. 2001. Hydrogen peroxide mediates the cell growth and transformation caused by the mitogenic oxidase Nox1. Proc Natl Acad Sci U S A 98: 5550-5555.

Bahar, G., R. Feinmesser, et al. 2007. Salivary analysis in oral cancer patients: DNA and protein oxidation, reactive nitrogen species, and antioxidant profile. Cancer 109: 54-59.

Barocas, D.A., S. Motley, et al. 2011. Oxidative stress measured by urine F2-isoprostane level is associated with prostate cancer. J Urol 185: 2102-2107.

Baynes, J.W. 1991. Role of oxidative stress in development of complications in diabetes. Diabetes 40: 405-412.

Beckman, J.S., H. Ischiropoulos, et al. 1992. Kinetics of superoxide dismutase- and iron-catalyzed nitration of phenolics by peroxynitrite. Arch Biochem Biophys 298: 438-445.

Beckman, K.B. and B.N. Ames. 1998. The free radical theory of aging matures. Physiol Rev 78: 547-581.

Beer, S.M., E.R. Taylor, et al. 2004. Glutaredoxin 2 catalyzes the reversible oxidation and glutathionylation of mitochondrial membrane thiol proteins: implications for mitochondrial redox regulation and antioxidant DEFENSE. J Biol Chem 279: 47939-47951.

Berlett, B.S. and E.R. Stadtman. 1997. Protein oxidation in aging, disease, and oxidative stress. J Biol Chem 272: 20313-20316.

Biteau, B., J. Labarre, et al. 2003. ATP-dependent reduction of cysteine-sulphinic acid by S. cerevisiae sulphiredoxin. Nature 425: 980-984.

Boschi-Muller, S., S. Azza, et al. 2000. A sulfenic acid enzyme intermediate is involved in the catalytic mechanism of peptide methionine sulfoxide reductase from Escherichia coli. J Biol Chem 275: 35908-35913.

Boyd, N.F. and V. McGuire. 1991. The possible role of lipid peroxidation in breast cancer risk. Free Radic Biol Med 10: 185-190.

Brandon, M., P. Baldi, et al. 2006. Mitochondrial mutations in cancer. Oncogene 25: 4647-4662.

Brennan, M.L., W. Wu, et al. 2002. A tale of two controversies: defining both the role of peroxidases in nitrotyrosine formation in vivo using eosinophil peroxidase and myeloperoxidase-deficient mice, and the nature of peroxidase-generated reactive nitrogen species. J Biol Chem 277: 17415-17427.

Bryk, R., P. Griffin, et al. 2000. Peroxynitrite reductase activity of bacterial peroxiredoxins. Nature 407: 211-215.

Cadet, J., M. Berger, et al. 1997. Oxidative damage to DNA: formation, measurement, and biological significance. Rev Physiol Biochem Pharmacol 131: 1-87.

Cadet, J., T. Douki, et al. 2002. Assessment of oxidative base damage to isolated and cellular DNA by HPLC-MS/MS measurement. Free Radic Biol Med 33: 441-449.

Cascorbi, I., S. Henning, et al. 2000. Substantially reduced risk of cancer of the aerodigestive tract in subjects with variant--463A of the myeloperoxidase gene. Cancer Res 60: 644-649.

Chae, H.Z. and S.G. Rhee. 1994. A thiol-specific antioxidant and sequence homology to various proteins of unknown function. Biofactors 4: 177-180.

Chang, J.W., H.B. Jeon, et al. 2001. Augmented expression of peroxiredoxin I in lung cancer. Biochem Biophys Res Commun 289: 507-512.

Chang, T.S., W. Jeong, et al. 2004. Characterization of mammalian sulfiredoxin and its reactivation of hyperoxidized peroxiredoxin through reduction of cysteine sulfinic acid in the active site to cysteine. J Biol Chem 279: 50994-51001.

Chang, X.Z., D.Q. Li, et al. 2007. Identification of the functional role of peroxiredoxin 6 in the progression of breast cancer. Breast Cancer Res 9: R76.

Chole, R.H., R.N. Patil, et al. 2010. Estimation of serum malondialdehyde in oral cancer and precancer and its association with healthy individuals, gender, alcohol, and tobacco abuse. Journal of cancer research and therapeutics 6: 487-491.

Chung, F.L., H.J.C. Chen, et al. 1996. Lipid peroxidation as a potential endogenous source for the formation of exocyclic DNA adducts. Carcinogenesis 17: 2105-2111.

Cobbs, C.S., D.S. Levi, et al. 1996. Manganese superoxide dismutase expression in human central nervous system tumors. Cancer Res 56: 3192-3195.

Collins, A., C. Gedik, et al. 2002. Comparative analysis of baseline 8-oxo-7, 8-dihydroguanine in mammalian cell DNA, by different methods in different laboratories: an approach to consensus. Carcinogenesis 23: 2129-2133.

Colombo, G., G. Aldini, et al. 2010. Water-Soluble alpha,beta-unsaturated aldehydes of cigarette smoke induce carbonylation of human serum albumin. Antioxid Redox Signal 12: 349-364.

Connor, K.M., N. Hempel, et al. 2007. Manganese superoxide dismutase enhances the invasive and migratory activity of tumor cells. Cancer Res 67: 10260-10267.

Costa, C., R. Soares, et al. 2002. Cyclo-oxygenase 2 expression is associated with angiogenesis and lymph node metastasis in human breast cancer. J Clin Pathol 55: 429-434.

Cross, J.V. and D.J. Templeton. 2004. Thiol oxidation of cell signaling proteins: Controlling an apoptotic equilibrium. J Cell Biochem 93: 104-111.

Dalle-Donne, I., D. Giustarini, et al. 2003. Protein carbonylation in human diseases. Trends Mol Med 9: 169-176.

Dalle-Donne, I., R. Rossi, et al. 2006. Biomarkers of oxidative damage in human disease. Clin Chem 52: 601-623.

Dalle-Donne, I., R. Rossi, et al. 2003. Protein carbonyl groups as biomarkers of oxidative stress. Clin Chim Acta 329: 23-38.

Davies, M.J. 2005. The oxidative environment and protein damage. Biochim Biophys Acta 1703: 93-109.

Dekkers, J.C., L.J. van Doornen, et al. 1996. The role of antioxidant vitamins and enzymes in the prevention of exercise-induced muscle damage. Sports Med 21: 213-238.

Dever, J.T. and A.A. Elfarra. 2006. In vivo metabolism of L-methionine in mice: evidence for stereoselective formation of methionine-d-sulfoxide and quantitation of other major metabolites. Drug Metab Dispos 34: 2036-2043.

Di Ilio, C., P. Sacchetta, et al. 1995. Glutathione peroxidase and glutathione reductase activities in cancerous and non-cancerous human kidney tissues. Cancer Lett 91: 19-23.

Dizdaroglu, M., P. Jaruga, et al. 2002. Free radical-induced damage to DNA: mechanisms and measurement. Free Radic Biol Med 32: 1102-1115.

Dizdaroglu, M., G. Kirkali, et al. 2008. Formamidopyrimidines in DNA: mechanisms of formation, repair, and biological effects. Free Radic Biol Med 45: 1610-1621.

el-Sharabasy, M.M., I. el-Dosoky, et al. 1993. Elevation of glutathione, glutathione-reductase and nucleic acids in both normal tissues and tumour of breast cancer patients. Cancer Lett 72: 11-15.

el Ghissassi, F., A. Barbin, et al. 1995. Formation of 1,N6-ethenoadenine and 3, N4-ethenocytosine by lipid peroxidation products and nucleic acid bases. Chem Res Toxicol 8: 278-283.

Ellis, H.R. and L.B. Poole. 1997. Novel application of 7-chloro-4-nitrobenzo-2-oxa-1,3-diazole to identify cysteine sulfenic acid in the AhpC component of alkyl hydroperoxide reductase. Biochemistry 36: 15013-15018.

Ergen, H.A., F. Narter, et al. 2007. Effects of manganase superoxide dismutase Ala-9Val polymorphism on prostate cancer: a case-control study. Anticancer Res 27: 1227-1230.

Esterbauer, H., R.J. Schaur, et al. 1991. Chemistry and biochemistry of 4-hydroxynonenal, malonaldehyde and related aldehydes. Free radical biology & medicine 11: 81-128.

Esterbauer, H., R.J. Schaur, et al. 1991. Chemistry and biochemistry of 4-hydroxynonenal, malonaldehyde and related aldehydes. Free Radic Biol Med 11: 81-128.

Estrela, J.M., A. Ortega, et al. 2006. Glutathione in cancer biology and therapy. Crit Rev Clin Lab Sci 43: 143-181.

Esworthy, R.S., M.A. Baker, et al. 1995. Expression of selenium-dependent glutathione peroxidase in human breast tumor cell lines. Cancer Res 55: 957-962.

Fraga, C.G., M.K. Shigenaga, et al. 1990. Oxidative damage to DNA during aging: 8-hydroxy-2'-deoxyguanosine in rat organ DNA and urine. Proc Natl Acad Sci USA 87: 4533-4537.

Friguet, B., E.R. Stadtman, et al. 1994. Modification of glucose-6-phosphate dehydrogenase by 4-hydroxy-2-nonenal. Formation of cross-linked protein that inhibits the multicatalytic protease. J Biol Chem 269: 21639-21643.

Fukumura, D., S. Kashiwagi, et al. 2006. The role of nitric oxide in tumour progression. Nat Rev Cancer 6: 521-534.

Garrison, W.M. 1987. Reaction mechanisms in radiolysis of peptides, polypeptides, and proteins. Chem Rev 87: 381-398.

Ghezzi, P. 2005. Regulation of protein function by glutathionylation. Free Radic Res 39: 573-580.

Giustarini, D., I. Dalle-Donne, et al. 2009. Oxidative stress and human diseases: Origin, link, measurement, mechanisms, and biomarkers. Crit Rev Clin Lab Sci 46: 241-281.

Grandhee, S.K. and V.M. Monnier. 1991. Mechanism of formation of the Maillard protein cross-link pentosidine. Glucose, fructose, and ascorbate as pentosidine precursors. J Biol Chem 266: 11649-11653.

Grimsrud, P.A., H. Xie, et al. 2008. Oxidative stress and covalent modification of protein with bioactive aldehydes. J Biol Chem 283: 21837-21841.

Gunther, M.R., L.C. Hsi, et al. 1997. Nitric oxide trapping of the tyrosyl radical of prostaglandin H synthase-2 leads to tyrosine iminoxyl radical and nitrotyrosine formation. J Biol Chem 272: 17086-17090.

Guzy, R.D., B. Sharma, et al. 2008. Loss of the SdhB, but Not the SdhA, subunit of complex II triggers reactive oxygen species-dependent hypoxia-inducible factor activation and tumorigenesis. Mol Cell Biol 28: 718-731.

Halliwell, B. and J.M. Gutteridge. 1990. Role of free radicals and catalytic metal ions in human disease: an overview. Methods Enzymol 186: 1-85.

Halliwell, B. and J.M.C. Gutteridge. 1985. Free radicals in biology and medicine. Oxford New York, Clarendon Press; Oxford University Press.

Hao, G., B. Derakhshan, et al. 2006. SNOSID, a proteomic method for identification of cysteine S-nitrosylation sites in complex protein mixtures. Proc Natl Acad Sci USA 103: 1012-1017.

Harrison, J.E. and J. Schultz. 1976. Studies on the chlorinating activity of myeloperoxidase. J Biol Chem 251: 1371-1374.

Hasegawa, Y., T. Takano, et al. 2002. Decreased expression of glutathione peroxidase mRNA in thyroid anaplastic carcinoma. Cancer Lett 182: 69-74.

Hawkins, C.L. and M.J. Davies. 1998. Hypochlorite-induced damage to proteins: formation of nitrogen-centred radicals from lysine residues and their role in protein fragmentation. Biochem J 332 (Pt 3): 617-625.

Hess, D.T., A. Matsumoto, et al. 2005. Protein S-nitrosylation: purview and parameters. Nat Rev Mol Cell Biol 6: 150-166.

Hirose, K., D.L. Longo, et al. 1993. Overexpression of mitochondrial manganese superoxide dismutase promotes the survival of tumor cells exposed to interleukin-1, tumor necrosis factor, selected anticancer drugs, and ionizing radiation. FASEB J 7: 361-368.

Hirotsu, S., Y. Abe, et al. 1999. Crystal structure of a multifunctional 2-Cys peroxiredoxin heme-binding protein 23 kDa/proliferation-associated gene product. Proc Natl Acad Sci U S A 96: 12333-12338.

Hoshi, T. and S. Heinemann. 2001. Regulation of cell function by methionine oxidation and reduction. J Physiol 531: 1-11.

Huang, P., L. Feng, et al. 2000. Superoxide dismutase as a target for the selective killing of cancer cells. Nature 407: 390-395.

Huang, Z., D. Komninou, et al. 2007. Enhanced levels of glutathione and protein glutathiolation in rat tongue epithelium during 4-NQO-induced carcinogenesis. Int J Cancer 120: 1396-1401.

Hulka, B.S. and T. Wilcosky. 1988. Biological markers in epidemiologic research. Arch Environ Health 43: 83-89.

Hur, G.C., S.J. Cho, et al. 2003. Manganese superoxide dismutase expression correlates with chemosensitivity in human gastric cancer cell lines. Clin Cancer Res 9: 5768-5775.

Ioannidis, J.P. and O.A. Panagiotou. 2011. Comparison of effect sizes associated with biomarkers reported in highly cited individual articles and in subsequent meta-analyses. JAMA 305: 2200-2210.

Ischiropoulos, H., L. Zhu, et al. 1992. Peroxynitrite-mediated tyrosine nitration catalyzed by superoxide dismutase. Arch Biochem Biophys 298: 431-437.

Ishii, N., T. Ishii, et al. 2007. The role of the electron transport SDHC gene on lifespan and cancer. Mitochondrion 7: 24-28.

Jarvela, S., I. Rantala, et al. 2010. Specific expression profile and prognostic significance of peroxiredoxins in grade II-IV astrocytic brain tumors. BMC Cancer 10: 104.

Jones, W.R. and P.C. Dedon. 1999. DNA oxidation as a source of endogenous electrophiles: Formation of ethenoadenine adducts in gamma-irradiated DNA. Journal of the American Chemical Society 121: 9231-9232.

Kadiiska, M.B., B.C. Gladen, et al. 2005a. Biomarkers of oxidative stress study II: are oxidation products of lipids, proteins, and DNA markers of CCl4 poisoning? Free Radic Biol Med 38: 698-710.

Kadiiska, M.B., B.C. Gladen, et al. 2005b. Biomarkers of oxidative stress study III. Effects of the nonsteroidal anti-inflammatory agents indomethacin and meclofenamic acid on measurements of oxidative products of lipids in CCl4 poisoning. Free Radic Biol Med 38: 711-718.

Kadirvel, R., K. Sundaram, et al. 2007. Supplementation of ascorbic acid and alpha-tocopherol prevents arsenic-induced protein oxidation and DNA damage induced by arsenic in rats. Hum Exp Toxicol 26: 939-946.

Karihtala, P., A. Mantyniemi, et al. 2003. Peroxiredoxins in breast carcinoma. Clin Cancer Res 9: 3418-3424.

Kasai, H. 1997. Analysis of a form of oxidative DNA damage, 8-hydroxy-2'-deoxyguanosine, as a marker of cellular oxidative stress during carcinogenesis. Mutation Research-Reviews in Mutation Research **387**: 147-163.

Kasai, H., P.F. Crain, et al. 1986. Formation of 8-hydroxyguanine moiety in cellular DNA by agents producing oxygen radicals and evidence for its repair. Carcinogenesis 7: 1849-1851.

Kasai, H., S. Nishimura, et al. 1987. Oral administration of the renal carcinogen, potassium bromate, specifically produces 8-hydroxydeoxyguanosine in rat target organ DNA. Carcinogenesis 8: 1959-1961.

Kim, J.H., P.N. Bogner, et al. 2008. Up-regulation of peroxiredoxin 1 in lung cancer and its implication as a prognostic and therapeutic target. Clin Cancer Res 14: 2326-2333.

Kim, K., M. Yu, et al. 2009. Expression of human peroxiredoxin isoforms in response to cervical carcinogenesis. Oncol Rep 21: 1391-1396.

Kinnula, V.L., S. Lehtonen, et al. 2002. Overexpression of peroxiredoxins I, II, III, V, and VI in malignant mesothelioma. J Pathol 196: 316-323.

Kinnula, V.L., T. Torkkeli, et al. 2004. Ultrastructural and chromosomal studies on manganese superoxide dismutase in malignant mesothelioma. Am J Respir Cell Mol Biol 31: 147-153.

Klatt, P. and S. Lamas. 2000. Regulation of protein function by S-glutathiolation in response to oxidative and nitrosative stress. Eur J Biochem 267: 4928-4944.

Klaunig, J.E., L.M. Kamendulis, et al. 2010. Oxidative stress and oxidative damage in carcinogenesis. Toxicol Pathol 38: 96-109.

Knox, W.E., P.K. Stumpf, et al. 1948. The inhibition of sulfhydryl enzymes as the basis of the bactericidal action of chlorine. J Bacteriol 55: 451-458.

Kristal, B.S. and B.P. Yu. 1992. An emerging hypothesis: synergistic induction of aging by free radicals and Maillard reactions. J Gerontol 47: B107-114.

Kryston, T.B., A.B. Georgiev, et al. 2011. Role of oxidative stress and DNA damage in human carcinogenesis. Mutat Res 711: 193-201.

Kumaraguruparan, R., R. Subapriya, et al. 2002. Antioxidant profile in the circulation of patients with fibroadenoma and adenocarcinoma of the breast. Clin Biochem 35: 275-279.

Kuninaka, S., Y. Ichinose, et al. 2000. Suppression of manganese superoxide dismutase augments sensitivity to radiation, hyperthermia and doxorubicin in colon cancer cell lines by inducing apoptosis. Br J Cancer 83: 928-934.

Lahdenranta, J., J. Hagendoorn, et al. 2009. Endothelial nitric oxide synthase mediates lymphangiogenesis and lymphatic metastasis. Cancer Res 69: 2801-2808.

Laurent, A., C. Nicco, et al. 2005. Controlling tumor growth by modulating endogenous production of reactive oxygen species. Cancer Res 65: 948-956.

Levine, R.L., L. Mosoni, et al. 1996. Methionine residues as endogenous antioxidants in proteins. Proc Natl Acad Sci U S A 93: 15036-15040.

Lin, C.C., T.H. Su, et al. 2009. Protein carbonylation in THP-1 cells induced by cigarette smoke extract via a copper-catalyzed pathway. Chem Res Toxicol 22: 1232-1238.

MacMillan-Crow, L.A., J.S. Greendorfer, et al. 2000. Tyrosine nitration of c-SRC tyrosine kinase in human pancreatic ductal adenocarcinoma. Arch Biochem Biophys 377: 350-356.

Mannello, F., G.A. Tonti, et al. 2009. Protein oxidation in breast microenvironment: Nipple aspirate fluid collected from breast cancer women contains increased protein carbonyl concentration. Cell Oncol 31: 383-392.

Mantymaa, P., T. Siitonen, et al. 2000. Induction of mitochondrial manganese superoxide dismutase confers resistance to apoptosis in acute myeloblastic leukaemia cells exposed to etoposide. Br J Haematol 108: 574-581.

Mattagajasingh, S.N., B.R. Misra, et al. 2008. Carcinogenic chromium(VI)-induced protein oxidation and lipid peroxidation: implications in DNA-protein crosslinking. J Appl Toxicol 28: 987-997.

Mayeux, R. 2004. Biomarkers: potential uses and limitations. NeuroRx 1: 182-188.

Menter, D.G., R.L. Schilsky, et al. 2010. Cyclooxygenase-2 and cancer treatment: understanding the risk should be worth the reward. Clin Cancer Res 16: 1384-1390.

Mieyal, J.J., M.M. Gallogly, et al. 2008. Molecular mechanisms and clinical implications of reversible protein S-glutathionylation. Antioxid Redox Signal 10: 1941-1988.

Milne, G.L., S.C. Sanchez, et al. 2007. Quantification of F2-isoprostanes as a biomarker of oxidative stress. Nat Protoc 2: 221-226.

Moore, K. and L.J. Roberts, 2nd 1998. Measurement of lipid peroxidation. Free Radic Res 28: 659-671.

Morabito, F., M. Cristani, et al. 2004. Lipid peroxidation and protein oxidation in patients affected by Hodgkin's lymphoma. Mediators Inflamm 13: 381-383.

Morrow, J.D., J.A. Awad, et al. 1992. Non-cyclooxygenase-derived prostanoids (F2-isoprostanes) are formed in situ on phospholipids. Proc Natl Acad Sci U S A 89: 10721-10725.

Morrow, J.D., K.E. Hill, et al. 1990. A series of prostaglandin F2-like compounds are produced in vivo in humans by a non-cyclooxygenase, free radical-catalyzed mechanism. Proc Natl Acad Sci U S A 87: 9383-9387.

Muscat, J.E., W. Kleinman, et al. 2004. Enhanced protein glutathiolation and oxidative stress in cigarette smokers. Free Radic Biol Med 36: 464-470.

Narko, K., A. Ristimaki, et al. 1997. Tumorigenic transformation of immortalized ECV endothelial cells by cyclooxygenase-1 overexpression. J Biol Chem 272: 21455-21460.

Navarro, J., E. Obrador, et al. 1999. Changes in glutathione status and the antioxidant system in blood and in cancer cells associate with tumour growth in vivo. Free Radic Biol Med 26: 410-418.

Naylor, S. 2003. Biomarkers: current perspectives and future prospects. Expert Rev Mol Diagn 3: 525-529.

Neeley, W.L. and J.M. Essigmann. 2006. Mechanisms of formation, genotoxicity, and mutation of guanine oxidation products. Chem Res Toxicol 19: 491-505.

Nelson, K.K., A.C. Ranganathan, et al. 2003. Elevated sod2 activity augments matrix metalloproteinase expression: evidence for the involvement of endogenous hydrogen peroxide in regulating metastasis. Clin Cancer Res 9: 424-432.

Noh, D.Y., S.J. Ahn, et al. 2001. Overexpression of peroxiredoxin in human breast cancer. Anticancer Res 21: 2085-2090.

Ober, M., U. Linne, et al. 2003. The two main DNA lesions 8-Oxo-7,8-dihydroguanine and 2,6-diamino-5-formamido-4-hydroxypyrimidine exhibit strongly different pairing properties. Angew Chem Int Ed Engl 42: 4947-4951.

Ogasawara, Y. and K. Ishii. 2010. Exposure to chrysotile asbestos causes carbonylation of glucose 6-phosphate dehydrogenase through a reaction with lipid peroxidation products in human lung epithelial cells. Toxicol Lett 195: 1-8.

Olinski, R., T. Zastawny, et al. 1992. DNA base modifications in chromatin of human cancerous tissues. FEBS Lett 309: 193-198.

Olson, S.Y. and H.J. Garban. 2008. Regulation of apoptosis-related genes by nitric oxide in cancer. Nitric Oxide 19: 170-176.

Oltra, A.M., F. Carbonell, et al. 2001. Antioxidant enzyme activities and the production of MDA and 8-oxo-dG in chronic lymphocytic leukemia. Free Radic Biol Med 30: 1286-1292.

Pattison, D.I. and M.J. Davies. 2001. Absolute rate constants for the reaction of hypochlorous acid with protein side chains and peptide bonds. Chem Res Toxicol 14: 1453-1464.

Pellegrini, N., D. Del Rio, et al. 2005. A review of recent studies on malondialdehyde as toxic molecule and biological marker of oxidative stress. Nutrition Metabolism and Cardiovascular Diseases 15: 316-328.

Perera, F.P. and I.B. Weinstein. 2000. Molecular epidemiology: recent advances and future directions. Carcinogenesis 21: 517-524.

Perluigi, M., F. Di Domenico, et al. 2010. Effects of UVB-induced oxidative stress on protein expression and specific protein oxidation in normal human epithelial keratinocytes: a proteomic approach. Proteome Sci 8: 13.

Pfeiffer, S. and B. Mayer. 1998. Lack of tyrosine nitration by peroxynitrite generated at physiological pH. J Biol Chem 273: 27280-27285.

Pfeiffer, S., K. Schmidt, et al. 2000. Dityrosine formation outcompetes tyrosine nitration at low steady-state concentrations of peroxynitrite. Implications for tyrosine modification by nitric oxide/superoxide in vivo. J Biol Chem 275: 6346-6352.

Pignatelli, B., C.Q. Li, et al. 2001. Nitrated and oxidized plasma proteins in smokers and lung cancer patients. Cancer Res 61: 778-784.

Poole, L.B., P.A. Karplus, et al. 2004. Protein sulfenic acids in redox signaling. Annu Rev Pharmacol Toxicol 44: 325-347.

Pouget, J.P., T. Douki, et al. 2000. DNA damage induced in cells by gamma and UVA radiation as measured by HPLC/GC-MS and HPLC-EC and Comet assay. Chem Res Toxicol 13: 541-549.

Ptolemy, A.S., R. Lee, et al. 2007. Strategies for comprehensive analysis of amino acid biomarkers of oxidative stress. Amino Acids 33: 3-18.

Pylvas, M., U. Puistola, et al. 2011. Elevated serum 8-OHdG is associated with poor prognosis in epithelial ovarian cancer. Anticancer Res 31: 1411-1415.

Qiu, H., F.W. Orr, et al. 2003. Arrest of B16 melanoma cells in the mouse pulmonary microcirculation induces endothelial nitric oxide synthase-dependent nitric oxide release that is cytotoxic to the tumor cells. Am J Pathol 162: 403-412.

Rabilloud, T., M. Heller, et al. 2002. Proteomics analysis of cellular response to oxidative stress. Evidence for in vivo overoxidation of peroxiredoxins at their active site. J Biol Chem 277: 19396-19401.

Radi, R. 2004. Nitric oxide, oxidants, and protein tyrosine nitration. Proc Natl Acad Sci U S A 101: 4003-4008.

Radicella, J.P., C. Dherin, et al. 1997. Cloning and characterization of hOGG1, a human homolog of the OGG1 gene of Saccharomyces cerevisiae. Proc Natl Acad Sci U S A 94: 8010-8015.

Raftery, M.J., Z. Yang, et al. 2001. Novel intra- and inter-molecular sulfinamide bonds in S100A8 produced by hypochlorite oxidation. J Biol Chem 276: 33393-33401.

Refsgaard, H.H., L. Tsai, et al. 2000. Modifications of proteins by polyunsaturated fatty acid peroxidation products. Proc Natl Acad Sci U S A 97: 611-616.

Remmer H., H.T., H Frańk and M. Müh-Zange. 1984. Cytochrome P-450 oxidation of alkanes originating as scission products during lipid peroxidation. Xenobiotica 14: 207-219.

Requena, J.R., C.C. Chao, et al. 2001. Glutamic and aminoadipic semialdehydes are the main carbonyl products of metal-catalyzed oxidation of proteins. Proc Natl Acad Sci U S A 98: 69-74.

Richter, C., J.W. Park, et al. 1988. Normal oxidative damage to mitochondrial and nuclear DNA is extensive. Proc Natl Acad Sci U S A 85: 6465-6467.

Ridnour, L.A., D.D. Thomas, et al. 2006. The biphasic nature of nitric oxide responses in tumor biology. Antioxid Redox Signal 8: 1329-1337.

Ripp, S.L., K. Itagaki, et al. 1999. Methionine S-oxidation in human and rabbit liver microsomes: evidence for a high-affinity methionine S-oxidase activity that is distinct from flavin-containing monooxygenase 3. Arch Biochem Biophys 367: 322-332.

Roberts, L.J. and J.D. Morrow. 2000. Measurement of F2-Isoprostanes as an index of oxidative stress in vivo. Free Radical Biology & Medicine 28: 505-513.

Roshchupkina, G.I., A.A. Bobko, et al. 2008. In vivo EPR measurement of glutathione in tumor-bearing mice using improved disulfide biradical probe. Free Radic Biol Med.

Rossner, P., Jr., M.B. Terry, et al. 2007. Plasma protein carbonyl levels and breast cancer risk. J Cell Mol Med 11: 1138-1148.

Salzman, R., L. Pacal, et al. 2009. Elevated malondialdehyde correlates with the extent of primary tumor and predicts poor prognosis of oropharyngeal cancer. Anticancer research 29: 4227-4231.

Schaur, R.J. 2003. Basic aspects of the biochemical reactivity of 4-hydroxynonenal. Mol Aspects Med 24: 149-159.

Schneider, C., K.A. Tallman, et al. 2001. Two distinct pathways of formation of 4-hydroxynonenal. Mechanisms of nonenzymatic transformation of the 9- and 13-hydroperoxides of linoleic acid to 4-hydroxyalkenals. J Biol Chem 276: 20831-20838.

Schroder, E., J.A. Littlechild, et al. 2000. Crystal structure of decameric 2-Cys peroxiredoxin from human erythrocytes at 1.7 A resolution. Structure 8: 605-615.

Schulte, P.A. 1992. Biomarkers in epidemiology: scientific issues and ethical implications. Environ Health Perspect 98: 143-147.

Sharov, V. S., D.A. Ferrington, et al. 1999. Diastereoselective reduction of protein-bound methionine sulfoxide by methionine sulfoxide reductase. FEBS Lett 455: 247-250.

Shigenaga, M.K., C.J. Gimeno, et al. 1989. Urinary 8-hydroxy-2'-deoxyguanosine as a biological marker of in vivo oxidative DNA damage. Proc Natl Acad Sci U S A 86: 9697-9701.

Slane, B.G., N. Aykin-Burns, et al. 2006. Mutation of succinate dehydrogenase subunit C results in increased O2.–, oxidative stress, and genomic instability. Cancer Res 66: 7615-7620.

Stadtman, E.R. and B.S. Berlett. 1998. Reactive oxygen-mediated protein oxidation in aging and disease. Drug Metab Rev 30: 225-243.

Stadtman, E.R. and R.L. Levine. 2003. Free radical-mediated oxidation of free amino acids and amino acid residues in proteins. Amino Acids 25: 207-218.

Steiner, C., W.H. Peters, et al. 2007. Genistein protects human mammary epithelial cells from benzo(a)pyrene-7,8-dihydrodiol-9,10-epoxide and 4-hydroxy-2-nonenal genotoxicity by modulating the glutathione/glutathione S-transferase system. Carcinogenesis 28: 738-748.

Sudarshan, S., C. Sourbier, et al. 2009. Fumarate hydratase deficiency in renal cancer induces glycolytic addiction and hypoxia-inducible transcription factor 1alpha stabilization by glucose-dependent generation of reactive oxygen species. Mol Cell Biol 29: 4080-4090.

Suresh, A., L. Guedez, et al. 2003. Overexpression of manganese superoxide dismutase promotes survival in cell lines after doxorubicin treatment. Br J Haematol 120: 457-463.

Szatrowski, T.P. and C.F. Nathan. 1991. Production of large amounts of hydrogen peroxide by human tumor cells. Cancer Res 51: 794-798.

Szeto, S.S., S.N. Reinke, et al. 2007. Ubiquinone-binding site mutations in the Saccharomyces cerevisiae succinate dehydrogenase generate superoxide and lead to the accumulation of succinate. J Biol Chem 282: 27518-27526.

Tamura, S., H. Tsukahara, et al. 2006. Evaluation of a urinary multi-parameter biomarker set for oxidative stress in children, adolescents and young adults. Free Radic Res 40: 1198-1205.

Tanriover, N., M.O. Ulu, et al. 2008. Neuronal nitric oxide synthase expression in glial tumors: correlation with malignancy and tumor proliferation. Neurol Res 30: 940-944.

Tanriverdi, T., H. Hanimoglu, et al. 2007. Glutathione peroxidase, glutathione reductase and protein oxidation in patients with glioblastoma multiforme and transitional meningioma. J Cancer Res Clin Oncol 133: 627-633.

Taysi, S., C. Uslu, et al. 2003. Malondialdehyde and nitric oxide levels in the plasma of patients with advanced laryngeal cancer. Surg Today 33: 651-654.

Thomas, D.D., M.G. Espey, et al. 2002. Protein nitration is mediated by heme and free metals through Fenton-type chemistry: an alternative to the NO/O2- reaction. Proc Natl Acad Sci U S A 99: 12691-12696.

Totter, J.R. 1980. Spontaneous cancer and its possible relationship to oxygen metabolism. Proc Natl Acad Sci U S A 77: 1763-1767.

Uchida, K. 2000. Role of reactive aldehyde in cardiovascular diseases. Free Radic Biol Med 28: 1685-1696.

Uchida, K. 2003. 4-Hydroxy-2-nonenal: a product and mediator of oxidative stress. Progress in lipid research 42: 318-343.

Uchida, K., Y. Kato, et al. 1990. A novel mechanism for oxidative cleavage of prolyl peptides induced by the hydroxyl radical. Biochem Biophys Res Commun 169: 265-271.

Uchida, K. and E.R. Stadtman. 1993. Covalent attachment of 4-hydroxynonenal to glyceraldehyde-3-phosphate dehydrogenase. A possible involvement of intra- and intermolecular cross-linking reaction. J Biol Chem 268: 6388-6393.

Valavanidis, A., T. Vlachogianni, et al. 2009. 8-hydroxy-2' -deoxyguanosine (8-OHdG): A critical biomarker of oxidative stress and carcinogenesis. J Environ Sci Health C Environ Carcinog Ecotoxicol Rev 27: 120-139.

van der Kemp, P.A., D. Thomas, et al. 1996. Cloning and expression in Escherichia coli of the OGG1 gene of Saccharomyces cerevisiae, which codes for a DNA glycosylase that excises 7,8-dihydro-8-oxoguanine and 2,6-diamino-4-hydroxy-5-N-methylformamidopyrimidine. Proc Natl Acad Sci U S A 93: 5197-5202.

Verzijl, N., J. DeGroot, et al. 2000. Age-related accumulation of Maillard reaction products in human articular cartilage collagen. Biochem J 350 Pt 2: 381-387.

Vineis, P. and K. Husgafvel-Pursiainen. 2005. Air pollution and cancer: biomarker studies in human populations. Carcinogenesis 26: 1846-1855.

Wang, L., G.G. Shi, et al. 2005. Expression of endothelial nitric oxide synthase correlates with the angiogenic phenotype of and predicts poor prognosis in human gastric cancer. Gastric Cancer 8: 18-28.

Wang, L.I., D.P. Miller, et al. 2001. Manganese superoxide dismutase alanine-to-valine polymorphism at codon 16 and lung cancer risk. J Natl Cancer Inst 93: 1818-1821.

Wang, T.C., K.Y. Jan, et al. 2007. Trivalent arsenicals induce lipid peroxidation, protein carbonylation, and oxidative DNA damage in human urothelial cells. Mutat Res 615: 75-86.

Wells-Knecht, K.J., D.V. Zyzak, et al. 1995. Mechanism of autoxidative glycosylation: identification of glyoxal and arabinose as intermediates in the autoxidative modification of proteins by glucose. Biochemistry 34: 3702-3709.

Wheatley-Price, P., K. Asomaning, et al. 2008. Myeloperoxidase and superoxide dismutase polymorphisms are associated with an increased risk of developing pancreatic adenocarcinoma. Cancer 112: 1037-1042.

Winterbourn, C.C. 1985. Comparative reactivities of various biological compounds with myeloperoxidase-hydrogen peroxide-chloride, and similarity of the oxidant to hypochlorite. Biochim Biophys Acta 840: 204-210.

Woo, H.A., W. Jeong, et al. 2005. Reduction of cysteine sulfinic acid by sulfiredoxin is specific to 2-cys peroxiredoxins. J Biol Chem 280: 3125-3128.

Xu, Y., F. Fang, et al. 2008. Mutations in the SOD2 promoter reveal a molecular basis for an activating protein 2-dependent dysregulation of manganese superoxide dismutase expression in cancer cells. Mol Cancer Res 6: 1881-1893.

Yanagawa, T., T. Ishikawa, et al. 1999. Peroxiredoxin I expression in human thyroid tumors. Cancer Lett 145: 127-132.

Yanagawa, T., S. Iwasa, et al. 2000. Peroxiredoxin I expression in oral cancer: a potential new tumor marker. Cancer Lett 156: 27-35.

Yano, T., F. Shoji, et al. 2009. Significance of the urinary 8-OHdG level as an oxidative stress marker in lung cancer patients. Lung Cancer 63: 111-114.

Yazdanpanah, M., X. Luo, et al. 1997. Cytotoxic aldehydes as possible markers for childhood cancer. Free Radic Biol Med 23: 870-878.

Yeh, C.C., C.Y. Lai, et al. 2010. Protein carbonyl levels, glutathione S-transferase polymorphisms and risk of colorectal cancer. Carcinogenesis 31: 228-233.

Yeung, B.H., K.Y. Wong, et al. 2008. Chemosensitisation by manganese superoxide dismutase inhibition is caspase-9 dependent and involves extracellular signal-regulated kinase 1/2. Br J Cancer 99: 283-293.

Yilmaz, I.A., T. Akcay, et al. 2003. Relation between bladder cancer and protein oxidation. Int Urol Nephrol 35: 345-350.

Yossepowitch, O., I. Pinchuk, et al. 2007. Advanced but not localized prostate cancer is associated with increased oxidative stress. J Urol 178: 1238-1243; discussion 1243-1234.

Zhang, B., Y. Wang, et al. 2009. Peroxiredoxins, a novel target in cancer radiotherapy. Cancer Lett 286: 154-160.

Zhang, H.J., T. Yan, et al. 1999. Comparison of effects of two polymorphic variants of manganese superoxide dismutase on human breast MCF-7 cancer cell phenotype. Cancer Res 59: 6276-6283.

Zhou, F.L., W.G. Zhang, et al. 2010. Involvement of oxidative stress in the relapse of acute myeloid leukemia. J Biol Chem 285: 15010-15015.

Ziech, D., R. Franco, et al. 2010. The role of reactive oxygen species and oxidative stress in environmental carcinogenesis and biomarker development. Chem Biol Interact 188: 334-339.

CHAPTER 3

β-Catenin: A Novel Biomarker and Therapeutic Target in Liver Cancer

Anupam Bishayee
Department of Pharmaceutical Sciences,
School of Pharmacy, American University of Health Sciences,
1600 East Hill Street, Signal Hill, CA 90755, USA
E-mail: abishayee@auhs.edu

ABSTRACT

Primary liver cancer, predominantly hepatocellular carcinoma (HCC), is the fifth most common cancer in the world with about one million new cases diagnosed every year with almost an equal number of deaths. The incidence and associated death rates of HCC has been increasing steadily in the United States. Chronic metabolic and alcoholic liver diseases, viral hepatitis (B and C) as well as dietary carcinogens, including aflatoxins and nitrosamines, significantly contribute to the pathogenesis of HCC. Sustained abnormal proliferation in the background of continued inflammation and oxidative stress represents the primary driving force for HCC. Identification of cellular pathways necessary for the proliferation and survival of tumor cells in HCC provides a valuable tool in designing effective prevention and intervention of this disease. The canonical Wnt/β-catenin signaling cascade represents one such relevant pathway. Accumulating evidence based on experimental as well as clinical observations implicates the activated Wnt/β-catenin signaling at various stages of hepatocellular neoplasia, making it an attractive and viable therapeutic target. This chapter reviews the current knowledge on β-catenin, a pivotal player in the Wnt/β-catenin pathway, as a potential biomarker in liver cancer. It also explores the possibility of targeting β-catenin-mediated signaling using a variety of synthetic and naturally occurring agents to achieve effective therapy as well as prevention of hepatic malignancy. Additionally, potential challenges associated with the inhibition of β-catenin as well as future avenues of research to enhance the translational and clinical impact of interfering with Wnt/β-catenin signaling pathway in HCC are presented and critically discussed.

Keywords: Biomarker, β-catenin, chemoprevention, liver cancer, therapy, Wnt signaling.

INTRODUCTION

Liver cancer is one of the most prevalent and life-threatening human cancers in the world. It ranks as the fifth most common cancer and the second leading cause of cancer-related death in men. In women, it represents the seventh most frequent cancer and sixth leading cause of cancer mortality (Jemal et al. 2011). Liver cancer incidence rates continue to increase in several low-risk regions of the world, including developed countries of Western Europe, North America and Oceania, while they are declining in certain highest-risk countries in Asia (Center and Jemal 2011). Hepatocellular carcinomas (HCC), malignant tumors arising from hepatic parenchymal cells, represent the majority (75-90%) of the primary liver cancers. Other less frequent forms of liver cancer include cholangiocarcinoma (originated from the ductal epithelium of the intrahepatic bile duct), hepatoblastoma (derived from immature hepatic precursor cells), and hepatocellular adenomas (benign tumors). The incidence of HCC has dramatically increased in the United States by more than 70% during the past 25 years (El-Serag 2004), with approximately 24,000 new cases and about 19,000 deaths expected to occur in 2010 alone (Jemal et al. 2010). Major risk factors for HCC include chronic infection with hepatitis B virus (HBV) and HCV; however, alcoholic cirrhosis, dietary carcinogens including aflatoxins and nitrosamines, non-alcoholic steatohepatitis and other metabolic liver diseases, obesity, diabetes and smoking also contribute to the development of HCC (Bartsch and Montesano 1984; Perz et al. 2006; Schütte et al. 2009; Blonski et al. 2010; Severi et al. 2010).

Lack of effective diagnostic tools for early detection coupled with limited treatment options available to patients with advanced stages of HCC contributes to a dismal prognosis and high mortality rate for this disease. Surgical resection remains the treatment of choice for patients with well-preserved hepatic functions. Nevertheless, it involves a high risk of post-operative complications as well as tumor recurrence. Liver transplantation is only a curative option provided the tumor is detected early, but this option is of limited value due to inadequate number of qualified donors as well as occurrence of the disease in the transplanted liver. Although a number of other strategies, including radio-frequency ablation, microwave ablation, trans-catheter arterial chemoembolization, cryoablation percutaneous ethanol injection, and yttrium-90 microspheres, are available for inoperable patients, difficulty in the management of patients and tumor recurrence are important limitations for these treatments. Currently, sorafenib (Nexavar^{TM}, Bayer) is the only drug approved by the United States Food and Drug Administration for the treatment of advanced HCC. However, only moderate improvement of survival, severe adverse side effects, and high costs call for other therapeutic as well as preventive approaches (Je et al. 2009; Lu 2010; Bishayee et al. 2010; Darvesh and Bishayee 2010; Thomas et al. 2010; Morgan 2011).

Chronic liver disease has been suggested to be a predisposing factor for at least 80% of HCC (Farazi and DePinho 2006). Essentially, any condition that is linked to the development of fibrosis and cirrhosis has been associated with subsequent occurrence of HCC. In response to hepatocyte injury as well as inflammation and oxidative stress, hepatic stellate cells and portal fibroblasts undergo activation and transformation resulting in fibrosis and ultimately cirrhosis (Wong and Ng 2008). Under these circumstances, the surviving hepatocytes proliferate as a means to regenerate the injured liver. This cellular proliferation in the background of continued inflammation and oxidative stress represents a driving force for liver tumorigenesis (Nejak-Bowen and Monga 2011). Identification of cellular pathways necessary for the proliferation and survival of tumor cells in HCC not only aids in understanding the pathophysiology, diagnosis and progression but also provides a valuable tool in designing effective prevention and intervention of this disease. One such relevant pathway is the Wnt/β-catenin signaling pathway. Emerging studies have shown that the canonical Wnt/β-catenin pathway is frequently activated in HCC and responsible for the initiation and progression of the disease (Lee et al. 2006; Takigawa and Brown 2008; Nejak-Bowen and Monga 2011; Fatima et al. 2011). This chapter reviews the current knowledge on β-catenin, a pivotal player in the Wnt/β-catenin pathway, as a potential biomarker of liver cancer, and explores the possibility of targeting this molecule to achieve effective therapy as well as prevention of this devastating disease.

Wnt/β-CATENIN PATHWAY

The term "Wnt" has been developed by combining the Drosophila segment polarity gene "*Wingless*" with the mouse proto-oncogene "*Int-1*" (Luu et al. 2004). The Wnt/β-catenin signaling pathway maintains a dual personality in liver biology; it plays an important role in hepatic development, regeneration, and growth, whereas it's aberrant activation is evident in several types of liver tumors. This pathway also contributes to liver physiology and pathology by regulating various fundamental cellular events, such as differentiation, proliferation, survival, inflammation, oxidative stress, morphogenesis, and regeneration (Lee et al. 2006a; Thompson and Monga 2007; Takigawa and Brown 2008; Lade and Monga 2011).

The canonical Wnt pathway is the most studied and best characterized mechanism of Wnt signaling in which the multifunctional protein β-catenin serves as the key signaling intermediate. Under normal resting conditions, most of the cellular β-catenin binds to E-cadherin and α-catenin to form a cytoskeleton in adherent junctions between the cells with cell-cell adhesion functions (Jamora and Fuchs 2002; Drees et al. 2005). In the absence of Wnt signaling, free cytoplasmic β-catenin is degraded through a coordinated action of a multi-protein destruction complex containing adenomatous polyposis coli (APC), glycogen synthase kinase-3β (GSK-3β), casein kinase 1α (CK1α) and Axin (Munemitsu et al. 1995; Sakanaka 2002) (Fig. 1A). β-Catenin is first phosphorylated at serine-45 by CK1α (Amit et al. 2002), which facilitates its subsequent phosphorylation at serine-33, serine-37 and threonine-41 by GSK-3β (Liu et al. 2002). Additionally, APC, a tumor suppressor protein, binds to β-catenin following its phosphorylation by GSK-3β (Rubinfeld et al. 1996). Phosphorylated β-catenin undergoes ubiquitination by cellular E3 ubiquitin ligase complex (β-TrCP) and subsequent degradation by 26S proteasome (Hart et al. 1999; Aberle et al. 1997). Thus, the cytoplasmic level of free β-catenin remains low in the absence of Wnt signaling. Under these circumstances, transcription factor T-cell factor (TCF)/lymphoid enhancer factor (LEF) represses gene expression through the recruitment of co-repressors, namely Groucho, HMG box repressor protein (HBP1) and C-terminal binding protein (CtBP) (Sampson et al. 2001; Hurlstone and Clevers 2002). The effect of Groucho is likely to be mediated by histone deacetylases (HDAC) (Arce et al. 2009).

When Wnt binds to the seven-pass transmembrane Frizzled (FZD) receptor and to the co-receptor low-density lipoprotein receptor-related protein 5 or 6 (LRP 5/6), activation of the canonical Wnt pathway is triggered (Fig. 1B). Consequently, FZD recruits cytoplasmic scaffolding protein Dishevelled (Dvl), which in turn, recruits Axin and GSK-3β to LRP 5/6 (Zeng et al. 2008). The LRP 5/6 is phosphorylated by GSK-3β and CK1α, and this event disrupts the formation of the destruction complex as

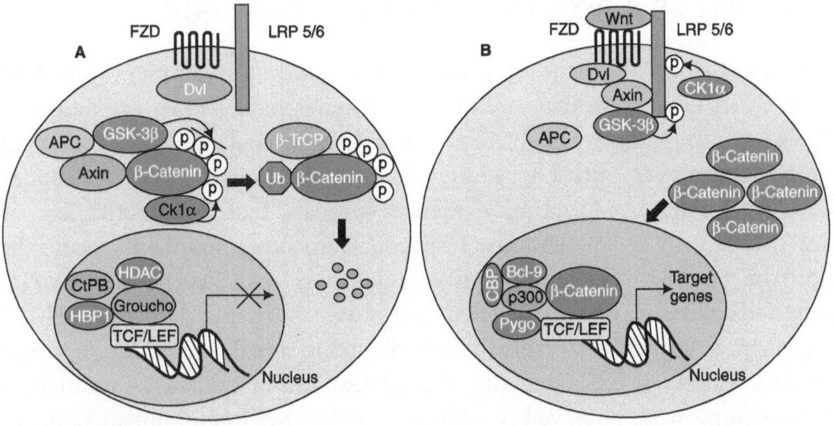

Fig. 1 Schematic representation of Wnt/β-catenin signaling cascade. A. In the absence of Wnt ligand, cytoplasmic β-catenin bound to its destruction complex consists of APC, Axin, CK1α and GSK-3β. β-Catenin is first phosphorylated by CK1α followed by further phosphorylation by GSK-3β. Phosphorylated β-catenin is recognized by β-TrCP followed by degradation by the 26S proteosomal system. In the nucleus, TCF-regulated transcription of target genes is repressed by the coordinated action of CtBP, Groucho, HBP1 and HDAC. B. Following binding of Wnt to FZD and simultaneously to LRP 5/6, Dvl is recruited to FZD followed by recruitment of Axin and GSK-3β to LRP 5/6. This results in the dissociation of β-catenin destruction complex. β-Catenin is stabilized and translocates to the nucleus, where it binds to TCF/LEF in the presence of Bcl-9, CBP/p300 and Pygogus, and activates the transcription of target genes.

Color image of this figure appears in the color plate section at the end of the book.

well as prevents the phosphorylation of β-catenin by GSK-3β. While non-phosphorylated β-catenin is not degraded, it accumulates into the cytoplasm and subsequently translocates to the nucleus. The nuclear translocation of β-catenin is facilitated by Bcl-9/legless adaptor protein that directs β-catenin to the nuclear protein Pygogus (Pygo) (Kramps et al. 2002). Once β-catenin is present in the nucleus, it displaces Groucho from TCF/LEF and recruits CREB-binding protein (CBP)/p300 (Hecht et al. 2000). This eventually activates the transcription of various target genes that regulate cell growth, survival, proliferation, anti-apoptosis, angiogenesis and tumor progression, such as cyclin D1, c-myc, survivin, cyclooxygenase-2(COX-2) and matrix metalloproteinase-7 (MMP-7) (Behrens et al. 1996; Crawford et al. 1999; Polakis 2000; Araki et al. 2003). Additionally, β-catenin-regulated genes include genes encoding for glutamine synthetase (GS), glutamate transporter, and cytochrome P450 (CYP) family members, including CYP1A2 and CYP2E1 (Cadoret et al. 2002; Loeppen et al. 2005).

β-CATENIN AS A POTENTIAL BIOMARKER IN HCC

Although the involvement of β-catenin in tumor development was first observed in colon cancer, a multitude of studies have shown activation of Wnt/β-catenin pathway in various other human malignancies, including breast, cervix, ovary, prostate, kidney, lung, skin, stomach and liver cancer (Luu et al. 2004). Mounting evidence suggests that hyperactivation of Wnt pathway or dysregulation of specific pathway components occur in approximately 50% of all human cancers (Goss and Khan 2011). In 1998, de La Coste and colleagues first observed the role of Wnt/β-catenin signaling in HCC. Since then, an aberrant activation of Wnt/β-catenin signaling has been reported in a wide variety of HCC patients. The nuclear and/or cellular accumulation of β-catenin, best observed in HCC samples by immunohistochemical staining, has been accepted as the reminiscent of activated Wnt/β-catenin signaling. In 33-67% of HCC patients, nuclear and/or cellular accumulation of β-catenin have been strongly associated with poorly differentiated tumor with high proliferative activity, vascular invasion and dismal prognosis (Endo et al. 2000; Wong et al. 2001; Inagawa et al. 2002; Coleman 2003). Additionally, a strong correlation has been established between nuclear accumulations of β-catenin with mutations (Wong et al. 2001). Accumulating studies reveal mutation of β-catenin gene (*CTNNB1*) leading to its nuclear and/or cytoplasmic localization in 20-40% of all HCC cases (Nejak-Bowen and Monga 2011). Interestingly, a higher frequency of β-catenin mutations in HCC associated with HCV infections have been observed than HBV infections or alcoholic risk factors (Hsu et al. 2000; Wong et al. 2001; Huang et al. 1999; Edamoto et al. 2003). Several studies indicate that β-catenin mutation and accumulation occur during the progression of HCC rather than early hepatocarcinogenic events (Inagawa et al. 2002; Kondo et al. 1999; Nhieu et al. 1999; Joo et al. 2003). Most of the mutations in HCC frequently occur within a single exon (exon 3) of *CTNNB1* gene, which is responsible for the phosphorylation and subsequent ubiquitination of β-catenin. Consequently, mutations in this region culminate in stabilization and nuclear accumulation. Immunohistochemically detected expression of GS has been found to be a sensitive and specific marker of β-catenin-mutated HCC (Cadoret et al. 2002; Zucman-Rossi et al. 2007). Moreover, alterations in the other components of the Wnt/β-catenin pathway, such as mutations in the components of the degradation complex of β-catenin, including *AXIN1* and *AXIN2*, inactivation of Axin and GSK-3β as well as overexpression of FZD7 receptor, lead to wild-type β-catenin stabilization (Zucman-Rossi et al. 2007; Taniguchi et al. 2002; Ban et al. 2003; Merle et al. 2004).

Zekri and colleagues (2011) have recently explored the possible diagnostic roles of various serum markers in Egyptian patients with HCC initiated by HCV infection. Serum β-catenin has been found to be the only marker that can significantly discriminate between patients with HCC and those with chronic hepatitis. Based upon this study, β-catenin could be considered as a potential marker for early diagnosis of HCV-associated HCC. Nevertheless, further studies with large-scale patient population are needed to confirm the diagnostic value of β-catenin in patients with HCC.

Very recently, Purcell and co-researchers (2011) have investigated an alternative activation of β-catenin by hepatocyte growth factor/c-Met signaling in a large cohort of 98 hepatoblastoma patients. Aberrant β-catenin activation has been observed in the cytoplasm and/or nucleus of 87% of tumor samples. The results also reveal a large subset of hepatoblastoma (83%) with cytoplasmic expression of tyrosine654-phosphorylated β-catenin and 30% showing additional nuclear accumulation. The sequence analysis shows mutations in 15% patients. Furthermore, an association between nuclear expression of c-Met-activated β-catenin and wild type *CTNNB1* has been established.

Unlike clinical situations, the role of β-catenin in the development and progression of liver cancer in pre-clinical experimental models of hepatocarcinogenesis is less clear. Several β-catenin transgenic mouse models have been developed to achieve overexpression of mutant or wild-type β-catenin. While most of these models have not been able to demonstrate spontaneous development of HCC (Tan et al. 2005; Cadoret et al. 2001; Harada et al. 2002; Nejak-Bowen et al. 2010), at least one recent study confirmed the association of mutant β-catenin with liver cancer (Stein et al. 2011). Based on several experimental studies, it can be hypothesized that mutations or activation of β-catenin may work in concert with other signaling pathways (such as Ha-ras or Lkb1) for the development of hepatic malignancy (Harada et al. 2004; Nejak-Bowen et al. 2010; Miyoshi et al. 2009). β-Catenin activation has been shown to confer additional growth and invasive advantage in a genetic model of liver cancer already accelerated by synergistic activity of the c-myc and transforming growth factor-α(TGF-α) transgenes (Calvisi et al. 2004). Similar effect was observed in the c-myc/E2F1 transgenic mouse model of hepatocarcinogenesis (Calvisi et al. 2005). In c-myc/TGF-β transgenic model, β-catenin activation has been observed in 25-33% tumors as an early event in hepatocarcinogenesis (Calvisi et al. 2001).

A high frequency of β-catenin activating mutations has also been reported in chemically-induced HCC models in rodents. Experimental liver cancer in rodents induced by diethylnitrosamine (DENA), an environmental and dietary hepatocarcinogen, has been established as one of the best-characterized experimental models of HCC (Bishayee and Dhir 2009; Bishayee et al. 2011a,b). DENA-initiated formation of preneoplastic foci, neoplastic nodules and ultimately HCC nodules in rodents closely resemble the development of human HCC (Peto et al. 1991; Verna et al. 1996). Moreover, DENA-initiated HCC has been shown to follow the sequential development of hepatic fibrosis and cirrhosis (Corpechot et al. 2002; Liu et al. 2009), which mimics human HCC. Yamada and colleagues (1999) reported *CTNNB1* point mutations in 31% adenomas and carcinomas induced in rats by administration of DENA, implying that the involvement of this pathway is related to an earlier stage of hepatocarcinogenesis. In another study in mice, almost 54% HCC samples from DENA-induced animals showed β-catenin mutations indicating importance of β-catenin alterations during the multi-stage hepatic carcinogenesis (Ogawa et al. 1999). Interestingly, Aydinlik and co-investigators (2001) have shown that β-catenin mutations are present in 80% of liver tumors from mice initiated with DENA and promoted by phenobarbital (PB). The same group (Loeppen et al. 2002) also observed that liver DENA/PB-induced liver tumors harboring β-catenin mutations were GS positive, whereas tumors without β-catenin mutations were GS negative. The accumulation of β-catenin in the cytoplasm and/or nucleus has been found to occur during stepwise progression of DENA-induced hepatocarcinogenesis in rats and this feature has been correlated with pathological features of atypical hyperplasia (Wang et al. 2006). Our laboratory has also investigated the expression of β-catenin by the immunohistochemical technique using an early stage, DENA-induced rat liver carcinogenesis model (Bishayee et al. 2011a). In the normal (untreated) livers, weak to moderate β-catenin immunoreactivity was observed on the cell membrane without cytoplasmic or nuclear accumulation (Fig. 2A). On the other hand, hepatocytes from DENA-exposed group showed an overexpression of β-catenin in cytoplasm and/or nucleus (Fig. 2B), confirming activated Wnt/β-catenin signaling pathway at an early stage of DENA-induced hepatocarcinogenic process in rats.

In an interesting study conducted by Xie et al. (2011), the promotional effect of 2-amino-3-methylimidazo[4,5-*f*]quinolone (IQ), a genotoxic carcinogen produced during cooking of protein-rich foods, has been investigated in mice initiated with DENA. Dietary IQ at a concentration of 100 or 300 ppm has been shown to increase altered hepatocyte foci, hepatocellular adenoma and carcinoma in DENA-initiated animals compared to DENA control. Moreover, IQ has been found to elevate the protein expression of Wnt1,

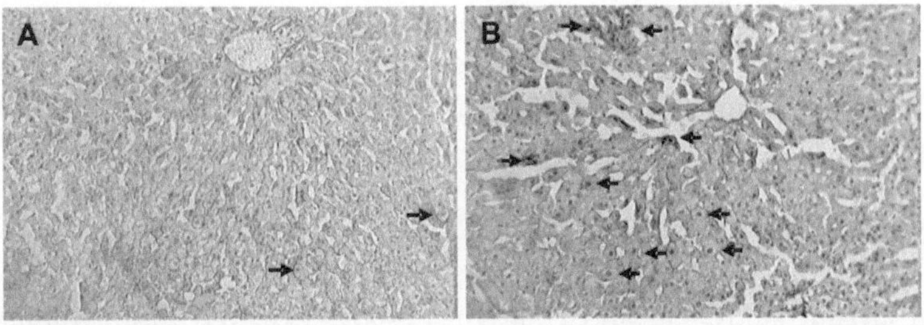

Fig. 2 Induction of hepatic β-catenin expression during DENA hepatocarcinogenesis in male Sprague-Dawley rats. Hepatocarcinogenesis was initiated with DENA (200 mg/kg) followed by promotion with PB in drinking water (0.05%, w/v) as per our published protocol (Bishayee et al. 2011a). Rats were sacrificed 18 weeks following DENA exposure, liver sections were prepared, and β-catenin expression was detected by immunohistochemistry. A. Normal hepatocyte show membrane-bound weak or moderate immunopositivity of β-catenin (black arrows). B. Accumulation of β-catenin in cytoplasm (green arrows) and nucleus (black arrows) during early phase of DENA-initiated liver cancer. Magnification: 100 × (unpublished laboratory observation).

Color image of this figure appears in the color plate section at the end of the book.

TGF-β, TGF-β receptor 1 and 2 as well as phosphorylated c-Jun, while suppressing E-cadherin and p21. Additionally, IQ-mediated translocation of β-catenin to the nucleus as well as upregulation of nuclear expression of c-Myc and cyclin D1 have been noticed.

The role of hepatocyte-specific β-catenin knockout (KO) on chemically-induced hepatocarcinogenesis has been investigated by at least two laboratories. In one study, the absence of β-catenin enhances DENA-induced hepatocarcinogenesis with pre-cirrhotic phenotype in mouse liver. However, functional β-catenin signaling was required to promote DENA hepatocarcinogenesis by PB (Rignall et al. 2011). In a separate study, the investigators showed a dramatic increase in HCC incidence in β-catenin KO mice exposed to DENA. Mechanistic studies reveal that activation of platelet-derived growth factor receptor α/phosphoinositide 3-kinase/Akt signaling partly due to aberrant activation of nuclear factor-κB in the background of increased liver oxidative stress, inflammation, injury, and fibrosis contributed to the accelerated liver tumorigenesis in β-catenin deficient mice (Zhang et al. 2010).

β-CATENIN AS A THERAPEUTIC TARGET IN LIVER CANCER

The final event result of sustained Wnt/β-catenin signaling is the induction of elevated expression of β-catenin-regulated genes that, in turn, impact the development and growth of liver tumors. Hence, interfering with Wnt/β-catenin signaling pathway may offer a potential opportunity for the development of novel drugs for the prevention as well as therapy of HCC. The following section presents studies showing innovative approaches of targeting Wnt/β-catenin signaling by a variety of synthetic as well as natural dietary compounds.

Synthetic Agents

In an elegant study, Hu and colleagues (2009) explored the possibility of blocking Wnt signaling by exogenous expression of Wnt antagonists to achieve antitumor therapeutic activity in HCC. Human Wnt inhibitory factor 1 (WIF1) and secreted frizzled-related protein 1 (sFRP1) were each fused with Fc fragment of human IgG1 to construct fusion proteins, namely WIF1-Fc and sFRP1-Fc. The fusion proteins inhibited the growth of Huh7, HepG2 and PLC/PRF/5 cells with downregulation of E2F1, cyclin D1 and c-myc at transcriptional level in Huh7 cells. Treatment of established tumors developed by implanting Huh7 cells in BALB/c nude mice with the fusion proteins inhibited tumor growth and prolonged survival accompanied by reduced β-catenin protein expression in tumor samples (Table 1).

In another study, an anti-Wnt-1 antibody decreased the viability and proliferation of Huh7 and Hep40 cells overexpressing Wnt-1 and harboring wild type β-catenin without any effect on normal hepatocytes with undetectable Wnt-1. The anti-Wnt-1 antibody also inhibited TCF4/β-catenin transcriptional activity and downregulated c-myc, cyclin D1 and survivin. Intratumoral injection of the antibody suppressed the growth of Huh7 tumor cells transplanted in nude mice with reduced expression of the genes studied *in vitro* (Wei et al. 2009).

The frequent overexpression of the FZD7 membrane receptor and subsequent activation of wild-type β-catenin protein have been reported in HCC (Merle et al. 2004; Bengochea et al. 2008). The pharmacological inhibition of the FZD7-Dvl protein-protein interaction could be a pertinent strategy for HCC therapy through disruption of β-catenin signaling. Nambotin and colleagues (2011) designed small interfering peptides (RHPDs) that are able to enter into cells and competitively antagonize the binding of FZD7 to the PDZ domain of Dvl. RHPDs have been found to decrease the viability of Huh7 cells through apoptosis depending on their affinity for PDZ. This effect was associated with the degradation of β-catenin. Similar results were achieved in transgenic mouse model in which intra-tumor injection of RHPDs inhibited HCC progression through β-catenin modulation.

Behari and coresearchers (2007) examined the effects of celecoxib, a COX-2 inhibitor, and R-etodolac, an enantiomer of the non-steroidal anti-inflammatory drug etodolac, which lacks COX-inhibitory activity, on the Wnt/β-catenin pathway in Hep3B and HepG2 cells. Celecoxib diminished tumor cell proliferation and survival through degradation of β-catenin protein via upregulation of GSK-3β and APC. R-etodolac caused a decrease in total and activated β-catenin protein, TCF/β-catenin transcriptional activity, GS expression, GSK-3β activation and an increase in β-catenin-E-cadherin association. The overall effect was R-etodolac-mediated induction of anti-hepatoma effects in both cell lines.

The chemopreventive potential of NIK-333, an acyclic retinoid, has been investigated in transgenic mice expressing the dominant negative form of retinoic receptor α (RARE mice) that generate reactive oxygen species and lead to the spontaneous development of liver tumors. A diet containing 0.06% NIK-333 completely abolished tumor appearance, significantly suppressed oxidative stress and diminished the expression of proliferating cell nuclear antigen, β-catenin and cyclin D1 (Sakabe et al. 2007).

Wei et al. (2010) have evaluated anticancer therapeutic effects of three small molecule antagonists, namely PKF118-310, PKF115-584 and CGP049090, on the TCF4/β-catenin complex against human HCC (HepG2, Hep40 and Huh7) *in vitro* as well as *in vivo*. All these agents displayed cytotoxicity against the aforementioned cells in a dose-responsive fashion. In HepG2 and Huh7 cells, treatment with the antagonists inhibited TCF4/β-catenin binding capacity and transcriptional activity with simultaneous downregulation of the endogenous TCF4/β-catenin target genes, such as c-myc, cyclin D1 and survivin. Interestingly, these effects were replicated in *in vivo* conditions as all three compounds suppressed the growth of xenografted HepG2 tumor, induced tumor apoptosis and reduced the expression of target genes.

Recently, Parody et al. (2010) assessed the status of Wnt/β-catenin/TCF pathway in a very early stage of rat hepatocarcinogenesis initiated by DENA and promoted by 2-acetylaminofluorene (2-AAF) as well as the impact of interferon-α2b (IFN-α2b) treatment. The investigators demonstrated that the canonical β-catenin signaling pathway is activated during the early stage of hepatocarcinogenesis even with a wild-type β-catenin. Moreover, the *in vivo* IFN-α2b treatment significantly reduced the binding of β-catenin to TCF4 transcription factor and consequently suppressed TCF transcriptional activity while stimulating the β-catenin-Forkhead box containing protein class O (FOXO) interaction. All these mechanisms have been linked to a diminished occurrence of rat glutathione S-transferase placenta form (rGST-P)-positive foci by an apoptotic process. Later study by the same group (Ceballos et al. 2011) reported that IFN-α2b decreased the proliferation and increased apoptosis in HepG2/C3A (a clonal derivative of HepG2) and Huh7 cells. Mechanistic study revealed that IFN-α2b attenuated Wnt/β-catenin signal by lowering β-catenin and FZD7 receptor proteins and the interaction of β-catenin with TCF4. Another interesting observation of this study is the augmentation of the effects of IFN-α2b by co-treatment of HCC cells with transforming growth factor β1.

Very recently, the effect of pegylated interferon-α2a (peg-IFN-α2a), a first line of therapy for HCV patients that also impacts the recurrence of HCC, on Wnt/β-catenin signaling has been studied in several models of HCC. In Hep3B, HepG2 and HuH7 cells, peg-IFN-α2a treatment decreased the transcriptional activity of TCF/β-catenin and increased the mRNA and protein expression of RanBP3, a known β-catenin nuclear export factor. Additionally, an elevated association between β-catenin and RanBP3 has been noticed following peg-IFN-α2a exposure. In transgenic mice overexpressing serine-45 mutant β-catenin in the liver, peg-IFN-α2a treatment led to increased nuclear RanBP3 and its association to β-catenin, decreased nuclear β-catenin and cyclin D1 as well as cytoplasmic GS and cell proliferation (Thompson et al. 2011).

Dietary Agents

Emerging studies strongly suggest that targeted disruption of β-catenin-mediated signaling is a promising strategy for early chemopreventive as well as therapeutic intervention, particularly with respect to intestinal tumorigenesis, by a number of bioactive food components (Clapper et al. 2004; Kundu et al. 2006; Sarkar et al. 2010; Teiten et al. 2011). As presented below, several dietary compounds have been also found to modulate β-catenin signaling, linked to either liver cancer preventive or therapeutic effects.

Auraptene, a natural bioactive coumarin derivative, is present in several citrus varieties, such as grapefruits and natsudaidai. It has been shown that feeding of rats with an auraptene-containing diet during the post-initiation period of DENA-induced hepatocarcinogenesis suppressed the development of HCC with β-catenin mutation, presumably by negative selection for neoplastic hepatocytes harboring the mutation (Hara et al. 2005).

Docosahexaenoic acid (DHA) and eicosapentaenoic acid (EPA) are long-chain omega-3 fatty acids predominantly present in oily fish, including salmon, cod, herring, mackerel and sardine. As per the study of Lim and coresearchers (2009), DHA and EPA treatment resulted in a dose-dependent reduction of cell viability and induction of apoptosis in several human HCC cell lines. Mechanistically, both DHA and EPA manifested dephosphorylation and subsequently activation of GSK-3β, leading to β-catenin degradation in Hep3B cells. Additional studies have shown that DHA induced the formation of β-catenin/Axin/GSK-3β binding complex, which represents a parallel mechanism of β-catenin degradation.

Chrysin is a naturally occurring flavone present in blue passion flower (*Passiflora caerulea*) as well as honeycomb. 5,7-Dimethoxyflavone (DMF), formed by methylation of chrysin, is present in pepper tree leaves. According to a recent study, oral administration of DMF exhibited a striking inhibitory effect on the development and multiplicity of hepatic nodules induced by DENA in male Wistar rats. An upregulation of the canonical Wnt/β-catenin signaling pathway has been observed as evidenced by an elevated expression of nuclear protein level of β-catenin, cytoplasmic Wnt 3a, CK2 (a positive regulator of Wnt pathway), Dvl2, and cyclin D1 with a concomitant decrease in the expression of GSK-3β. The chemopreventive effect of DMF has been associated with a reversal of the alterations of the components of the Wnt/β-catenin pathway (Khan et al. 2011).

CONCLUSIONS AND FUTURE PERSPECTIVES

While the precise role of the canonical Wnt/β-catenin signaling pathway in normal as well as abnormal hepatic functions has not yet been fully understood, emerging studies provide convincing evidence that aberrant activation of Wnt/β-catenin signaling plays a vital role in the pathogenesis and progression of HCC. It is also apparent that there are several mechanisms which contribute to the activation of β-catenin. There also exists the possibility of crosstalk between Wnt/β-catenin signaling and other signaling pathways in HCC. As Wnt/β-catenin pathway is tightly regulated at multiple cellular and subcellular levels, the pathway itself represents a valuable target to inhibit the development and growth of liver tumors. Several preclinical studies as highlighted in this review, indeed, show that several agents from dietary as well as non-dietary origin have been able to disrupt activated Wnt/β-catenin signaling to halt the growth of a variety of HCC cells *in vitro* and/or *in vivo* or inhibit the development and progression of chemically-induced liver tumors in animals. What is truly fascinating that all these structurally diverse molecules target nearly all components of Wnt/β-catenin pathway as they are able to strengthen β-catenin-E-cadherin cooperation, antagonize Wnt function, intensify β-catenin-destruction complex interaction, promote β-catenin degradation, facilitate extranuclear β-catenin transport as well as downregulate β-catenin transcriptional activity. As activation of the Wnt/β-catenin pathway occurs during the late as well as early phases of multi-stage hepatocarcinogenesis, the inhibitors of canonical Wnt/β-catenin signaling may be effective in both the treatment and chemoprevention of HCC.

Although the impressive numbers of preclinical studies highlighted in this chapter provide the *"proof-of-principle"* of blocking or diminishing liver cancer by interference of the Wnt/β-catenin signaling by various natural and synthetic compounds, the efficacy and safety of these agents need to be validated by well-controlled clinical studies. Nevertheless, there are certain limitations as well as challenges of targeting Wnt/β-catenin signaling. For example, there is a caveat of blocking multifunctional β-catenin. In addition to being an indispensable component of Wnt/β-catenin pathway, β-catenin is also involved in cell-cell adhesion and plays a pivotal role in tissue homeostasis and renewal. Thus, diminishing merely the level of β-catenin, but not its transcriptional activity, may constitute severe adverse physiological consequences. Again, Wnt/β-catenin signaling is implicated in the maintenance and self-renewal of normal stem cells as well as proliferation or differentiation of progenitor cells. Thus, the systemic inhibition of Wnt/β-catenin signaling could produce adverse effects. Moreover, it is conceivable that the Wnt/β-catenin pathway is connected to other signaling pathways. While this could intensify the effects of Wnt/β-catenin inhibitors on tumor cells, this may also give neoplastic cells the opportunity to develop therapy resistance. Furthermore, it is very important to select appropriate patient populations to achieve maximum benefit of targeted therapy with β-catenin inhibitors. Activation of β-catenin through mutations in *CTNNB1* gene can be detected by positive immunostaining of GS in tumor samples. For these patients, use of RNAi would selectively target tumor cells and spare normal healthy hepatocytes. This may prevent the adverse effects of universal β-catenin suppression. Nevertheless, there would be still challenges for patients with mutant-negative tumors with aberrant Wnt/β-catenin signaling which warrants for further studies. Another potential challenge of indiscriminate downregulation of β-catenin-mediated signaling could be the loss of expression of several CYP isozymes responsible for activation or deactivation of commonly used chemotherapeutic drugs. Hence, the use of

Table 1 Modulation of β-catenin signaling in liver cancer by agents from dietary and non-dietary sources.

Agents	*Experimental models*	*Anti-hepatocarcinogenic effects*	*Effects on β-catenin signaling*	*References*
Synthetic agents				
WIF1-Fc, sFRP1-Fc	Huh7, HepG2 and PLC/PRF/5 cells	↓cell proliferation	↓TCF activity; ↓c-myc; ↓cyclin D1; ↓E2F1	Hu et al. 2009
	Huh7 and Hep3B xenografts in nude mice	↓tumor volume	↓β-catenin	
Anti-Wnt-1 antibody	Huh7 and Hep40 cells	↓cell proliferation	↓TCF4/β-catenin transcriptional activity; ↓β-catenin; ↓c-myc; ↓cyclin D1; ↓survivin	Wei et al. 2009
	Huh7 xenograft in nude mice	↓tumor volume	↓c-myc; ↓cyclin D1; ↓survivin	
	Huh7 and Hep3B xenografts in nude mice	↓tumor volume	↓β-catenin	
RHPDs	Huh7 cells	↓cell viability	↑β-catenin degradation	Nambotin et al. 2011
	SV40-TAg transgenic mice	↓tumor growth	↑β-catenin degradation	
Celecoxib	Hep3B and HepG2 cells	↓cell proliferation	↓β-catenin; ↑APC; ↑GSK-3β	Behari et al. 2007
R-Etodolac	Hep3B and HepG2 cells	↓cell proliferation	↓β-catenin; ↓TCF4/β-catenin transcriptional activity; ↓cyclin D1; ↓GS; ↑GSK-3β activation; ↑β-catenin-E-cadherin interaction	Behari et al. 2007
NIK-333	RARE transgenic mice	↓tumor incidence	↓β-catenin; ↓cyclin D1	Sakabe et al. 2007
PKF118-310, PKF115-584, CGP049090	HepG2, Hep40 and Huh7 cells	↑cytotoxicity	↓TCF4/β-catenin binding and transcriptional activity; ↓c-myc; ↓cyclin D1; ↓survivin	Wei et al. 2010
	HepG2 xenograft in nude mice	↓tumor volume	↓c-myc; ↓cyclin D1; ↓survivin	

Table 1 Contd. ...

Table 1 Contd. ...

IFN-α2b	DENA-initiated and 2-AAF-promoted hepatocarcinogenesis in male Wistar rats	↓GST-P foci	↓TCF4/β-catenin binding; ↓cyclin D1; ↓MMP-7; ↓Axin 2; ↓SP5; ↓FZD7; ↑β-catenin-FOXO3a interaction	Parody et al. 2010
	HepG2/C3A and Huh7 cells	↓cell proliferation	↓β-catenin; ↓FZD7; ↓TCF4/β-catenin binding	Ceballos et al. 2011
Peg-IFN-α2a	HepG2, Hep3B and HuH-7 cells		↓β-catenin/TCF transcriptional activity; ↑RanBP3; ↑Dkk-1	Thompson et al. 2011
	Serine-45 mutant β-catenin transgenic mice treated with DENA	↓cell proliferation	↓β-catenin; ↓cyclin D1; ↓GS; ↑RanBP3; ↑β-catenin-RanBP3 association	
Dietary compounds				
Auraptene	DENA-induced hepatocarcinogenesis in male F344 rats	↓HCC	↓β-catenin gene mutation	Hara et al. 2005
DHA, EPA	Hep3B, Huh7 and HepG2 cells	↓cell viability	↓β-catenin; ↓c-Met; ↓β-catenin activity; ↓p-GSK-3β; ↑β-catenin degradation	Lim et al. 2009
DMF	DENA-induced hepatocarcinogenesis in male Wistar rats	↓hepatic nodules	↓Wnt3a; ↓β-catenin; ↓β-arrestin 2; ↓CK 2; ↓Dlv 2; ↑GSK-3β; ↓cyclin D1	Khan et al. 2011

β-catenin inhibitory agents should be performed with caution in conjunction with other drug therapy.

In conclusion, substantial experimental and clinical evidence as presented here strongly suggests that inappropriate activation of β-catenin could be a valuable biomarker in liver cancer and this biomolecule may represent a novel target for the prevention and therapy of this incurable disease.

ACKNOWLEDGEMENTS

A part of our research on β-catenin and liver cancer as presented in this review was carried out at the Northeast Ohio Medical University (Rootstown, OH). I would like to thank Dr. Alexandros Georgakilas for providing me with the opportunity to review this emerging area of research, Dr. Lois Garland-Patterson for her vision, mentorship, and inspiration, and Dr. Altaf S. Darvesh for critical reading of the manuscript.

REFERENCES

Aberle, H., A. Bauer, J. Stappert, A. Kispert and R. Kemler. 1997. Beta-catenin is a target for the ubiquitin-proteasome pathway. EMBO J. 16: 3797-3804.

Amit, S., A. Hatzubai, Y. Birman, J.S. Andersen, E. Ben-Shushan, M. Mann, Y. Ben-Neriah and I. Alkalay. 2002. Axin-mediated CKI phosphorylation of beta-catenin at Ser 45: a molecular switch for the Wnt pathway. Genes Dev. 16: 1066-1076.

Araki, Y., S. Okamura, S.P. Hussain, M. Nagashima, P. He, M. Shiseki, K. Miura and C.C. Harris. 2003. Regulation of cyclooxygenase-2 expression by the Wnt and ras pathway. Cancer Res. 63: 728-734.

Arce, L., K.T. Pate and M.L. Waterman. 2009. Groucho binds two conserved regions of LEF-1 for HDAC-dependent repression. BMC Cancer 9: 159.

Aydinlik, H., T.D. Nguyen, O. Moennikes, A. Buchmann and M. Schwarz. 2001. Selective pressure during tumor promotion by phenobarbital leads to clonal outgrowth of beta-catenin-mutated mouse liver tumors. Oncogene 20: 7812-7816.

Ban, K.C., H. Singh, R. Krishnan and H.F. Seow. 2003. GSK-3beta phosphorylation and alteration of beta-catenin in hepatocellular carcinoma. Cancer Lett. 199: 201-208.

Bartsch, H. and R. Montesano. 1984. Relevance of nitrosamines to human cancer. Carcinogenesis 5: 1381-1393.

Behari, J., G. Zeng, W. Otruba, M.D. Thompson, P. Muller, A. Micsenyi, S.S. Sekhon, L. Leoni and S.P. Monga. 2007. R-Etodolac decreases beta-catenin levels along with survival and proliferation of hepatoma cells. J. Hepatol. 46: 849-857.

Behrens, J., J.P. von Kries, M. Kühl, L. Bruhn, D. Wedlich, R. Grosschedl and W. Birchmeier. 1996. Functional interaction of beta-catenin with the transcription factor LEF-1. Nature 382: 638-642.

Bengochea, A., M.M. de Souza, L. Lefrançois, E. Le Roux, O. Galy, I. Chemin, M. Kim, J.R. Wands, C. Trepo, P. Hainaut, J.Y. Scoazec, L. Vitvitski and P. Merle. 2008. Common dysregulation of Wnt/Frizzled receptor elements in human hepatocellular carcinoma. Br. J. Cancer 99: 143-150.

Bishayee, A. and N. Dhir. 2009. Resveratrol-mediated chemoprevention of diethylnitrosamine-initiated hepatocarcinogenesis: inhibition of cell proliferation and induction of apoptosis. Chem. Biol. Interact. 179: 131-144.

Bishayee, A., T. Politis and A.S. Darvesh. 2010. Resveratrol in the chemoprevention and treatment of hepatocellular carcinoma. Cancer Treat. Rev 36: 43-53.

Bishayee, A., D. Bhatia, R.J. Thoppil, A.S. Darvesh, E. Nevo and E.P. Lansky. 2011a. Pomegranate-mediated chemoprevention of experimental hepatocarcinogenesis involves Nrf2-regulated antioxidant mechanisms. Carcinogenesis 32: 888-896.

Bishayee, A., T. Mbimba, R.J. Thoppil, E. Háznagy-Radnai, P. Sipos, A.S. Darvesh, H. Folkesson and J. Hohmann. 2011b. Anthocyanin-rich black currant (*Ribes nigrum* L.) extract affords chemoprevention against diethylnitrosamine-induced hepatocellular carcinogenesis in rats. J. Nutr. Biochem. 22: 1035-1046.

Blonski, W., D.S. Kotlyar and K.A. Forde. 2010. Non-viral causes of hepatocellular carcinoma.World J. Gastroenterol. 16: 3603-3615.

Cadoret, A., C. Ovejero, S. Saadi-Kheddouci, E. Souil, M. Fabre, B. Romagnolo, A. Kahn and C. Perret. 2001. Hepatomegaly in transgenic mice expressing an oncogenic form of beta-catenin. Cancer Res. 61: 3245-3249.

Cadoret, A., C. Ovejero, B. Terris, E. Souil, L. Lévy, W.H. Lamers, J. Kitajewski, A. Kahn and C. Perret. 2002. New targets of beta-catenin signaling in the liver are involved in the glutamine metabolism. Oncogene 21: 8293-8301.

Calvisi, D.F., V.M. Factor, R. Loi and S.S. Thorgeirsson. 2001. Activation of beta-catenin during hepatocarcinogenesis in transgenic mouse models: relationship to phenotype and tumor grade. Cancer Res. 61: 2085-2091.

Calvisi, D.F., S. Ladu, V.M. Factor and S.S. Thorgeirsson. 2004. Activation of beta-catenin provides proliferative and invasive advantages in c-myc/TGF-alpha hepatocarcinogenesis promoted by phenobarbital. Carcinogenesis 25: 901-908.

Calvisi, D.F., E.A. Conner, S. Ladu, E.R. Lemmer, V.M. Factor and S.S. Thorgeirsson. 2005. Activation of the canonical Wnt/beta-catenin pathway confers growth advantages in c-Myc/E2F1 transgenic mouse model of liver cancer. J. Hepatol. 42: 842-859.

Ceballos, M.P., J.P. Parody, M.D. Alvarez, P.I. Ingaramo, C.E. Carnovale and M.C. Carrillo. 2011. Interferon-α2b and transforming growth factor-β1 treatments on HCC cell lines: Are Wnt/β-catenin pathway and Smads signaling connected in hepatocellular carcinoma? Biochem. Pharmacol. 82: 1682-1691.

Center, M.M. and A. Jemal. 2011. International trends in liver cancer incidence rates. Cancer Epidemiol. Biomarkers Prev. 20: 2362-2368.

Clapper, M.L., J. Coudry and W.C. Chang. 2004. β-Catenin-mediated signaling: a molecular target for early chemopreventive intervention. Mutat. Res. 555: 97-105.

Coleman, W.B. 2003. Mechanisms of human hepatocarcinogenesis. Curr. Mol. Med. 3: 573-588.

Corpechot, C., V. Barbu, D. Wendum, N. Kinnman, C. Rey, R. Poupon, C. Housset and O. Rosmorduc. 2002. Hypoxia-induced VEGF and collagen I expressions are associated with angiogenesis and fibrogenesis in experimental cirrhosis. Hepatology 35: 1010-1021.

Crawford, H.C., B.M. Fingleton, L.A. Rudolph-Owen, K.J. Goss, B. Rubinfeld, P. Polakis and L.M. Matrisian. 1999. The metalloproteinase matrilysin is a target of beta-catenin transactivation in intestinal tumors. Oncogene 18: 2883-2891.

Darvesh, A.S. and A. Bishayee. 2010. Selenium in the prevention and treatment of hepatocellular carcinoma. Anticancer Agents Med. Chem. 10: 338-345.

de La Coste, A., B. Romagnolo, P. Billuart, C.A. Renard, M.A. Buendia, O. Soubrane, M. Fabre, J. Chelly, C. Beldjord, A. Kahn and C. Perret. 1998. Somatic mutations of the beta-catenin gene are frequent in mouse and human hepatocellular carcinomas. Proc. Natl. Acad. Sci. USA 95: 8847-8851.

Drees, F., S. Pokutta, S. Yamada, W.J. Nelson and W.I. Weis. 2005. Alpha-catenin is a molecular switch that binds E-cadherin-beta-catenin and regulates actin-filament assembly. Cell 123: 903-915.

Edamoto, Y., A. Hara, W. Biernat, L. Terracciano, G. Cathomas, H.M. Riehle, M. Matsuda, H. Fujii, J.Y. Scoazec and H. Ohgaki. 2003. Alterations of RB1, p53 and Wnt pathways in hepatocellular carcinomas associated with hepatitis C, hepatitis B and alcoholic liver cirrhosis. Int. J. Cancer 106: 334-341.

El-Serag, H.B. 2004. Hepatocellular carcinoma: recent trends in the United States. Gastroenterology 127: S27-S34.

Endo, K., T. Ueda, J. Ueyama, T. Ohta and T. Terada. 2000. Immunoreactive E-cadherin, alpha-catenin, beta-catenin, and gamma-catenin proteins in hepatocellular carcinoma: relationships with tumor grade, clinicopathologic parameters, and patients' survival. Hum. Pathol. 31: 558-565.

Farazi, P.A. and R.A. DePinho. 2006. Hepatocellular carcinoma pathogenesis: from genes to environment. Nat. Rev. Cancer 6: 674-687.

Fatima, S., N.P. Lee and J.M. Luk. 2011. Dickkopfs and Wnt/β-catenin signaling in liver cancer. World J. Clin. Oncol. 2: 311-325.

Goss, K.H. and M. Khan. 2011. Targeting the Wnt Pathway in Cancer. Springer, New York, USA.

Hara, A., K. Sakata, Y. Yamada, T. Kuno, N. Kitaori, T. Oyama, Y. Hirose, A. Murakami, T. Tanaka and H. Mori. 2005. Suppression of the β-catenin mutation by dietary exposure of auraptene, a citrus antioxidant, in N,N-diethylnitrosamine-induced hepatocellular carcinomas in rats. Oncol. Rep. 14: 345-351.

Harada, N., H. Miyoshi, N. Murai, H. Oshima, Y. Tamai, M. Oshima and M.M. Taketo. 2002. Lack of tumorigenesis in the mouse liver after adenovirus-mediated expression of a dominant stable mutant of beta-catenin. Cancer Res. 62: 1971-1977.

Harada, N., H. Oshima, M. Katoh, Y. Tamai, M. Oshima and M.M. Taketo. 2004. Hepatocarcinogenesis in mice with beta-catenin and Ha-ras gene mutations. Cancer Res. 64: 48-54.

Hart, M., J.P. Concordet, I. Lassot, I. Albert, R. del los Santos, H. Durand, C. Perret, B. Rubinfeld, F. Margottin, R. Benarous and P. Polakis. 1999. The F-box protein beta-TrCP associates with phosphorylated beta-catenin and regulates its activity in the cell. Curr. Biol. 9: 207-210.

Hecht, A., K. Vleminckx, M.P. Stemmler, F. van Roy and R. Kemler. 2000. The p300/CBP acetyltransferases function as transcriptional coactivators of beta-catenin in vertebrates. EMBO J. 19: 1839-1850.

Hsu, H.C., Y.M. Jeng, T.L. Mao, J.S. Chu, P.L. Lai and S.Y. Peng. 2000. Beta-catenin mutations are associated with a subset of low-stage hepatocellular carcinoma negative for hepatitis B virus and with favorable prognosis. Am. J. Pathol. 157: 763-770.

Hu, J., A. Dong, V. Fernandez-Ruiz, J. Shan, M. Kawa, E. Martínez-Ansó, J. Prieto and C. Qian. 2009. Blockade of Wnt signaling inhibits angiogenesis and tumor growth in hepatocellular carcinoma. Cancer Res. 69: 6951-6959.

Huang, H., H. Fujii, A. Sankila, B.M. Mahler-Araujo, M. Matsuda, G. Cathoma and H. Ohgaki. 1999. Beta-catenin mutations are frequent in human hepatocellular carcinomas associated with hepatitis C virus infection. Am. J. Pathol. 155: 1795-1801.

Hurlstone, A. and H. Clevers. 2002. T-cell factors: turn-ons and turn-offs. EMBO J. 21: 2303-2311.

Inagawa, S., M. Itabashi, S. Adachi, T. Kawamoto, M. Hori, J. Shimazaki, F. Yoshimi and K. Fukao. 2002. Expression and prognostic roles of beta-catenin in hepatocellular carcinoma: correlation with tumor progression and postoperative survival. Clin. Cancer Res. 8: 450-456.

Jamora, C. and E. Fuchs. 2002. Intercellular adhesion, signalling and the cytoskeleton. Nat. Cell Biol. 4: E101-E108.

Je, Y., F.A.B. Schutz and T.K. Choueiri. 2009. Risk of bleeding with vascular endothelial growth factor receptor tyrosine-kinase inhibitors sunitinib and sorafenib: a systematic review and mata-analysis of clinical trials. Lancet Oncol. 10: 967-974.

Jemal, A., F. Bray, M.M. Center, J. Ferlay, E. Ward and D. Forman. 2011. Global cancer statistics. CA Cancer J. Clin. 61: 69-90.

Jemal, A., A. Siegel, J. Xu and E. Ward. 2010. Cancer statistics. CA Cancer J. Clin. 60: 1-24.

Joo, M., H.K. Lee and Y.K. Kang. 2003. Expression of beta-catenin in hepatocellular carcinoma in relation to tumor cell proliferation and cyclin D1 expression. J. Korean Med. Sci. 18: 211-217.

Khan, M.S., D. Halagowder and S.N. Devaraj. 2011. Methylated chrysin induces co-ordinated attenuation of the canonical Wnt and NF-κB signaling pathway and upregulates apoptotic gene expression in the early hepatocarcinogenesis rat model. Chem. Biol. Interact. 193: 12-21.

Kondo, Y., Y. Kanai, M. Sakamoto, T. Genda, M. Mizokami, R. Ueda and S. Hirohashi. 1999. Beta-catenin accumulation and mutation of exon 3 of the beta-catenin gene in hepatocellular carcinoma. Jpn. J. Cancer Res. 9: 1301-1309.

Kramps, T., O. Peter, E. Brunner, D. Nellen, B. Froesch, S. Chatterjee, M. Murone, S. Züllig and K. Basler. 2002. Wnt/wingless signaling requires BCL9/legless-mediated recruitment of pygopus to the nuclear beta-catenin-TCF complex. Cell 109: 47-60.

Kundu, J.K., K.Y. Choi and Y.J. Surh. 2006. β-Catenin-mediated signaling: a novel molecular target for chemoprevention with anti-inflammatory substances. Biochim. Biophys. Acta 1765: 14-24.

Lade, A.G. and S.P. Monga. 2011. Beta-catenin signaling in hepatic development and progenitors: which way does the WNT blow? Dev. Dyn. 240: 486-500.

Lee, H.C., M. Kim and J.R. Wands. 2006. Wnt/frizzled signaling in hepatocellular carcinoma. Front. Biosci. 11: 1901-1915.

Lee, T.H., D.I. Tai, C.J. Cheng, C.S. Sun, C.Y. Lin, M.J. Sheu, W.P. Lee, C.Y. Peng, A.H. Wang and S.L. Tsai. 2006a. Enhanced nuclear factor-kappa B-associated Wnt-1 expression in hepatitis B- and C-related hepatocarcinogenesis: identification by functional proteomics. J. Biomed. Sci. 13: 27-39.

Lim, K., C. Han, Y. Dai, M. Shen and T. Wu. 2009. Omega-3 polyunsaturated fatty acids inhibit hepatocellular carcinoma cell growth through blocking β-catenin and cyclooxygenase-2. Mol. Cancer Ther. 8: 3046-3055.

Liu, C.,Y. Li, M. Semenov, C. Han, G.H. Baeg, Y. Tan, Z. Zhang, X. Lin and X. He. 2002. Control of beta-catenin phosphorylation/degradation by a dual-kinase mechanism. Cell 108: 837-847.

Liu, Y.F., B.S. Zha, H.L. Zhang, X.J. Zhu, Y.H. Li, J. Zhu, X.H. Guan, Z.Q. Feng and J.P. Zhang. 2009. Characteristic gene expression profiles in the progression from liver cirrhosis to carcinoma induced by diethylnitrosamine in a rat model. J. Exp. Clin. Cancer Res 28: 107.

Loeppen, S., D. Schneider, F. Gaunitz, R. Gebhardt, R. Kurek, A. Buchmann and M. Schwarz. 2002. Overexpression of glutamine synthetase is associated with beta-catenin-mutations in mouse liver tumors during promotion of hepatocarcinogenesis by phenobarbital. Cancer Res. 62: 5685-5688.

Loeppen, S., C. Koehle, A. Buchmann and M. Schwarz. 2005. A beta-catenin-dependent pathway regulates expression of cytochrome P450 isoforms in mouse liver tumors. Carcinogenesis 26: 239-248.

Lu, S.C. 2010. Where are we in the chemoprevention of hepatocellular carcinoma? Hepatology 51: 734-736.

Luu, H.H., R. Zhang, R.C. Haydon, E. Rayburn, Q. Kang, W. Si, J.K. Park, H. Wang, Y. Peng, W. Jiang and T.C. He. 2004. Wnt/β-catenin signaling pathway as a novel cancer drug target. Curr. Cancer Drug Targets 4: 653-671.

Merle, P., S. de la Monte, M. Kim, M. Herrmann, S. Tanaka, A. Von Dem Bussche, M.C. Kew, C. Trepo and J.R. Wands. 2004. Functional consequences of frizzled-7 receptor overexpression in human hepatocellular carcinoma. Gastroenterology 127: 1110-1112.

Miyoshi, H., A. Deguchi, M. Nakau, Y. Kojima, A. Mori, M. Oshima, M. Aoki and M.M. Taketo. 2009. Hepatocellular carcinoma development induced by conditional beta-catenin activation in Lkb1+/- mice. Cancer Sci. 100: 2046-2053.

Morgan, T.R. 2011. Chemoprevention of hepatocellular carcinoma in chronic hepatitis C. Recent Results Cancer Res. 188: 85-99.

Munemitsu, S., I. Albert, B. Souza, B. Rubinfeld and P. Polakis. 1995. Regulation of intracellular beta-catenin levels by the adenomatous polyposis coli (APC) tumor-suppressor protein. Proc. Natl. Acad. Sci. USA 92: 3046-3050.

Nambotin, S.B., L. Lefrancois, X. Sainsily, P. Berthillon, M. Kim, J.R. Wands, M. Chevallier, P. Jalinot, J.Y. Scoazec, C. Trepo, F. Zoulim and P. Merle. 2011. Pharmacological inhibition of Frizzled-7 displays anti-tumor properties in hepatocellular carcinoma. J. Hepatol. 54: 288-299.

Nejak-Bowen, K.N., M.D. Thompson, S. Singh, W.C. Bowen Jr, M.J. Dar, J. Khillan, C. Dai and S.P. Monga. 2010. Accelerated liver regeneration and hepatocarcinogenesis in mice overexpressing serine-45 mutant beta-catenin. Hepatology 51: 1603-1613.

Nejak-Bowen, K.N. and S.P. Monga. 2011. Beta-catenin signaling, liver regeneration and hepatocellular cancer: sorting the good from the bad. Semin. Cancer Biol. 21: 44-58.

Nhieu, J.T., C.A. Renard, Y. Wei, D. Cherqui, E.S. Zafrani and M.A. Buendia. 1999. Nuclear accumulation of mutated beta-catenin in hepatocellular carcinoma is associated with increased cell proliferation. Am. J. Pathol. 155: 703-710.

Ogawa, K.,Y. Yamada, K. Kishibe, K. Ishizaki and Y. Tokusashi. 1999. Beta-catenin mutations are frequent in hepatocellular carcinomas but absent in adenomas induced by diethylnitrosamine in B6C3F1 mice. Cancer Res. 59: 1830-1833.

Parody, J.P., M.L. Alvarez, A.D. Quiroga, M.P. Ceballos, D.E. Frances, G.B. Pisani, J.M. Pellegrino, C.E. Carnovale and M.C. Carrillo. 2010. Attenuation of the Wnt/β-catenin/TCF pathway by *in vivo* interferon-α2b (IFN-α2b) treatment in preneoplastic rat livers. Growth Factors 28: 166-177.

Perz, J.F., G.L. Armstrong, L.A. Farrington, Y.J. Hutin and B.P. Bell. 2006. The contributions of hepatitis B virus and hepatitis C virus infections to cirrhosis and primary liver cancer worldwide. J. Hepatol. 45: 529-538.

Peto, R., R. Gary, P. Brantom and P. Grasso. 1991. Dose and time relationships for tumor induction in the liver and esophagus of 4080 inbred rats by chronic ingestion of *N*-nitrosodiethylamine or *N*-nitrosodimethylamine. Cancer Res. 51: 6452-6469.

Polakis, P. 2000. Wnt signaling and cancer. Genes Dev. 14: 1837-1851.

Purcell, R., M. Childs, R. Miabach, C. Miles, C. Turner, A. Zimmermann and M. Sullivan. 2011. HGF/c-Met related activation of β-catenin in hepatoblastoma. J. Expt. Clin. Cancer Res. 30: 96.

Rignall, B., A. Braeuning, A. Buchmann and M. Schwarz. 2011. Tumor formation in liver of conditional β-catenin-deficient mice exposed to a diethylnitrosamine/phenobarbital tumor promotion regimen. Carcinogenesis 32: 52-57.

Rubinfeld, B., I. Albert, E. Porfiri, C. Fiol, S. Munemitsu and P. Polakis. 1996. Binding of GSK3β to the APC-β-catenin complex and regulation of complex assembly. Science 272: 1023-1026.

Sakanaka, C. 2002. Phosphorylation and regulation of beta-catenin by casein kinase I epsilon. J. Biochem. 132: 697-703.

Sampson, E.M., Z.K. Haque, M.C. Ku, S.G. Tevosian, C. Albanese, R.G. Pestell, K.E. Paulson and A.S. Yee. 2001. Negative regulation of the Wnt-beta-catenin pathway by the transcriptional repressor HBP1. EMBO J. 20: 4500-4511.

Sarkar, F.H., Y. Li, Z. Wang and D. Kong. 2010. The role of nutraceuticals in the regulation of Wnt and Hedgehog signaling in cancer. Cancer Metastasis Rev. 29: 383-394.

Sakabe, T., H. Tsuchiya, M. Endo, A. Tomita, K. Ishii, K. Gonda, R. Murai, K. Takubo, Y. Hoshikawa, A. Kurimasa, N. Ishibashi, S. Yanagida and G. Shiota. 2007. An antioxidant effect by acyclic retinoid suppresses liver tumor in mice. Biochem. Pharmacol. 73: 1405-1411.

Schütte, K., J. Bornschein and P. Malfertheiner. 2009. Hepatocellular carcinoma – epidemiological trends and risk factors. Dig. Dis. 27: 80-92.

Severi, T., H. van Malenstein, C.Verslype and J.F. van Pelt. 2010. Tumor initiation and progression in hepatocellular carcinoma: risk factors, classification, and therapeutic targets. Acta Pharmacol. Sin. 31: 1409-1420.

Stein, T.J., A. Jochem, K.E. Holmes and E.P. Sandgren. 2011. Effect of mutant β-catenin on liver growth homeostasis and hepatocarcinogenesis in transgenic mice. Liver Int. 31: 303-312.

Takigawa, Y. and A.M.C. Brown. 2008. Wnt signaling in liver cancer. Curr. Drug Targets 9: 1013-1024.

Tan, X., U. Apte, A. Micsenyi, E. Kotsagrelos, J.H. Luo, S. Ranganathan, D.K. Monga, A. Bell, G.K. Michalopoulos and S.P. Monga. 2005. Epidermal growth factor receptor: a novel target of the Wnt/beta-catenin pathway in liver. Gastroenterology 129: 285-302.

Taniguchi, K., L.R. Roberts, I.N. Aderca, X. Dong, C. Qian, L.M. Murphy, D.M. Nagorney, L.J. Burgart, P.C. Roche, D.I. Smith, J.A. Ross and W. Liu. 2002. Mutational spectrum of beta-catenin, AXIN1, and AXIN2 in hepatocellular carcinomas and hepatoblastomas. Oncogene 21: 4863-4871.

Teiten, M.H., F. Gaascht, M. Dicato and M. Diederich. 2011. Targeting the Wingless pathway with natural compounds as chemopreventive or chemotherapeutic agents. Curr. Pharm. Biotechnol. 13: 245-254.

Thomas, M.B., D. Jaffe, M.M. Choti, J. Belghiti, S. Curley, Y. Fong, G. Gores, R. Kerlan, P. Merle, B. O'Neil, R. Poon, L. Schwartz, J. Tepper, F. Yao, D. Haller, M. Mooney and A. Venook. 2010. Hepatocellular carcinoma: consensus recommendations of the National Cancer Institute Clinical Trials Planning Meeting. J. Clin. Oncol. 28: 3994-4005.

Thompson, M.D. and S.P. Monga. 2007. WNT/β-catenin signaling in liver health and disease. Hepatology 45: 1298-1305.

Thompson, M.D., M.J. Dar and S.P. Monga. 2011. Pegylated interferon alpha targets Wnt signaling by inducing nuclear export of β-catenin. J. Hepatol. 54: 506-512.

Verna, L., J. Whysner and G.M. Williams. 1996. *N*-nitrosodiethylamine mechanistic data and risk assessment: bioactivation, DNA-adduct formation, mutagenicity, and tumor initiation. Pharmacol. Ther. 71: 57-81.

Wang, Q.M., K.M. Yang, H.Y. Zhou, Z.H. Yu, X. Li and H.J. Yang. 2006. Role of beta-catenin in hepatocarcinogenesis of rats. Hepatobiliary Pancreat. Dis. Int. 5: 85-89.

Wei, W., M.S. Chua, S. Grepper and S.K. So. 2009. Blockade of Wnt-1 signaling leads to anti-tumor effects in hepatocellular carcinoma cells. Mol. Cancer 8: 76.

Wei, W., M.S. Chua, S. Grepper and S. So. 2010. Small molecule antagonists of Tcf4/β-catenin complex inhibit the growth of HCC cells *in vitro* and *in vivo*. Int. J. Cancer 126: 2426-2436.

Wong, C.M., S.T. Fan and I.O. Ng. 2001. β-Catenin mutation and overexpression in hepatocellular carcinoma: clinicopathologic and prognostic significance. Cancer 92: 136-145.

Wong C.M and O.L. Ng. 2008. Molecular pathogenesis of hepatocellular carcinoma. Liver Int. 28: 160-174.

Xie, X.L., M. Wei, A. Kakehashi, S. Yamano, M. Tajiri and H. Wanibuchi. 2011. 2-Amino-3-methylimidazo[4,5-f]quinolone (IQ) promotes mouse hepatocarcinogenesis by activating transforming growth factor-β and Wnt/β-catenin signaling pathway. Toxicol. Sci. 125: 392-400.

Yamada, Y., N. Yoshimi, S. Sugie, M. Suzui, K. Matsunaga, K. Kawabata, A. Hara and H. Mori. 1999. Beta-catenin (Ctnnb1) gene mutations in diethylnitrosamine (DEN)-induced liver tumors in male F344 rats. Jpn. J. Cancer Res. 90: 824-828.

Zekri, A.R., A.A. Bahnassy, H.M. Alam El-Din, H.M. Morsy, S. Shaarawy, N.Z. Moharram and S.S. Daoud. 2011. Serum levels of β-catenin as a potential marker for genotype 4/hepatitis C-associated hepatocellular carcinoma. Oncol. Rep. 26: 825-831.

Zeng, X., H. Huang, K. Tamai, X. Zhang, Y. Harada, C. Yokota, K. Almeida, J. Wang, B. Doble, J. Woodgett, A. Wynshaw-Boris, J.C. Hsieh and X. He. 2008. Initiation of Wnt signaling: control of Wnt coreceptor Lrp6 phosphorylation/activation via frizzled, dishevelled and axin functions. Development 135: 367-375.

Zhang, X.F., X. Tan, G. Zeng, A. Misse, S. Singh, Y. Kim, J.E. Klaunig and S.P. Monga. 2010. Conditional beta-catenin loss in mice promotes chemical hepatocarcinogenesis: role of oxidative stress and platelet-derived growth factor receptor alpha/phosphoinositide 3-kinase signaling. Hepatology 52: 954-965.

Zucman-Rossi, J., S. Benhamouche, C. Godard, S. Boyault, G. Grimber, C. Balabaud, A.S. Cunha, P. Bioulac-Sage and C. Perret. 2007. Differential effects of inactivated Axin1 and activated beta-catenin mutations in human hepatocellular carcinomas. Oncogene 26: 774-780.

CHAPTER 4

Urine Biomarkers for Prostate Cancer Detection

Yan-Hua Chen[1,2], Art A. Rodriguez[3] and Qun Lu[1,2*]

[1]Department of Anatomy and Cell Biology, 7N100 Brody Medical Sciences Building;
[2]Leo Jenkins Cancer Center, The Brody School of Medicine, 600 Moye Boulevard, East Carolina University, Greenville, North Carolina 27834, U.S.A
[3]Department of Chemistry, 300 Science and Technology Building, Thomas Harriot College of Arts and Sciences, East Carolina University, Greenville, North Carolina 27858, U.S.A

ABSTRACT

Prostate cancer (PCa) is the most common solid tumor malignancy in men in the US. Even though PCa is responsible for over 30,000 deaths per year in the US alone, effective screening and diagnosis are a matter of controversy.

The approach to screening has been widely adopted clinically by the discoveries of *prostate specific antigen* (PSA) as a serum biomarker, digital rectal examination (DRE), and transrectal ultrasound (TRUS) as live imaging of the prostate. Urine biomarkers are also attractive because they can offer simple, non-invasive early detection of PCa and allow repeated sampling as well as broader population access to the screening. However, their reliability is debatable because the test results are often subject to changes by patient diet, behavior, ways of urine collection, and storage.

There has been significant progress in recent years on the discovery and development of urine biomarkers for PCa detection. Notably, new tests using PCA3 noncoding mRNA and TMPRSS2:ERG gene fusion are being validated as potential urine PCa biomarkers. Additional urine biomarkers currently under investigation include PCADM-1, sarcosine, engrailed, δ-catenin/NPRAP/Neurojugin, minichromosome maintenance 5 protein, AGR2, and prostate specific membrane antigen, among many others. An ideal urine biomarker should fulfill important criteria such as its

*Corresponding author: E-mail: luq@ecu.edu

ability to distinguish between normal and cancerous tissues, scientific support of its functional relevance to PCa progression, and its ease of interpretation by clinicians. It is likely that a single biomarker by itself may not be able to fulfill all of the criteria. Therefore, a combination of multiple biomarkers with existing tools would provide the best possibility of reliable PCa detection.

ADVANCES AND CHALLENGES IN PCa MANAGEMENT

American Cancer Society statistical data shows that new cases of cancer in the United States in 2010 reached 1,529,560 (excluding nonmelanoma skin cancers), from which 569,460 died. In the near future, cancer is expected to pass heart disease as the number one killer. Cancer has a huge economic impact on our society with estimated total costs over $200 billion in recent years. Prostate cancer (PCa) is the most prevalent noncutaneous cancer and the second leading cause of cancer death in men in the United States (Jemal et al. 2010).

As numerous studies have shown, early detection saves lives and increases treatment options. For example, when PCa is detected early, the five-year survival rate is estimated almost at 100 percent, which underscores the importance of early screening and diagnosis. While there is insufficient data to support for or argue against PCa testing in men with an average risk of developing the disease, the American Cancer Society initially recommended that the PSA test and the DRE should be offered to men with average risk beginning at the age of 50. Individuals who are at high risk of developing PCa (e.g., African Americans or men with a strong family history of PCa) should begin the screenings at age 45.

Over the past several decades, the five-year survival rate for PCa has increased significantly due in large part to the availability of PSA screening. This diagnostic option has led to the early detection of many cases of PCa at both the local and regional stages. However, accumulating data has shown that the serum PSA level could potentially be elevated as a result of conditions other than PCa, such as benign prostatic hyperplasia (BPH). Consequently, false positives are a major problem for the PSA test and can lead to unnecessary biopsies or other uncomfortable interventions (for review, see Tricoli et al. 2004). In addition, preoperative serum PSA levels do not always correlate with cancer volume or the Gleason score of radical prostatectomy specimens (Stamey et al. 2002). Many studies estimate that PSA tests can miss

PCa by more than 20 to 30 percent. Recently, several US national reviews of the clinical data led to the conclusion against PSA screening for PCa.

Therefore, it is important to develop more reliable biomarkers to complement the PSA test for PCa risk assessment and the prediction of aggressiveness. The discovery and validation of additional PCa biomarkers independent of the PSA test will not only improve early detection and prognosis, but also provide better guidance for treatment options.

URINE BIOMARKERS FOR PCa DETECTION

Many methodologies have been exploited for PCa detection. Clinically, DRE during office visits and TRUS can reveal signs of prostate enlargement and PCa. Currently, the laboratory serum test on PSA levels, often presented as the ratio of free to total PSA, would provide the rationale for biopsies if it surpassed an age-tailored, scientifically still-questionable threshold. This article will discuss other potential new biomarkers under investigations with the focus on urine biomarkers.

Urine is naturally an appealing source when looking into urinary tract diseases. For PCa detection, viable urinary tests are not available clinically at present. Urine test results are often subject to changes based on patient diet, behavior, ways of collection, and storage. However, urinary biomarkers are attractive because they are inexpensive, easy to collect in large quantities, amenable for repeated samplings, and simple to process for storage and transportation. Most importantly, they are non-invasive to patients, so a viable urine PCa biomarker makes early detection screening more convenient and accessible. Here, we will consider the potential urine biomarker types ranging from DNA, RNA, protein, metabolites, and cellular organelles from urine sediments.

GENETIC COMPONENTS INVOLVING DNA OR RNA

Risk Assessment and the Prediction of PCa Progression

Currently, it is estimated that over 35 single nucleotide polymorphisms (SNPs) have been identified and found to be associated with PCa. There are over 19,000 publications that cover the topic of biomarkers and PCa, but surprisingly only one biomarker, i.e. serum PSA, is currently used in the clinic (Aly and Grönberg 2011).

Genetic testing can provide valuable guidance for disease risk assessment. For example, the presence of apolipoprotein E4 allele has been linked to an eight times higher risk for Alzheimer's disease (Roses 2006). For colorectal cancer, adenomatosis polyposis coli (APC) mutations account for over 90 percent of the colon cancer cases in families carrying the mutations (Hegde and Roa 2006). For Alzheimer's disease, it is unfortunate that the ApoE4 identification is not able to provide the disease treatment recommendation because there is no effective therapy for treating Alzheimer's disease. However, treating PCa during the early stages can improve survival tremendously. Therefore, genetic tests showing a high correlation to disease occurrence and aggressiveness will offer a powerful tool for providing guidance for PCa detection and treatment strategies.

Recent genome-wide association (GWA) studies of PCa have detected numerous variants associated with the disease as well as common gene fusions and expression 'signatures' in prostate tumors (Witte 2009; Catalona et al. 2011). In the past few years, multiple GWA studies have detected over a dozen germline SNPs that are each associated with a modest increase in the risk of PCa (Witte 2007; Haiman et al. 2007; Thomas et al. 2008; Eeles et al. 2008). One of the striking findings shows that at least three distinct loci within a ~1 Mb segment of chromosome 8q24 harbor germline variants that are associated with PCa (e.g. see Haiman et al. 2007). These combined SNPs could cause a substantially higher risk among the limited proportion of men who carry large numbers of genetic variants.

In addition to germline variants detected by GWA studies, somatic mutations can now be identified by expression array analyses of PCa specimens. One of the most exciting studies that used expression arrays found common gene fusions in PCa tumors (Tomlins et al. 2005). The bioinformatics approach for cancer outlier profile analysis (COPA) has been successfully used to search for genes that are overexpressed in a subset of prostate tumors. This type of approach has revealed that the 5'-untranslated region (5'UTR) of the androgen-regulated gene TMPRSS2 is fused to ETS family genes, leading to the androgenic induction of oncogenic transcription factor overexpression (Tomlins et al. 2005).

However, the small effect size for each associated GWA SNP (with most of the odds ratios less than 1.30) suggests that these SNPs collectively might only account for approximately 15 percent of the familial risk of PCa (Eeles et al. 2008).

In contrast to using genetic markers to screen non-diseased men, it is clinically important to identify markers that can tell which prostate tumors will progress rapidly and be life-threatening versus those that are relatively latent and might not impact a man's health status. This is especially important since the current standard treatment options for PCa result in considerable co-morbidity and detrimental effects on a patient's quality of life.

PCA3 Noncoding mRNA Expression

There are encouraging recent studies that show potential utility of nucleotide sequence variations and expressions of PCA3 and TMPRSS2 detected in the urines of PCa patients.

A urine test based on the prostate cancer antigen 3 (PCA3) gene, originally called DD3 (Differential Display clone 3), is probably the best adjunct to serum PSA for predicting biopsy outcomes, and has proven its clinical relevance by surpassing the predictive abilities of traditional serum biomarkers (Ploussard and De la Taille 2010).

PCA3 was first identified in 1999 using differential display analysis to compare the mRNA expression patterns with the normal and tumor tissues of human prostate (Bussemakers et al. 1999). Studies show that PCA3 is a segment of non-coding messenger RNA from chromosome 9q21-22 and is overexpressed in 95 percent of PCa, although the functions of PCA3 is unclear at present (Bourdoumis et al. 2010; Hessels and Schalken 2009). PCA3 RNA is extracted and determined in voided urine samples after attentive DRE of the prostate, which allows cells to exfoliate in urine (Bourdoumis et al. 2010). This PCA3-based urine test initially used the quantitative RT-PCR method (Hessels et al. 2003). Later, the quantitative APTIMA PCA3 assay was developed for clinical studies to support the diagnosis of PCa (Groskopf et al. 2006). The procedures of this analysis include the purification of target mRNA (PCA3 and PSA), amplification of target RNA sequences, and specific detection of amplified products. The readout is the ratio of PCA3/PSA mRNA copy numbers after normalization.

Although the sensitivity of the PCA3 test was less than that of PSA, its specificity appeared to be improved, particularly in patients with a previously negative biopsy (Vlaeminck-Guillem et al. 2010). Groskopf et al. (2006) reported that, while using APTIMA PCA3 assay, the sensitivity was 69 percent and specificity was 79 percent at a PCA3/PSA mRNA ratio cutoff of 50. This specificity was much better than that of the serum PSA assay used for the same study group (28 percent at a cutoff of 2.5µg/L). In another clinical trial involving 463 European men reported by Haese et al. (2008), the sensitivity and specificity were 46 percent and 73 percent, respectively, with the PCA3 score at a cutoff of 35. Ochiai et al. (2011) conducted the PCA3 urine test on 105 Japanese men and reported that the probability of PCa was 69 percent at a PCA3 score >50 and 5 percent at a PCA3 score <20. These combined data indicate that PCA3 urine test can be potentially supplemented as a biomarker to improve the prediction for PCa.

TMPRSS2-ETS Fusion

Another potential urine biomarker for PCa is TMPRSS2-ETS fusion. The fusion of TMPRSS2 with ETS transcription factor genes in PCa was initially reported by Tomlins et al. (2005) and later confirmed by several other groups. The ETS-related gene (*ERG*) is the most common fusion partner for the androgen regulated gene *TMPRSS2*. Both genes are located within three Mb of chromosome 21, and the most common mechanism for fusion is through an interstitial deletion (Demichelis and Rubin 2007). TMPRSS2:ERG fusion has been identified as a common molecular event in PCa development and is reported to being present in 54 percent of PCa (Mehra et al. 2007). Because the androgen hormones activate the TMPRSS2 gene promoter, the prostate expresses the fusion gene, promoting cell growth and invasion and reducing cell death (Tuma 2010).

By using fluorescence in situ hybridization (FISH) method, Rajput et al. (2007) reported that *TMPRSS2:ERG* fusion is more often found in moderately to poorly differentiated tumors (35/86, 40.7%) than in well-differentiated tumors (1/15, 6.7%). To evaluate the specificity and sensitivity of the *TMPRSS2:ERG* fusion in urine, multiple laboratories have performed the test to determine the amount of TMPRSS2:ERG mRNA in urine sediments after attentive DRE procedure by quantitative PCR method. Nguyen et al. (2011) reported a lack of fusion-positive samples in PCa-free subjects and a clinical sensitivity of 45.4 percent in PCa patients prior to treatments. In the study described above from University of Michigan in Ann Arbor, TMPRSS2:ERG score from the urine test corresponded to Gleason score, tumor volume, and Epstein criteria for significant PCa (Mehra et al. 2007). Another study, reported by Dr. James B. Amberson from DIANON Systems in Connecticut, showed that TMPRSS2:ERG urine test has a specificity of 89 percent and a sensitivity of 39 percent (Lau 2010).These clinical results indicate that TMPRSS2:ERG urine test may eventually be used to supplement the PSA score to screen PCa in the primary care setting. It was also indicated that it has the potential to help patients avoid unnecessary biopsies.

δ-Catenin/NPRAP/Neurojungin mutations

δ-Catenin/NPRAP/Neurojungin (gene designation: *CTNND2*) is an adhesive junction protein of the β-catenin superfamily and was initially identified as a neural specific protein in the brain (Paffenholz and Franke 1997; Zhou et al. 1997; Lu et al. 1999). However, examination of human EST data bank revealed *δ-catenin* mRNA sequences in kidney, lung, ovarian, brain, breast, prostate, and esophageal tumors. Other studies demonstrated that *δ-catenin* mRNA was overexpressed in PCa compared to that in BPH (Burger et al. 2002). Tissue microarray (TMA) studies showed that *δ*-catenin protein expression begins to rise in prostatic intraepithelial neoplasia (PIN) and is correlated with increasing Gleason scores and tapers when Gleason score reaches 10, indicating that δ-catenin is a potential indicator for PCa progression (Lu et al. 2005).

The frequently increased expression of the brain protein δ-catenin in peripheral tissues of cancer raised important questions, such as what mechanisms are used in cancer that result in a high level of δ-catenin expression. Recent studies showed that E2F1 and Pax6 positively upregulates δ-catenin expression in PCa cells in culture (Kim et al. 2008). While there are no changes in methylation status and gene amplification, *δ-catenin* displayed increased nucleotide sequence variations in the promoter and 5'UTR of PCa patients (Wang et al. 2009).

The unique feature of SNPs and mutations in *δ-catenin* gene in relation to PCa is that there are not always recurrent SNPs in the same site. Rather, there is an increased incidence of SNPs in *δ-catenin* 5'UTR of prostatic adenocarcinomas. Some nucleotide changes reflect true somatic mutations. For example, -9G-A was identified only in the tumor clusters that were isolated using laser capture micro-dissection (LCM) technology, but not in the adjacent benign glands (Table 1). To verify the authenticity of the variations, two sets of PCR primers were designed to detect SNPs in two PCR products of different sizes (Wang et al. 2009). These analyses indicated that δ-*catenin* gene promoter and 5'UTR are subject to substantial genetic variations. PCa patients showing an increased presence of any single or multiple variations in these regions may overcome the barriers and increase the odds of linking the occurrence of SNPs to PCa. However, these studies were performed on PCa tissues. Although *δ-catenin* gene can be amplified from urine, further studies are needed to determine whether its mutations can be identified and correlated with the disease at urine level.

Table 1 The polymorphic changes in the nucleotide sequence of *δ-catenin* gene promoter in PCa.

Mutation sites	*Mutation/SNP*	*Function*
–9 G-A	True mutation	Increase expression
–13 C-T	True recurrent mutation	Not known
–17 T-A	True mutation	Not known
–58 C-T	SNP	Not known
–97 A-C or insert C	Recurrent mutations	Not known
–98 G-C/T	Recurrent mutations	Not known

NON-NUCLEIC ACID COMPONENTS INVOLVING PROTEINS

PCADM-1

By using DNA-protein binding assays for the identification of novel transcriptional regulatory proteins associated with the development of PCa, one gene in particular was identified: a novel 33 kDa chromosomal protein, termed prostate cancer diagnostic marker -1 or PCADM-1 (Ohkia et al. 2004). Sequencing analysis revealed that the gene had six specific point mutations, which appeared to provide the mutant protein with DNA binding capability. RT-PCR assays indicated that the mRNA was expressed in human PCa cells and tissues but was not present in BPH, HGPIN, or seminal vesicle tissue. Western blots, in situ hybridization, and immunolabeling studies of human PCa tissues also indicated that PCADM-1 expression is specifically associated with the glandular epithelial cells. The levels of expression increased from +1 to +5 with an increased Gleason score of 4, 6 and 8, respectively. The study showed that the protein was not expressed by other cancers, indicating that it was PCa-specific.

Ohkia et al. (2004) applied the ELISA method to show that the protein was a highly sensitive urine marker for PCa. A beta test of the urine of 43 patients with known PCa and 40 patients with BPH yielded a sensitivity of 75 percent and a specificity of approximately 78 percent. They subsequently carried out a blinded study with 533 patients with either PCa (n=48), BPH (n=88), urinary problems (n=108), no symptoms (n=289) or normal people (n=40). Their data showed that the PCADM-1 urine assay had a sensitivity of 79 percent and a specificity of 83 percent overall (NPV=79%; PPV=78%). A second study of 157 patients diagnosed with PCa (n=76) and BPH (n=81) yielded a sensitivity of 83 percent and specificity of 74 percent. While there is no further literature available for this gene in association with PCa, a biotechnology company (MacroArray Technologies, LLC) reportedly planned trials to develop an FDA-approved test for PCa detection.

Engrailed

Many genes involved in early embryonic development are turned off after birth but are subsequently re-expressed in cancer. The *HOX* genes are a family of homeodomain-containing transcription factors that specify the early identities of cells and tissues (Shah and Sukumar 2010). Studies have shown that *HOX* genes are often dysregulated in cancers, and accumulating evidence suggests that targeting HOX/PBX binding may have therapeutic value (Daniels et al. 2010; Morgan et al. 2007; Morgan et al. 2010; Shears et al. 2008; Plowright et al. 2009). *Engrailed-2* (*En2*) is a member of *Hox* gene family, showing a very high degree of functional conservation during development (Morgan 2006). *En2* is a transcriptional repressor, but it also has a regulatory role at the translational level. In addition to its roles in embryonic development, *En2* displays oncogenic properties in breast cancer. For example, forced overexpression of *En2* in the nonmalignant mammary cells induces malignant phenotypes including increased cell proliferation and a loss of contact inhibition (Martin et al. 2005). Furthermore, downregulation of En2 expression causes a dramatic decrease in PCa cell proliferation (Bose et al. 2008).

Recent studies show that *En2* is expressed in and secreted by PCa, but not in normal prostatic tissue. Since EN2 localization is close to the cellular membrane structure and shows the properties of secretory proteins in embryonic development (Maizel et al. 1999), it is interesting that EN2-positive blebs can be found within the lumen of malignant prostatic ducts.

These features of EN2 prompted two studies in the United Kingdom (Morgan et al. 2011). One study was performed at Surrey University and used ELISA assay to screen representative populations of PCa patients and relevant controls. It was very interesting that the stability of EN2 protein in urine lasted at least four days at room temperature, allowing for postal collection of samples with the mean age of men in the study ranging from 57 to 67 years. Median PSA (6.3 to 7.6 ng/mL) was similar in men suspected of PCa versus the two low PSA control groups (0.9 to 1.1 ng/mL). Using a cutoff of 42.5 µg/L, EN2 protein was detected in 54 of the 82 (66%) men with PCa confirmed by biopsy. Notably, in this EN2 positive group the PSA was <2.5 µg/L in nine men. In men with high PSA and no cancer on biopsy, EN2 was detected in six of 58 (10.3%). EN2 detection was also infrequent in the control groups.

A receiver operating characteristic (ROC) analysis for this dataset revealed that the area under the curve (AUC) was 0.8021 ($P < 0.001$), indicating that EN2 may have diagnostic potential for PCa. The mean concentration of EN2 protein was 10.4-fold higher in PCa patients compared to that in all the men that were not known to have cancer (Morgan et al. 2011). There was no significant correlation between EN2 expression and combined Gleason score, although the majority of patients in this cohort had diseases with a combined Gleason score of 6 or 7. There was also no significant correlation between serum PSA level and the presence or absence of urinary EN2 protein. Additionally, EN2 was found in the urine of three of 10 men with high-grade prostatic intraepithelial neoplasia (PIN). Interestingly, two of these three men had confirmed PCa within six months upon re-biopsy.

To validate these findings on urinary EN2 secretion in PCa, a similar study was completed independently at Cambridge University following the same collection and test protocol as the Surrey University study (Morgan et al. 2011). This analysis found the results to be similar to those obtained for the patients at Surrey University. Of 13 control patients, two had detectable levels of EN2 in their urine (15%), whereas 47 of 81 patients with PCa tested positive for EN2 (58%). These two studies published in the same article collectively suggest the potential of EN2 as a urine PCa biomarker and warrant further investigations.

δ-Catenin/NPRAP/Neurojungin

The surprising discovery of a brain-enriched, neuronal gene δ-*catenin* (*CTNND2*) being overexpressed in a large percentage of primary human prostatic adenocarcinomas has also suggested a potential PCa biomarker candidate. Because δ-catenin is primarily expressed in the brain, its expression background is low in normal prostatic glands (Burger et al. 2002; Lu et al. 2005).

Using TMA analysis, 90 cases of PCa patient prostatic tissues were obtained from prostatectomy and then compared with 90 cases of normal prostates (Lu et al. 2005). 92 percent of the PCa samples showed strong (46/72 cases) or moderate (20/72 cases) staining (immunoscore ≥2) for δ-catenin; only six (8%) of the 72 cases of specimens were either negative or had an immunoscore of less than two. Although 49 (75%) of the 65 cases of benign specimens showed an immunoscore of less than two, six (9%) of the 65 cases of benign samples showed an immunoscore equal to two, and 10 (15%) of the 65 cases of benign specimens showed an immunoscore of higher than two. These data indicated a clinical sensitivity of 91.7 percent and a specificity of 75.4 percent if specimens are evaluated using δ-catenin immunoscore alone.

TMA analysis also showed that δ-catenin expression is increased with the down-regulation of cell-cell junction proteins such as E-cadherin and $p120^{ctn}$ in PCa epithelial cells. Increased δ-catenin expression with increasing Gleason scores provided a potential biomarker candidate for molecular diagnostics based on junctional complex proteins (Lu et al. 2005). A recent study of isolated human urine sediments showed that δ-catenin, caveolin-1, and CD59 were all detected in cell-free human voided urine prostasomes. δ-Catenin immunoreactivity, which is associated with membrane blebs when it is overexpressed, was significantly increased in the urine of PCa patients ($P < 0.0005$). These studies raised the intriguing possibility for its potential utility for non-invasive PCa detection (Lu et al. 2009).

The development of a biosensor for ELISA and quantum dot (QD) application has been underway and suggested the feasibility to detect human δ-catenin in voided urine. Besides the traditional ELISA approach, the QD bioconjugate method requires only 500 uL of urine in a pre-prepared sample cuvette that contains the bioconjugate QD for fluorescence measurement. A small shift from a standard sample (in a double-beam instrument) can quickly establish the presence of a protein such as δ-catenin.

QDs are tiny inorganic materials or composites, usually made of Au (gold), Ag (silver), CdSe/ZnS (cadmium selenide/zinc sulfide), or PbSe (lead selenide), with a range in size from 2-12 nanometers in diameter (Bruchez et al. 1998). Unlike organic-based biosensors, the nano-scale QD-biosensors possess very narrow absorption or emission bands and do not degrade due to light source exposure. Furthermore, QDs possess higher sensitivities and broader wavelength selectivity. Although the development of QD-based biosensors is a relatively recent technology, the extreme versatility, selectivity and robustness of these biosensors have created a high demand in the biosensor market of the development of sensors for a variety of applications.

As an initial step, we explored whether the QD/antibody bioconjugates can recognize and interact with δ-catenin. In the presence of δ-catenin protein, the fluorescence spectra of QD/anti-δ-catenin+δ-catenin protein is obtained (Fig. 1). The emission peaks of the QD-antibody bioconjugates can experience a 2~6 nm shift when binding to the δ-catenin antigen (QD-anti-δ-catenin+δ-catenin). These measurements support the concept of developing and utilizing QD biosensors to determine the presence of δ-catenin in body fluids.

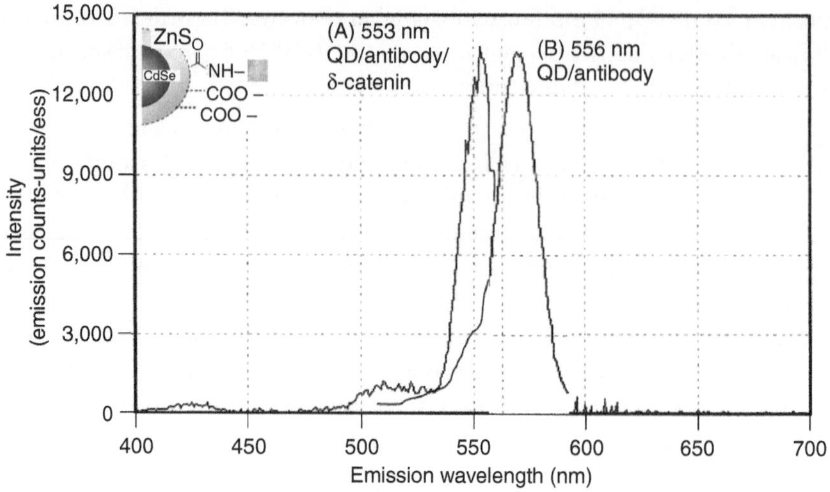

Fig. 1 Fluorescence spectrum of QD-antibody+δ-catenin (a) versus QD-antibody (b). Note the shift of emission peak due to QD-antibody interaction with δ-catenin protein in solution. To the upper left is the QD structure illustration.

Minichromosome Maintenance 5 Protein

The minichromosome maintenance (Mcm) proteins are highly conserved proteins that are essential for initiating and regulating eukaryotic DNA replication. Recent studies have demonstrated the potential use of Mcm family (Mcm2-7) proteins as markers of proliferation. Meng et al. (2001) characterized the pattern of Mcm2 staining in benign and malignant prostate tissues and examined the role of Mcm2 expression in disease-free survival after surgery in men with localized PCa. Immunohistochemistry with a Cox proportional hazards analysis was performed and determined that Mcm2 expression is consistently increased in malignant glands and significantly associated with disease-free survival in univariate (P = 0.002) and multivariate (P = 0.01) analyses. Patients with high Mcm2 expression exhibited shorter disease-free survival. Mcm2 expression was not associated with any traditional, clinical, or pathological factors and therefore is an independent predictor of survival in these patients with PCa.

A subsequent study investigated the use of Mcm7 as a proliferation marker in 79 lymph node negative PCa and found that Mcm7 had a significantly higher proliferation index than Ki-67 at PIN or adenocarcinoma stages (Padmanabhan et al. 2004). On the other hand, Stoeber et al. (2002) used an immunofluorometric assay and measured Mcm5 levels in urine sediments of 353 patients who either had hematuria or lower urinary tract symptoms, or were undergoing follow-up cystoscopy for urothelial neoplasia (Stoeber et al. 2002). A two-sided analysis with McNemar's test and Fisher's exact test found that Mcm5 was detected in primary and recurrent bladder cancers with 87 percent (95% confidence interval [CI] = 77% to 94%) sensitivity and 87 percent (95% CI = 83% to 91%) specificity. Patients with PCa had higher levels of Mcm5 in their urine sediments than did men without malignancy (P<.001).

Whereas these studies have shown that Mcm proteins are useful as a proliferation marker, recent studies suggest that deregulated MCM expression actively contributes to tumor formation and progression as well as malignant conversion, which all may be therapeutic targets (Honeycutt et al. 2006; Majid et al. 2010).

Because the Mcm family proteins are potential cancer diagnostic markers, Dudderidge et al. (2010) directly assessed whether urinary Mcm5 level is associated with PCa. Studies from 88 men with PCa and from two control groups negative for malignancy detected PCa with 82 percent sensitivity (confidence interval (CI)= 72-89%) and with a specificity ranging from 73 percent (CI=68-78%) to 93 percent (CI=76-99%). Attentive prostate massage led to increased Mcm5 signals compared with pre-massage samples (median 3440 (interquartile range (IQR) 2280 to 5220) vs 2360 (IQR <1800 to 4360); P=0.009), and was associated with significantly increased diagnostic sensitivity (82 vs 60%; P=0.012). Therefore, urinary Mcm5 detection seems to be a simple, accurate, and non-invasive method for identifying patients with PCa. These studies suggest that large-scale prospective trials are now required to evaluate this test further in diagnosis and screening.

Prostate-specific Membrane Antigen

Prostate-specific membrane antigen (PSMA) is a 750-residue integral membrane glycoprotein and the target of an in-vivo imaging agent for metastatic PCa. By using a dual monoclonal antibody (mAb) sandwich assay, which detects PSMA at a sensitivity <1 ng/mL, Sokoloff et al. (2000) reported that prostate tissue membranes expressed the highest levels of PSMA than that of other tissues did.

Interestingly, several recent studies suggest that a combination of PSMA with PSA and PCA3 may serve as a better urinary biomarker for PCa. Talesa et al. (2009) investigated the diagnostic value of five different genes that are associated with PCa carcinogenesis, encoding serine protease Hepsin, PCA3,

UDP-N-acetyl-alpha-D-galatosamine transferase (GalNAC-T3), PSA, and PSMA. 44 patients with previously untreated, histologically verified PCa and 46 patients with BPH were enrolled in the study. Using logistic regression analysis, it was demonstrated that the diagnostic potential of the combined urinary PSA and PSMA levels was significantly better than the individual potentials of each marker.

An independent study by Rigau et al. (2011) indicated that a single marker may not necessarily reflect the multifactorial nature of PCa. This study analyzed urine samples from 154 patients following prostate massage, which showed an elevated serum PSA (>4 ng/ml) and/or abnormal DRE results. This study found that PSMA, PSA, and PCA3 were all significant predictors of PCa. Furthermore, the use of a multiplex model showed that the AUC was 0.74 versus 0.82 in the diagnostic "gray zone" (4-10ng/ml). Fixing the sensitivity at 96 percent, a specificity of 34 percent and 50 percent in the gray zone was obtained. These results suggest a strategy for the development of a more accurate model for PCa diagnosis by using a multiplexed, urine-based diagnostic test for PCa with a higher specificity but the same sensitivity as that of the serum-PSA test.

AGR2

Anterior gradient 2 (AGR2) is another interesting protein involving neural tissues in development but is associated with PCa progression. AGR2 is a secreted protein of 17~19kDa and a human homologue of the Xenopus laevis protein xenopus anterior gradient-2 (XAG-2). XAG-2 is expressed in the cement gland and the neural plate during the formation of the anterior-posterior axis in embryogenesis (Aberger et al. 1998). Its homologue in the newt (nAG) is expressed and secreted during regeneration of amputated limbs and is involved in the nerve-dependent limb regrowth (Kumar et al. 2007). In human, AGR2 has been shown to be expressed in prostate, breast, small intestine, colon, lung, and pancreas. Several studies have reported its association with breast cancer and a variety of other tumors, including cancers of pancreatic, lung, colon, esophageal, and prostatic tissues.

AGR is a member of the endoplasmatic reticulum (ER) protein disulphide isomerase (PDI) gene family (Park et al. 2009). PDI proteins facilitate the isomerization of their substrates into bioactive conformations as they transport through the ER for secretion or membrane association. The failure to isomerize can lead to the accumulation of misfolded membrane and secreted proteins in the ER, initiating ER stress, cell-cycle arrest, and apoptosis.

AGR2 shows positive expression in high grade PIN and PCa but there was a trend towards lower levels of AGR2 with increasing Gleason score. Its expression level was associated with poor survival of prostate cancer patients (Zhang et al. 2007). There are two recent studies which demonstrated its presence in voided urine of PCa patients. Bu et al. (2011) used nucleic acid amplification method to detect AGR2 transcript in urine. From gene expression analysis of $CD26^+$ cancer cells, AGR2 transcript appears to be a better marker than that of PCA3. The transcript levels based on array signal intensity values (σ) is 170 for PCA3 but is 4,200 for AGR2. Another study focused on the development of a sandwich ELISA using paired monoclonal antibodies. AGR2 protein is stable in urine at 4°C, so that immediate processing and special care against RNA degradation are not necessary. The sample preparation in this study was unique in that they were desalted and concentrated by filters. Wayner et al. (2012) showed that the developed ELISA detected AGR2 protein, ranging from 3.6 to 181 pg/ml in an initial cohort of 12 samples. AGR2 was not detected in the urine of non-cancer and a bladder cancer patient. Because AGR2 is specific to cancer cells, and apparently secreted into the urine, further studies may be needed to determine whether it can be a viable biomarker using larger sample size.

OTHER MARKERS

Sarcosine

While gene and protein expression have been extensively profiled in human cancers, a 2009 University of Michigan study explored global metabolomic alterations that characterize neoplastic progression of PCa (Sreekumar et al. 2009). By using a combination of high-throughput liquid-and-gas chromatography-based mass spectrometry, this study reported the profiling of more than 1,126 metabolites across 262 clinical samples related to PCa (42 tissues and 110 each of urine and plasma). The unbiased metabolomic profiling was able to distinguish benign prostate, clinically localized prostate tumors, and metastatic PCa. From the profiles, sarcosine was identified as a differential metabolite that was highly increased during PCa progression to metastasis and can be detected in urine.

Sarcosine is an intermediate and byproduct in glycine synthesis and degradation. Glycine-*N*-methyl transferase generates sarcosine from glycine, while sarcosine can also be metabolized to glycine by the enzyme sarcosine dehydrogenase. Sarcosine is a natural amino acid found in muscles and other body tissues. It can also be found naturally as an intermediate in the metabolism of choline to glycine. When it is formed from the dietary intake of choline and from the metabolism of methionine, sarcosine is rapidly degraded into glycine. Besides the importance of being a constituent of protein, glycine plays a significant role in many physiological processes.

The University of Michigan study also found that sarcosine levels were increased in invasive PCa cell lines compared to benign prostate epithelial cells (Sreekumar et al. 2009). It was found that knockdown of glycine-N-methyl transferase attenuated PCa invasion. Interestingly, the addition of exogenous sarcosine or suppressing sarcosine dehydrogenase induced an invasive phenotype in benign prostate epithelial cells. Since androgen receptor and the ERG gene fusion product coordinately regulate components of the sarcosine pathway, the authors concluded that, by profiling the metabolomic alterations of PCa progression, sarcosine is a potentially important metabolic intermediary of cancer cell invasion.

Interests in sarcosine and other potential metabolites and their relationships with PCa have spread rapidly since the 2009 Michigan study with over 35 articles in the literature today. However, the studies reached controversial conclusions. One recent study reported a different result in 2010. After measuring sarcosine levels in urine samples from 106 PCa patients following DRE, they found that sarcosine in urine failed as a marker in detecting PCa and identifying aggressive tumors (Jentzmik et al. 2010). A follow-up study by the same group further suggested that sarcosine in PCa tissue is not a differential metabolite for disease aggressiveness and biochemical progression (Jentzmik et al. 2011). Another study on serum sarcosine also did not support its application as a PCa biomarker (Struys et al. 2010). Therefore, while sarcosine may play important roles in PCa progression, its urine biomarker potential awaits further clarification. Meyer et al. (2011) suggested that the inconsistent results may be due to the fact that sarcosine and alanine co-elute on a HPLC column; therefore, an improved separation method may help determine whether sarcosine in serum or urine is a viable biomarker for PCa detection.

Exosomes and Cellular Shedding

Many molecules present in PCa patient urines are secreted from prostate cells through conventional secretory pathways or excreted through the shedding of small (50-100 nm) membrane-containing vesicles called prostasomes (Duijvesz et al. 2011). For example, RNA markers such as PCA3 are recognized for being associated with exosomes (Jansen et al. 2009).

Pan et al. (1983) first described exosome biogenesis in the reticulocyte maturation of sheep. At the early endosome stage, cytoplasmic content is taken up by the inward budding of endosomal membranes, forming exosomes individually called the *multivesicular body* (MVB). When the MVB fuses with the plasma membrane, the vesicles are secreted with various cellular components including DNAs, RNAs, proteins, lipids, and others.

Based on this principle, a number of approaches have been employed to investigate exosomes isolated from urine samples. Recently, we have shown that PCa cells both in culture and in human voided urines contain prostatesomes of ~100 nm in size, indistinguishable from exosomes described by others (Lu et al. 2009). These exosome-like MVBs excreted by PCa cells express caveolin-1, CD59, and δ-catenin/NPRAP/neurojungin. Furthermore, as a proof-of-concept, a transcriptome analysis revealed the presence of two potential PCa biomarkers described earlier: *PCA3* and *TMPRSS2:ERG* in exosomes isolated from the urine of patients (Nilsson et al. 2009). These studies provide evidence that, at least partially, the basis of these PCa markers detectable in urine is due to their association with exosomes or cellular shedding materials. Finally, the emerging technologies of enumerating circulating tumor cells (CTC) in blood samples may advance the studies of their application in PCa patient urine (Danila et al. 2011).

CONCLUSION

Significant progress has been made in recent years on the discovery and development of urine biomarkers for PCa detection. New tests using PCA3 noncoding mRNA and TMPRSS2:ERG gene fusion are being validated in many clinical studies as potential urine biomarkers. There are a number of additional urine biomarkers under investigation including PCADM-1, sarcosine, engrailed, δ-catenin/NPRAP/Neurojugin, minichromosome maintenance 5 protein, AGR2, and prostate specific membrane antigen, as well as many others (Table 2).

Table 2 Potential urine biomarkers under study for PCa detection.

Maker candidate	*Partition in urine*	*Function*	*PCa relevance*
PCA3	Sediment and supernatant	Non-coding mRNA	Overexpression
TMPRSS2:ERG	Sediment	Gene regulation	Gene fusion
PCADM-1	Not determined	DNA binding Protein	Not known

Table 2 Contd....

Table 2 Contd....

Engrailed	Not determined	Homeodomain transcription factor	Overexpression
δ-Catenin	Sediment and supernatant	Neuronal synaptic adhesion	Overexpression
AGR2	Filter concentrated	ER protein isomerization	Association with Gleason
Mcm5	Not determined	Eukaryotic DNA replication	Overexpression
PSMA	Not determined	Type II membrane glycoprotein	Imaging of prostate
Sarcosine	Not determined	Metabolism	Increased level

There are several features of these urine markers that are worth noting. First, the most reliable urine markers of PCa appear to be associated with urinary sediments and/or cellular shedding materials such as exosomes. This feature may explain the relatively lower sensitivity but higher specificity of PCA3 assay than those of the serum PSA test. Second, it is interesting that some of these genes, such as engrailed-2 and δ-catenin/NPRAP/Neurojungin, are neutrally enriched proteins. Third, the relatively high sensitivity and specificity of some of these urine markers indicate that the commonly believed variable urinary components due to diet and lifestyle could be overcome in a viable urine test. Fourth, while the profiling of gene or protein signatures is often advocated as potentially better biomarker options, they have yet to show promise. Finally, an ideal urine biomarker should scientifically support the functional relevance to PCa progression, and some of these potential markers display this quality. It is likely that there is no absolute boundary between a pre- versus cancerous lesion, so the combination of multiple biomarkers with existing clinical tools will have the best possibility of reliable PCa detection.

ACKNOWLEDGEMENT

This study is supported in part by the National Cancer Institute CA111891 (Q.L.) and National Institute for Environmental Health Sciences ES016888 (Y.H.C).

REFERENCES

Aberger, F., G. Weidinger, H. Grunz, and K. Richter. 1998. Anterior specification of embryonic ectoderm: the role of the Xenopus cement gland-specific gene XAG-2. Mech Dev. 72(1-2): 115-130.

Aly, M., F. Wiklund, and H. Grönberg. 2011. Early detection of prostate cancer with emphasis on genetic markers. Acta Oncol. 50 Suppl 1: 18-23.

Bose, S.K., R.S. Bullard, and C.D. Donald. 2008. Oncogenic role of engrailed-2 (en-2) in prostate cancer cell growth and survival. Transl Oncogenomics. 3: 37-43.

Bourdoumis, A., A.G. Papatsoris, M. Chrisofos, E. Efstathiou, A. Skolarikos, and C. Deliveliotis. 2010. The novel prostate cancer antigen 3 (PCA3) biomarker. Int Braz J Urol. 36(6): 665-668.

Bruchez, M. Jr., M. Moronne, P. Gin, S. Weiss, and A.P. Alivisatos. 1998. Semiconductor nanocrystals as fluorescent biological labels. Science. 281(5385): 2013-2016.

Burger, M.J., M.A. Tebay, P.A. Keith, H.M. Samaratunga, J. Clements, M.F. Lavin, and R.A. Gardiner. 2002. Expression analysis of delta-catenin and prostate-specific membrane antigen: their potential as diagnostic markers for prostate cancer. Int J Cancer. 100(2): 228-237.

Bu, H., S. Bormann, G. Schäfer, W. Horninger, P. Massoner, A. Neeb, V.K. Lakshmanan, D. Maddalo, A. Nestl, H. Sültmann, A.C. Cato, and H. Klocker. 2011. The anterior gradient 2 (AGR2) gene is overexpressed in prostate cancer and may be useful as a urine sediment marker for prostate cancer detection. Prostate. 71(6): 575-587.

Bussemakers, M.J., A. van Bokhoven, G.W. Verhaegh, F.P. Smit, H.F. Karthaus, J.A. Schalken, F.M. Debruyne, N. Ru, and W.B. Isaacs. 1999. DD3: a new prostate specific gene, highly overexpressed in prostate cancer. Cancer Res. 59(23): 5975-5979.

Catalona, W.J., J.E. Bailey-Wilson, N.J. Camp, S.J. Chanock, K.A. Cooney, D.F. Easton, R.A. Eeles, L.M. Fitzgerald, M.L. Freedman, J. Gudmundsson, R.A. Kittles, E.H. Margulies, B.B. McGuire, E.A. Ostrander, T.R. Rebbeck, J.L. Stanford, S.N. Thibodeau, J.S. Witte, and W.B. Isaacs. 2011. National Cancer Institute Prostate Cancer Genetics Workshop. Cancer. 71(10): 3442-3446.

Danila, D.C., M. Fleisher, and H.I. Scher. 2011. Circulating tumor cells as biomarkers in prostate cancer. Clin Cancer Res. 17(12): 3903-3912.

Daniels, T.R., I.I. Neacato, J.A. Rodríguez, H.S. Pandha, R. Morgan, and M.L. Penichet. 2010. Disruption of HOX activity leads to cell death that can be enhanced by the interference of iron uptake in malignant B cells. Leukemia. 24(9): 1555-1565.

Demichelis, F., and M.A. Rubin. 2007. TMPRSS2-ETS fusion prostate cancer: biological and clinical implications. J Clin Pathol. 60(11): 1185-1186.

Dudderidge, T.J., J.D. Kelly, A. Wollenschlaeger, O. Okoturo, T. Prevost, W. Robson, H.Y. Leung, G.H. Williams, and K. Stoeber. 2010. Diagnosis of prostate cancer by detection of minichromosome maintenance 5 protein in urine sediments. Br J Cancer. 103(5): 701-707.

Duijvesz, D., T. Luider, C.H. Bangma, and G. Jenster. 2011. Exosomes as biomarker treasure chests for prostate cancer. Eur Urol. 59(5): 823-831.

Eeles, R.A., Z. Kote-Jarai, G.G. Giles, A.A. Olama, M. Guy, S.K. Jugurnauth, S. Mulholland, D.A. Leongarmornlert, S.M. Edwards, J. Morrison, H.I. Field,

M.C. Southey, G. Severi, J.L. Donovan, F.C. Hamdy, D.P. Dearnaley, K.R. Muir, C. Smith, M. Bagnato, A.T. Ardern-Jones, A.L. Hall, L.T. O'Brien, B.N. Gehr-Swain, R.A. Wilkinson, A. Cox, S. Lewis, P.M. Brown, S.G. Jhavar, M. Tymrakiewicz, A. Lophatananon; UK Genetic Prostate Cancer Study Collaborators; British Association of Urological Surgeons' Section of Oncology; UK ProtecT Study Collaborators, A. Horwich, R.A. Huddart, V.S. Khoo, C.C. Parker, C.J. Woodhouse, A. Thompson, T. Christmas, C. Ogden, C. Fisher, C. Jamieson, C.S. Cooper, D.R. English, J.L. Hopper, D.E. Neal, and D.F. Easton. 2008. Multiple newly identified loci associated with prostate cancer susceptibility. Nat Genet. 40(3): 316-321.

Groskopf, J., S.M. Aubin, I.L. Deras, A. Blase, S. Bodrug, C. Clark, S. Brentano, J. Mathis, J. Pham, T. Meyer, M. Cass, P. Hodge, M.L. Macairan, L.S. Marks, and H. Rittenhuose. 2006. APTIMA PCA3 molecular urine test: development of a method to aid in the diagnosis of prostate cancer. Clin Chem. 52(6): 1089-1095.

Haese, A., A. de la Taille, H. van Poppel, M. Marberger, A. Stenzl, P.F. Mulders, H. Huland, C.C. Abbou, M. Remzi, M. Tinzl, S. Feyerabend, A.B. Stillebroer, M.P. van Gils, and J.A. Schalken. 2008. Clinical utility of the PCA3 urine assay in European men scheduled for repeat biopsy. Eur Urol. 54(5): 1081-1088.

Haiman, C.A., N. Patterson, M.L. Freedman, S.R. Myers, M.C. Pike, A. Waliszewska, J. Neubauer, A. Tandon, C. Schirmer, G.J. McDonald, S.C. Greenway, D.O. Stram, L. Le Marchand, L.N. Kolonel, M. Frasco, D. Wong, L.C. Pooler, K. Ardlie, I. Oakley-Girvan, A.S. Whittemore, K.A. Cooney, E.M. John, S.A. Ingles, D. Altshuler, B.E. Henderson, and D. Reich. 2007. Multiple regions within 8q24 independently affect risk for prostate cancer. Nat Genet. 39(5): 638-644.

Hedge, M.R. and B.B. Roa. 2006. Detecting mutations in the APC gene in familial adenomatous polyposis (FAP). Curr Protoc Hum Genet. Chapter 10: Unit 10.8.

Hessels, D. and J.A. Schalken. 2009. The use of PCA3 in the diagnosis of prostate cancer. Nat Rev Urol. 6(5): 255-261.

Hessels, D., J.M. Klein Gunnewiek, I. van Oort, H.F. Karthaus, G.J. van Leenders, B. van Balken, L.A. Kiemeney, J.A. Witjes, and J.A. Schalken. 2003. DD3(PCA3)-based molecular urine analysis for the diagnosis of prostate cancer. Eur Urol. 44(1): 8-15.

Honeycutt, K.A., Z. Chen, M.I. Koster, M. Miers, J. Nuchtern, J. Hicks, D.R. Roop, and J.M. Shohet. 2006. Deregulated minichromosomal maintenance protein MCM7 contributes to oncogene driven tumorigenesis. Oncogen. 25(29): 4027-4032.

Jansen, F.H., J. Krijgsveld, A. van Rijswijk, G.J. van den Bemd, M.S. van den Berg, W.M. van Weerden, R. Willemsen, L.J. Dekker, T.M. Luider, and G. Jenster. 2009. Exosomal secretion of cytoplasmic prostate cancer xenograft-derived proteins. Mol Cell Proteomics. 8(6): 1192-1205.

Jemal, A., R. Siegel, J. Xu, and E. Ward. 2010. Cancer statistics, 2010. CA Cancer J Clin. 60(5): 277-300.

Jentzmik, F., C. Stephan, K. Miller, M. Schrader, A. Erbersdobler, G. Kristiansen, M. Lein, and K. Jung. 2010. Sarcosine in urine after digital rectal examination fails as a marker in prostate cancer detection and identification of aggressive tumours. Eur Urol. 58(1): 12-18.

Jentzmik, F., C. Stephan, M. Lein, K. Miller, B. Kamlage, B. Bethan, G. Kristiansen, and K. Jung. 2011. Sarcosine in prostate cancer tissue is not a differential metabolite for prostate cancer aggressiveness and biochemical progression. J Urol. 185(2): 706-711.

Kim, K., M. Oh, H. Ki, T. Wang, S. Bareiss, M.E. Fini, D. Li, and Q. Lu. 2008. Identification of E2F1 as a positive transcriptional regulator for delta-catenin. Biochem Biophys Res Commun. 369(2): 414-420.

Kumar, A., J.W. Godwin, P.B. Gates, A.A. Garza-Garcia, and J.P. Brockes. 2007. Molecular basis for the nerve dependence of limb regeneration in an adult vertebrate. Science. 318(5851): 772-777.

Lau, M. 2010. Genitourinary Cancers Symposium Highlights for Internists. N A J Med Sci. 3(2): 106.

Lu, Q., M. Paredes, M. Medina, J. Zhou, R. Cavallo, M. Peifer, L. Orecchio, and K.S. Kosik. 1999. Delta-catenin, an adhesive junction-associated protein which promotes cell scattering. J Cell Biol. 144(3): 519-532.

Lu, Q., L.J. Dobbs, C.W. Gregory, G.W. Lanford, M.P. Revelo, S. Shappell, and Y.H. Chen. 2005. Increased expression of delta-catenin/neural plakophilin-related armadillo protein is associated with the down-regulation and redistribution of E-cadherin and p120ctn in human prostate cancer. Hum Pathol. 36(10): 1037-1048.

Lu, Q., J. Zhang, R. Allison, H. Gay, W.X. Yang, N.A. Bhowmick, G. Frelix, S. Shappell, and Y.H. Chen. 2009. Identification of extracellular delta-catenin accumulation for prostate cancer detection. Prostate. 69(4): 411-418.

Majid, S., A.A. Dar, S. Saini, Y. Chen, V. Shahryari, J. Liu, M.S. Zaman, H. Hirata, S. Yamamura, K. Ueno, Y. Tanaka, and R. Dahiya. 2010. Regulation of minichromosome maintenance gene family by microRNA-1296 and genistein in prostate cancer. Cancer Res. 70(7): 2809-2818.

Maizel, A., O. Bensaude, A. Prochiantz, and A. Joliot. 1999. A short region of its homeodomain is necessary for engrailed nuclear export and secretion. Development. 126(14): 3183-3190.

Martin, N.L., M.K. Saba-El-Leil, S. Sadekova, S. Meloche, and G. Sauvageau. 2005. EN2 is a candidate oncogene in human breast cancer. Oncogene 24(46): 6890-6901.

Mehra, R., S.A. Tomlins, R. Shen, O. Nadeem, L. Wang, J.T. Wei, K.J. Pienta, D. Ghosh, M.A. Rubin, A.M. Chinnaiyan, and R.B. Shah. 2007. Comprehensive assessment of TMPRSS2 and ETS family gene aberrations in clinically localized prostate cancer. Mod Pathol. 20(5): 538-544.

Meng, M.V., G.D. Grossfeld, G.H. Williams, S. Dilworth, K. Stoeber, T.W. Mulley, V. Weinberg, P.R. Carroll, and T.D. Tisty. 2001. Minichromosome maintenance protein 2 expression in prostate: characterization and association with outcome after therapy for cancer. Clin Cancer Res. 7(9): 2712-2718.

Meyer, T.E., S.D. Fox, H.J. Issaq, X. Xu, L.W. Chu, T.D. Veenstra, and A.W. Hsing. 2011. A reproducible and high-throughput HPLC/MS method to separate sarcosine from α- and β-alanine and to quantify sarcosine in human serum and urine. Anal Chem. 83(14): 5735-5740.

Morgan, R. 2006. Engrailed: complexity and economy of a multi-functional transcription factor. FEBS Lett. 580(11): 2531-2533.

Morgan, R., P.M. Pirard, L. Shears, J. Sohal, R. Pettengell, and H.S. Pandha. 2007. Antagonism of HOX/PBS dimer formation blocks the in vivo proliferation of melanoma. Cancer Res. 67(12): 5806-5813.

Morgan, R., L. Plowright, K.J. Harrington, A. Michael, and H.S. Pandha. 2010. Targeting HOX and PBX transcription factors in ovarian cancer. BMC Cancer. 10: 89.

Morgan, R., A. Boxall, A. Bhatt, M. Bailey, R. Hindley, S. Langley, H.C. Whitaker, D.E. Neal, M. Ismail, H. Whitaker, N. Annels, A. Michael, and H. Pandha. 2011. Engrailed-2 (EN2): a tumor specific urinary biomarker for the early diagnosis of prostate cancer. Clin Cancer Res. 17(5): 1090-1098.

Nguyen, P.N., P. Violette, S. Chan, S. Tanguay, W. Kassouf, A. Aprikian, and J.Z. Chen. 2011. A panel of TMPRSS2:ERG fusion transcript markers for urine-based prostate cancer detection with high specificity and sensitivity. Eur Urol. 59(3): 407-414.

Nilsson, J., J. Skog, A. Nordstrand, V. Baranov, L. Mincheva-Nilsson, X.O. Breakefield, and A. Widmark. 2009. Prostate cancer-derived urine exosomes: a novel approach to biomarkers for prostate cancer. Br J Cancer. 100(10): 1603-1607.

Ochiai, A., K. Okihara, K. Kamoi, T. Iwata, A. Kawauchi, T. Miki, and Z. Fors. 2011. Prostate cancer gene 3 urine assay for prostate cancer in Japanese men undergoing prostate biopsy. Int J Urol. 18(3): 200-205.

Ohkia, A., Y. Hu, M. Wang, F.U. Garcia, and M.E. Stearns. 2004. Evidence for prostate cancer-associated diagnostic marker-1: immunohistochemistry and in situ hybridization studies. Clin Cancer Res. 10(7): 2452-2458.

Padmanabhan, V., P. Callas, G. Philips, T.D. Trainer, and B.G. Beatty. 2004. DNA replication regulation protein Mcm7 as a marker of proliferation in prostate cancer. J Clin Pathol. 57(10): 1057-1062.

Paffenholz, R., and W.W. Franke. 1997. Identification and localization of a neurally expressed member of the plakoglobin/armadillo multigene family. Differentiation. 61(5): 293-304.

Pan, B.T., R. Blostein, and R.M. Johnstone. 1983. Loss of the transferrin receptor during the maturation of sheep reticulocytes in vitro. An immunological approach. Biochem J. 210(1): 37-47.

Park, S.W., G. Zhen, C. Verhaeghe, Y. Nakagami, L.T. Nguyenvu, A.J. Barczak, N. Killeen, and D.J. Erle. 2009. The protein disulfide isomerase AGR2 is essential for production of intestinal mucus. Proc Natl Acad Sci U. S. A. 106(17): 6950-6955.

Ploussard, G. and A. de la Taille. 2010. Urine biomarkers in prostate cancer. Nat Rev Urol. 7(2): 101-109.

Plowright, L., K.J. Harrington, H.S. Pandha, and R. Morgan. 2009. HOX transcription factors are potential therapeutic targets in non-small-cell lung cancer (targeting HOX genes in lung cancer). Br J Cancer. 100(3): 470-475.

Roses, A.D. 2006. On the discovery of the genetic association of Apolipoprotein E genotypes and common late-onset Alzheimer disease. J Alzherimers Dis. 9(3 Suppl): 361-366.

Rajput, A.B., M.A. Miller, A. De Luca, N. Boyd, S. Leung, A. Hurtado-Coll, L. Fazli, E.C. Jones, J.B. Palmer, M.E. Gleave, M.E. Cox, and D.G. Huntsman. 2007. Frequency of the TMPRSS2:ERG gene fusion is increased in moderate to poorly differentiated prostate cancers. J Clin Pathol. 60(11): 1238-1243.

Rigau, M., I. Ortega, M.C. Mir, C. Ballesteros, M. Garcia, M. Llauradó, E. Colás, N. Pedrola, M. Montes, T. Sequeiros, T. Ertekin, B. Majem, J. Planas, A. Ruiz, M. Abal, A. Sánchez, J. Morote, J. Reventós, and A. Doll. 2011. A Three-Gene panel on urine increases PSA specificity in the detection of prostate cancer. Prostate doi: 10.1002/pros.21390. [Epub ahead of print].

Shah, N., and S. Sukumar. 2010. The Hox genes and their roles in oncogenesis. Nat Rev Cancer. 10(5): 361-371.

Shears, L., L. Plowright, K. Harrington, H.S. Pandha, and R. Morgan. 2008. Disrupting the interaction between HOX and PBX causes necrotic and apoptotic cell death in the renal cancer lines CaKi-2 and 769-P. J Urol. 180(5): 2196-2201.

Sokoloff, R.L., K.C. Norton, C.L. Gasior, K.M. Marker, and L.S. Grauer. 2000. A dual-monoclonal sandwich assay for prostate-specific membrane antigen: levels in tissues, seminal fluid and urine. Prostate. 43(2): 150-157.

Sreekumar, A., L.M. Poisson, T.M. Rajendiran, A.P. Khan, Q. Cao, J. Yu, B. Laxman, R. Mehra, R.J. Lonigro, Y. Li, M.K. Nyati, A. Ahsan, S. Kalyana-Sundaram, B. Han, X. Cao, J. Byun, G.S. Omenn, D. Ghosh, S. Pennathur, D.C. Alexander, A. Berger, J.R. Shuster, J.T. Wei, S. Varambally, C. Beecher, and A.M. Chinnaiyan. 2009. Metabolomic profiles delineate potential role for sarcosine in prostate cancer progression. Nature. 457(7231): 910-914.

Stamey, T.A., I.M. Johnstone, J.E. McNeal, A.Y. Lu, and C.M. Yemoto. 2002. Preoperative serum prostate specific antigen levels between 2 and 22 ng./ml. correlate poorly with post-radical prostatectomy cancer morphology: prostate specific antigen cure rates appear constant between 2 and 9 ng./ml. J Urol. 167(1): 103-111.

Stoeber, K., R. Swinn, A.T. Prevost, P. de Clive-Lowe, I. Halsall, S.M. Dilworth, J. Marr, W.H. Turner, N. Bullock, A. Doble, C.N. Hales, and G.H. Williams. 2002. Diagnosis of genito-urinary tract cancer by detection of minichromosome maintenance 5 protein in urine sediments. J Natl Cancer Inst. 19(14): 1071-1079.

Struys, E.A., A.C. Heijboer, J. van Moorselaar, C. Jakobs, and M.A. Blankenstein. 2010. Serum sarcosine is not a marker for prostate cancer. Ann Clin Biochem. 47(Pt 3): 282.

Talesa, V.N., C. Antognelli, C. Del Buono, F. Stracci, M.R. Serva, E. Cottini, and E. Mearini. 2009. Diagnostic potential in prostate cancer of a panel of urinary molecular tumor markers. Cancer Biomark. 5(6): 241-251.

Thomas, A., N.J. Camp, J.M. Farnham, K. Allen-Brady, and L.A. Cannon-Albright. 2008. Shared genomic segment analysis. Mapping disease predisposition genes in extended pedigrees using SNP genotype assays. Ann Hum Genet. 72(Pt 2): 279-287.

Tomlins, S.A., D.R. Rhodes, S. Perner, S.M. Dhanasekaran, R. Mehra, X.W. Sun, S. Varambally, X. Cao, J. Tchinda, R. Kuefer, C. Lee, J.E. Montie, R.B. Shah, K.J. Pienta, M.A. Rubin, and A.M. Chinnaiyan. 2005. Recurrent fusion of TMPRSS2 and ETS transcription factor genes in prostate cancer. Science. 310(5748): 644-648.

Tricoli, J.V., M. Schoenfeldt, and B.A. Conley. 2004. Detection of prostate cancer and predicting progression: current and future diagnostic markers. Clin Cancer Res. 10(12 Pt 1): 3943-3953.

Tuma, R.S. 2010. New tests for prostate cancer may be nearing the clinic. J Natl Cancer Inst. 102(11): 752-754.

Vlaeminck-Guillem, V., A. Rufflon, J. André, M. Devonec, and P. Paparel. 2010. Urinary prostate cancer 3 test: toward the age of reason? Urology. 75(2): 447-453.

Wang, T., Y.H. Chen, H. Hong, Y. Zeng, J. Zhang, J.P. Lu, B. Jeansonne, and Q. Lu. 2009. Increased nucleotide polymorphic changes in the 5'-untranslated region of delta-catenin (CTNND2) gene in prostate cancer. Oncogene. 28(4): 555-564.

Wayner, E.A., S.I. Quek, R. Ahmad, M.E. Ho, M.A. Loprieno, Y. Zhou, W.J. Ellis, L.D. True, and A.Y. Liu. 2011. Development of an ELISA to detect the secreted prostate cancer biomarker AGR2 in voided urine. Prostate. Doi: 10.1002/pros.21508. [Epub ahead of print].

Witte, J.S. 2009. Prostate cancer genomics: towards a new understanding. Nat Rev Genet. 10(2): 77-82.

Witte, J.S. 2007. Multiple prostate cancer risk variants on 8q24. Nat Genet. 39(5): 579-580.

Zhang, Y., S.S. Forootan, D. Liu, R. Barraclough, C.S. Foster, P.S. Rudland, and Y. Ke. 2007. Increased expression of anterior gradient-2 is significantly associated with poor survival of prostate cancer patients. Prostate Cancer Prostatic Dis. 10(3): 293-300.

Zhou, J., U. Liyanage, M. Medina, C. Ho, A.D. Simmons, M. Lovett, and K.S. Kosik. 1997. Presenilin 1 interaction in the brain with a novel member of the Armadillo family. Neuroreport. 8(8): 2085-2090.

CHAPTER 5

Ectodomain Shedding as a Mechanism to Generate Cancer Biomarkers (Ectomarkers)

Maria J. Ruiz-Echevarria[1, 2, 3, *], Xiaofei Chen[4] and Ryan F. Overcash[4]

[1]Department of Oncology and [2]Department of Anatomy and Cell Biology, Brody School of Medicine, East Carolina University, 600 Moye Blvd., Greenville, NC 27834, USA; and [3]Lineberger Comprehensive Cancer Center, University of North Carolina at Chapel Hill, Chapel Hill, NC 27599, USA

[4]Department of Biochemistry and Molecular Biology, Brody School of Medicine, East Carolina University, 600 Moye Blvd., Greenville, NC 27834, USA

ABSTRACT

Progress in the understanding of tumor biology has propelled the development of new targeted therapies and the identification of new biomarkers for aggressiveness and response to treatment. Because tumor tissue is not always easy to obtain and may require invasive procedures, the main efforts on current biomarker studies are directed to identifying soluble biomarkers on nontumor biospecimens such as serum, urine and other body fluids. Protein shedding promotes the irreversible release of cell surface protein domains (ectodomains) that can accumulate in body fluids and function as specific biomarkers and in some cases play a role in the tumorigenic process. Cleavage of the transmembrane proteins is mediated by a family of multidomain zinc metalloproteinases known as ADAM (a disintegrin and metalloproteinase) or by matrix metalloproteinases (MMPs). While the participation of these sheddases in ectodomain release and cancer has been recognized, little is known about the molecular mechanisms that lead to their activation or about the role that the ectodomain may play in tumorigenesis. This chapter will explore mechanisms that lead to sheddase activation and subsequent

*Corresponding author: E-mail: ruizechevarriam@ecu.edu

ectodomain shedding as well as possible mechanisms by which ectodomain shedding promotes tumorigenesis and will highlight the value of some of the known soluble biomarkers.

INTRODUCTION

As much as 10 percent of the proteins on the cell surface are subjected to proteolytic cleavage and release of their extracellular–ecto–domains (Fong et al. 2011). This process, known as ectodomain shedding, is responsible for activation of many cytokines and growth factors that are synthesized as membrane bound proforms and need to undergo proteolytic shedding for activation. In addition, it is the first step in the cleavage/activation of growth factor receptors, adhesion molecules, and ectoenzymes. Therefore, ectodomain shedding may also affect the number of receptors available for ligand binding, or produce soluble molecules with new functions that can behave as paracrine or autocrine effectors or as competitive inhibitors of the ligand by binding to the membrane receptor (Hayashida et al. 2010). Ectodomain shedding is therefore a switch that regulates the activity of the cleaved substrate.

Ectodomain shedding affects membrane proteins of different topologies – type I, type II and GPI anchored proteins – however, aside from their localization at the plasma membrane, no similarities can be found between the various proteins that are cleaved or the primary sequences at the cleavage sites (Wong et al. 1989; Pandiella et al. 1992; Sahasrabudhe et al. 1992; Sisodia 1992). While this could reflect the existence of many different sheddases, it is well known that membrane proteins are cleaved by a common mechanism that is sensitive to metalloproteinase inhibitors. The main family of sheddases is the ADAMs (a disintegrin and metalloproteinases) family, although several matrix metalloproteinases (MMPs 2, 7, 9, and 14) have also been shown to promote ectodomain shedding of different substrates.

Due to the wide variety of substrates affected by ectodomain shedding and the various roles of the soluble molecules produced, protein shedding is involved in many different vital biological processes such as proliferation, apoptosis, adhesion, migration and differentiation. In fact, some MMPs and ADAMs are highly expressed in cancer and are implicated in multiple stages of this disease including the acquisition of invasive and metastatic properties. Importantly, the increase in the shedding activity in cancer cells promotes an increased release of ectodomain fragments with potential diagnostic value. Proteins that are shed from cell membranes have been identified in blood or other bodily fluids of patients afflicted with a variety of cancers. As sampling of bodily fluids is minimally invasive, these ectodomains are potentially very useful as biomarkers, thus we coined the term "ectomarkers".

The goal of this review is to discuss recent advances in the study of ectodomain shedding in cancer and its potential value for biomarker discovery. Specific examples of potential shed biomarkers will be discussed.

THE SHEDDASES: STRUCTURE AND FUNCTION

Sheddases are proteases that cleave transmembrane proteins at extracellular membrane-proximal sites, releasing soluble ectodomains from the cell surface. A family of multidomain zinc metalloproteinases known as ADAM (a disintegrin and metalloproteinase) proteins has emerged as major sheddases, although matrix metalloproteinases (MMPs) may also participate in ectodomain shedding. Evidence of ADAMs having a role in cancer is accumulating (Duffy et al. 2009; Mochizuki and Okada 2007; Rocks et al. 2006) and therapeutic ADAM inhibitors are being tested as anti-cancer agents (Kataoka 2009; Lu et al. 2008).

ADAMs and MMPs are members of the metzincin family of zinc-dependent proteases. To date, 22 ADAMs (ADAM1-3, 6-12, 15, 17-22, 28-30, 32, 33) have been identified in the human genome, 12 of which (ADAM1, 8-10, 12, 15, 17, 19, 21, 28, 30, 33) contain a catalytic site consensus sequence in the metalloproteinase domain and are therefore predicted to display protease activity. Please refer to the White Lab (http://people.virginia.edu/~jw7g/) for an up-to-date list of all ADAM family members. Besides the proteolytic function in cleaving transmembrane and extracellular matrix (ECM) proteins, ADAMs also bind to integrins through their disintegrin domain (reviewed in White 2003). Binding to integrins may reduce cell surface expression of the integrin and/or allow activation of the ADAM (reviewed in Murphy 2008). The role of ADAMs has been documented in a number of biological processes such as fertilization, angiogenesis, immunity, and development, as well as diseases including Alzheimer's disease, inflammation, asthma, and cancer (reviewed in Duffy et al. 2003; Edwards et al. 2008; Seals and Courtneidge 2003).

The following section will provide a brief description of the domain structure and biological functions of ADAMs. We will focus on those ADAMs with a known function in cancer.

ADAMs Structure

A typical ADAM is a type I transmembrane protein of about 750 amino acids. The domain structure of ADAMs consists of a prodomain, a metalloproteinase domain, a disintegrin domain, a cysteine-rich domain, an EGF-like domain, a transmembrane domain, and a cytoplasmic domain (Wolfsberg et al. 1995; Seals and Courtneidge 2003; Becherer and Blobel 2003). The prodomain is at the N-terminus of ADAMs, following a signal sequence that directs the protein to the secretory pathway. The function of the prodomain is two-fold: to keep the metalloproteinase site inactive through a cysteine switch (Van Wart and Birkedal-Hansen 1990), and to chaperone the proper folding of the protein (Milla et al. 1999; Anders et al. 2001). The removal of the prodomain is carried out by furin-like proprotein convertases (Lum et al. 1998) or by an autocatalytic mechanism (Howard et al. 2000; Schlomann et al. 2002) in the trans-Golgi network. C-terminal to the prodomain there is a conserved metalloproteinase domain present in all members of the ADAM family. Only about half of them contain a catalytic consensus sequence --HEXXHXXGXXG...M-- that is the signature of the metzincin family (Bode et al. 1993). The presence of this sequence correlates with protease activity (White 2003). Although ADAMs' major substrates are the ectodomain of transmembrane proteins, they are also capable of degrading ECM proteins. The disintegrin domain shares a high sequence similarity with snake venom disintegrins, which are small soluble proteins that compete for binding to platelet integrin $\alpha_{IIb}\beta_3$ complex with its natural ligand fibrinogen thereby blocking platelet aggregation (Blobel and White 1992; Niewiarowski et al. 1994). However, none of the ADAMs, except for human ADAM15 (Nath et al. 1999; Zhang et al. 1998), contain a classical integrin-binding tripeptide Arg-Gly-Asp (RGD) and instead, they interact with integrins in an RGD-independent manner via a conserved RX_6DEVF sequence in the disintegrin domain (Eto et al. 2002). The cysteine-rich domain has been shown to play a role in cell adhesion by binding to syndecan (Iba et al. 2000) and fibronectin (Gaultier et al. 2002), and in substrate recognition (Reddy et al. 2000; Janes et al. 2005). In addition, it regulates ADAM protease function (Smith et al. 2002) and facilitates the removal of the prodomain (Milla et al. 1999). The function of the ADAM EGF-like domain is not clear yet. It has been suggested that the EGF-like domain of fertilin β (ADAM2) is involved in sperm-egg binding and fusion (Chen et al. 1998). The cytoplasmic domain is highly variable among different ADAM family members. The presence of motifs such as phosphorylation sites and proline-rich SH3 binding sites in this domain suggests it may be involved in intracellular signal transduction (Seals and Courtneidge 2003). Finally, it is worth mentioning that as a result of alternative splicing some ADAMs (ADAM 9, 11, 12, 28) also encode a soluble, secreted form lacking the transmembrane and cytoplasmic domains (Gilpin et al. 1998; Katagiri et al. 1995; Mazzocca et al. 2005; Roberts et al. 1999).

ADAMs Function in Cancer

Twelve human ADAMs have potential protease activity. Their substrates include cytokines, growth factors and their cognate receptors. As a result of ectodomain shedding, activated, shed cytokines or growth factors can engage in paracrine or autocrine signaling, truncated receptors can be inactivated or the bound fragment can undergo regulated intramembrane proteolysis (RIP), releasing intracellular fragments that can enter the nucleus and participate in transcriptional regulation. Since these molecules play very important biological roles, their deregulation can have profound consequences. Indeed, changes in the expression of nearly every proteolytic active ADAM are associated with various types of cancer (Table 1). The function of these ADAMs in cancer has been extensively reviewed (Crawford et al. 2009; Moss et al. 2008; Pruessmeyer and Ludwig 2009; Saftig and Reiss 2011; Lucas and Day 2009; Arribas and Esselens 2009; Blobel 2005; Mochizuki and Okada 2009) and is summarized in Table 1.

SIGNALING PATHWAYS AND MOLECULAR MECHANISMS INVOLVED IN SHEDDASE ACTIVATION

While the participation of ADAMs in cancer has been recognized, little is known about the molecular mechanisms that lead to their activation. In the following sections, we will discuss the signaling pathways that lead to, and the modifications that promote sheddase activation and subsequent ectodomain shedding as well as possible mechanisms by which ectodomain shedding promotes tumorigenesis. In the last section of the chapter we will review the use of some of the known shed ectodomains as potential biomarkers (Fig. 1).

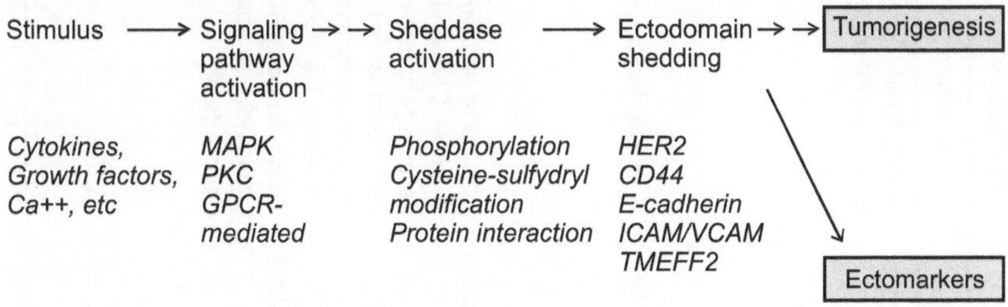

Fig. 1 Mechanisms leading to ectodomain shedding and tumorigenesis.

Table 1 ADAM proteases involved in cancer.

	*Phenotype of knockout mice**	*Tumor expression***	*Potential role in tumorigenesis***
ADAM8	Viable, fertile, no pathology (reduced CHL-1 shedding in brain)	Brain[1], lung[2], kidney[3, 33] pancreas[4], prostate[5]	Protein expression levels correlate with glioma cell invasiveness[1] and unfavorable patient outcome in prostate cancer, pancreatic ductal adenocarcinoma and renal cell carcinoma[3-5, 33]. Increases cell invasion *in vitro*[2, 4]. Autoproteolytic shed ectodomain[6] is a potential biomarker for lung cancer[2]. A soluble isoform contributes to the bone metastatic phenotype of lung cancer[7].
ADAM9	Viable, fertile, no pathology	Breast[8, 9], liver[10-12], kidney[13], pancreas[14, 15], prostate[16, 17], adenoid cystic carcinoma[18], stomach[19], melanoma[20], multiple myeloma[21]	Affects cell adhesion through interaction with integrins [20-26]. Sheddase for EGF and FGFR2iiib[27]. The membrane bound and a secreted isoform have opposing roles as migration suppressors and enhancers in breast cancer cells respectively[28].
ADAM10	Embryonic lethal (E9.5), defective CNS and heart development, somitegenesis and vasculogenesis, phenocopies Notch deficiency (more severe than presenilin 1 and 2 double knockouts)	Leukemia[29], liver[30], oral cavity[31], ovary[32], prostate[16], pancreas[34], adenoid cystic carcinoma[37], melanoma[36], stomach[35], uterus[32]	Promotes cell growth, migration, invasion and anchorage-independent growth *in vitro*[31, 34, 36, 38-40]. Sheddase for L1[32, 41-44], MICA[30, 47] and HER2[46] in various cancers. Ectodomain shedding of L1 by ADAM10 promotes liver metastasis in colon cancer[45].
ADAM12	Viable, fertile, 30% embryonic lethality, brown adipose abnormalities, no muscle defect	Bladder[48], brain[49], breast[8, 51-54], liver[10], lung[56], stomach[19]	Deficiency in mice delays prostate tumor progression[58]. Expression of a secreted isoform promotes metastasis in breast cancer[54]. Binds to syndecans through the cysteine-rich domain to support tumor cell adhesion[59, 60]. Cleaves HB-EGF to promote glioblastoma proliferation[49]. Involved in ECM protein degradation and thus cancer invasion and metastasis[53].

Table 1 Contd. ...

Table 1 Contd. ...

ADAM15	Viable, fertile, tumor neovascularization reduced, age onset osteoarthritis	Breast[61], pancreas[15], prostate[61], stomach[19], lung[63]	Downregulation attenuates many of the malignant characteristics of the prostate cancer cell line PC3[64]. Enhances cell proliferation and migration in breast cancer cell lines by shedding of E-cadherin and subsequent activation of HER2 and HER3[65].
ADAM17	Perinatal lethality, defective epithelia, heart and lung development, phenocopies EGFR, TGFα, HB-EGF and amphiregulin knockout mice	Breast[79], colon[78], kidney[33], ovary[50], prostate[16]	Plays a pivotal role in promoting tumor growth by releasing EGFR family ligands, including amphiregulin[66, 67], epigin[68], HB-EGF[69-71], heregulin[72], and TGFα[40, 66, 73-75]. Regulates shedding of MICA to facilitate tumor immune escape[47]. Increases motility of prostate and ovarian cancer cells by virtue of cleaving TMEFF2 and ALCAM respectively[76, 77].
ADAM19	80% postnatal lethality, multiple cardiovascular defects	Brain[1], kidney[3, 33]	Protein expression levels and protease activity correlate with glioma cell invasiveness[1].
ADAM28		Bladder[83], breast[81], chondrosacoma[82], lung[62, 80]	Expression correlates with cell proliferation and lymph node metastasis in NSCLC patients[80] and histological grade in chondrosarcoma[82]. Serum levels gradually increase with the staging of NSCLC[84]. Cleaves and releases IGF-1 from IGF-1/IGFBP-3 complex to promote cell proliferation[81]. Cleaves CTGF and releases $VEGF_{165}$ from CTGF/$VEGF_{165}$ complex to promote angiogenesis[57].
ADAM33	Viable, fertile, no pathology		Increases gastric cancer cell migration and proliferation by upregulating IL-18 secretion[55].

Table 1 Contd. ...

Table 1 Contd. ...

*Adapted from Edwards et al. 2008.

**(1)Wildeboer et al. 2006, (2)Ishikawa et al. 2004, (3)Roemer et al. 2004a, (4)Valkovskaya et al. 2007, (5)Fritzsche et al. 2006, (6)Schlomann et al. 2002, (7)Hernández et al. 2010, (8)Lendeckel et al. 2005, (9)O'Shea et al. 2003, (10)Le Pabic et al. 2003, (11)Tannapfel et al. 2003, (12)Tao et al. 2010, (13)Fritzsche et al. 2008b, (14) Grutzmann et al. 2004, (15)Yamada et al. 2007, (16)Karan et al. 2003, (17)Fritzsche et al. 2008a, (18)Xu et al. 2010a, (19)Carl-McGrath et al. 2005, (20)Zigrino et al. 2005, (21)Karadag et al. 2006, (22)Cominetti et al. 2009, (23)Mazzocca et al. 2005, (24)Mongaret et al. 2011, (25)Nath et al. 2000, (26)Zhou et al. 2001, (27)Peduto et al. 2005, (28)Fry and Toker 2010, (29)Wu et al. 1997, (30)Kohga et al. 2009, (31)Ko et al. 2007, (32)Fogel et al. 2003, (33)Roemer et al. 2004b, (34)Gaida et al. 2010, (35)Wang et al. 2011, (36)Lee et al. 2010, (37)Xu et al. 2010b, (38)Arima et al. 2007, (39)Armanious et al. 2011, (40)Xu and Derynck 2010, (41)Gutwein et al. 2003, (42)Maretzky et al. 2005, (43)Mechtersheimer et al. 2001, (44)Yang et al. 2011a, (45)Gavert et al. 2007, (46)Liu et al. 2006, (47)Waldhauer et al. 2008, (48)Frohlich et al. 2006, (49)Kodama et al. 2004, (50)Tanaka et al. 2005, (51)Kveiborg et al. 2005, (52)Narita et al. 2010, (53)Roy et al. 2004, (54)Roy et al. 2011, (55) Kim et al. 2009a, (56)Rocks et al. 2006, (57)Mochizuki et al. 2010, (58)Peduto et al. 2006, (59)Iba et al. 1999, (60)Iba et al. 2000, (61)Kuefer et al. 2006, (62)Wright et al. 2010, (63)Schütz et al. 2005, (64)Najy et al. 2008a, (65)Najy et al. 2008b, (66)Kenny and Bissell 2007, (67)Lemjabbar et al. 2003, (68)Sahin and Blobel 2007, (69)Ebi et al. 2010, (70) Schäfer et al. 2004, (71)Maretzky et al. 2011a, (72)Zhou et al. 2006, (73)Borrell-Pagès et al. 2003, (74)Maretzky et al. 2011b, (75)Lin et al. 2012, (76)Ali and Knaüper 2007, (77)Rosso et al. 2007, (78)Blanchot-Jossic et al. 2005, (79)McGowan et al. 2008, (80)Ohtsuka et al. 2006, (81)Mitsui et al. 2006, (82)Matsuura et al. 2010, (83) Yang et al. 2011b, (84)Kuroda et al. 2010.

ADAM17 and ADAM10 appear to be the most critical proteases in cancer because of their role in epidermal growth factor receptor (EGFR) activation and in ectodomain shedding of other transmembrane proteins with an important role in this disease, i.e. the shedding of the Notch receptor (Brou et al. 2000; Bozkulak and Weinmaster 2009; van Tetering et al. 2009) and Notch ligands (Six et al. 2003; Parr-Sturgess et al. 2010). In addition, ADAM17 is also known to contribute to angiogenesis, necessary for tumor growth and metastasis, and to the inflammatory responses often observed in cancer. These phenotypes are in part due to its role in shedding of the vascular endothelial growth factor receptor (VEGFR; Swendeman et al. 2008; Rahimi et al. 2009; Van Schaeybroeck et al. 2011) and the adhesion proteins L-selectin, ICAM, VCAM, CD-44 and N-cadherin among others (Smalley and Ley 2005; Killock and Ivetic 2010; Wang et al. 2010; Scheller et al. 2011). Due to its predominant role in cancer, in this section, we will primarily focus on the molecular mechanisms that lead up to ADAM17 activation.

Signaling Pathways that Lead to Sheddase Activation

ADAM17-mediated ectodomain shedding occurs in response to a variety of agonists including protein kinase C activators, calcium ionophores, cytokines, growth factors and ligands of G protein-coupled receptors (GPCR). These inducers can trigger different signal cascades that commonly converge in the activation of EGFR and lead to cancer. The link between ADAM17 and EGFR is illustrated by the fact that mice lacking ADAM17 resemble mice lacking the EGFR or certain EGFR ligands (Luetteke et al. 1993; Sibilia and Wagner 1995; Peschon et al. 1998; Mine et al. 2005). Three main signaling pathways have been implicated in the regulated shedding of membrane proteins: mitogen-activated protein kinase (MAPK), protein kinase C (PKC) and GPCR-mediated EGFR transactivation pathways (Fan and Derynck 1999; Gechtman et al. 1999; Xu and Derynck 2010). Their role in sheddase activation is presented below.

MAPK-Dependent ADAM17-Regulated Ectodomain Shedding

In mammals there are three major MAPK pathways, the extracellular signal regulated kinase 1 and 2 (ERK1 and 2), the p38 and the c-jun N-terminal kinase (JNK) and all of them play a role in ectodomain shedding. ERK1/2, which play a central role in survival and mitogenic signaling, are mainly activated by growth factors, certain tyrosine kinases, oncogenes, stress and intracellular pathways like PKC and Ca^{2+} messenger systems (Kubota et al. 2011; Yang et al. 2011c), while p38 and JNK, which are involved in various stress responses – i.e. cell death, survival, differentiation—are mainly triggered by cytokines and stress stimuli (Keshet and Seger 2010).

Stimulation of ERK1/2 can result in ADAM17 activation (Fan and Derynck 1999; Getchman et al. 1999; Díaz-Rodríguez et al. 2002; Weskamp et al. 2004; Swendeman et al. 2008; Mendelson et al. 2010; Van Schaeybroeck et al. 2011; Maretzky et al. 2011b). In fact, inhibition of the ERK pathway by specific inhibitors or dominant negative mutants of MEK1 and 2, required for ERK1/2 activation, completely blocks growth factor induced ADAM17-mediated shedding, while constitutively activated forms of ERK or MEK induce ectodomain shedding to the level observed by addition of growth factors (Fan and Derynck 1999). These results suggest that the ERK pathway can function as a general mechanism to induce ADAM17-mediated ectodomain shedding. ERK signaling has also been described as a regulator of constitutive (basal) shedding of kidney injury molecule 1 (KIM-1; Zhang et al. 2007), a protein upregulated in renal cell carcinoma. In several cases, a synergistic or additive effect of the p38 and ERK pathways in ectodomain shedding has been suggested (Fan and Derynck 1999; Singh et al. 2005).

The p38 MAPK signaling cascade has been implicated as a mechanism in constitutive shedding --i.e KIM-1 described above-- and in ADAM17-mediated ectodomain shedding induced by growth factors, anisomycin, UV light, interleukin 1β and stress (Fan and Derynck 1999; Takenobu et al. 2003; Fischer et al. 2004; Weskamp et al. 2004; Xu and Derynck 2010). Specific inhibitors of p38 or siRNA specific for the α isoform of p38 can block basal or induced ADAM17-dependent shedding while constitutively active forms of the MKK6, an upstream activator of p38, promote shedding (Fan and Derynck 1999; Xu and Derynck 2010). However, it has also been suggested that activation of the p38 MAPK can stimulate ADAM17-independent shedding of certain tyrosine kinase receptors or membrane anchored growth factors, although the sheddase involved has not been identified. (Rizoli et al. 1999; Díaz-Rodríguez et al. 2002; Montero et al. 2002). Finally, FGF-7 stimulation of the FGFR-2, promotes ADAM17 activation by a mechanism that requires p38, Src and PI3 kinase (Maretzky et al. 2011a).

The participation of the JNK MAPK in ADAM17-mediated ectodomain shedding has also been documented, i.e. for the endothelial protein C receptor

(EPCR), a protein with a role in tumor metastasis (Qu et al. 2007; Bezuhly et al. 2009). JNK activity is required for EPCR shedding in response to thrombin, phorbol 12-myristate-13-acetate (PMA), anisomycin, ionomycin and H_2O_2. However, p38 and ERK1/2 also play a role in ECPR shedding in response to different stimuli suggesting a possible synergism between the MAPKs in the regulation of EPCR shedding (Menschikowski et al. 2009). Finally, a recent report demonstrated that activation of JNK3 as a result of BDNF growth factor binding to the p75NTR receptor in sympathetic neurons mediates upregulation of ADAM17 and subsequent receptor cleavage which then triggers further activation of JNK3 (Kenchappa et al. 2010). This mechanism contributes to apoptosis during neuronal development and it is conceivable that its deregulation may lead to tumorigenicity.

Protein Kinase C (PKC)-Dependent ADAM17-Regulated Ectodomain Shedding

Induction of PKC by phorbol esters, specifically by PMA or 12-O-tetradecanoyl-phorbol-13-acetate (TPA), is a hallmark feature of ectodomain shedding. Different proteins are shed in response to PKC activation –i.e. EGF, TGFα, heparin-binding EGF (HB-EGF), interleukins (ILs), IL-6 receptor α (IL6Rα), L-selectin, amyloid precursor protein, neuregulin and VEGFR-1 (Thabard et al. 2001; Wheeler et al. 2003; Rahimi et al. 2009; Hayashida et al. 2010), and using specific inhibitors or cells lacking expression of ADAM17 (ADAM17^{-}/$^{-}$), it has been demonstrated that PKC-mediated ectodomain cleavage occurs in many instances via the activation of ADAM17 (Montero et al. 2002; Sahin et al. 2004; Rahimi et al. 2009; Kveiborg et al. 2011).

As is the case for other MAPKs, different PKC isoforms are known to regulate shedding in a cell type specific manner. For example, activation of the PKCδ and η isoforms controls the shedding of IL6Rα in myeloma cells, while PKCδ regulates the basal and induced shedding of HB-EGF in fibrosarcoma cells (Kveiborg et al. 2011) and neuregulin shedding in Chinese hamster ovary (CHO) cells, PC12 and primary neuronal cultures (Esper and Loeb 2009). Moreover, shedding of the same substrate can be regulated by different PKC isoforms depending on the stimulus (Herrlich et al. 2008; Dang et al. 2011). Even the same stimulus can activate different PKC isoforms as it is described for HB-EGF shedding after PMA stimulation in fibrosarcoma cells which results in PCKα and PKCδ activation (Kveiborg et al. 2011; Dang et al. 2011). In this case, it has been suggested that activation of both PKC isoforms acts in parallel to promote regulated and basal shedding of HB-EGF.

G-Protein-Coupled Receptor (GPCR)-Dependent ADAM-17 Regulated Ectodomain Shedding

Stimulation of several GPCRs can promote EGFR activation and therefore its downstream signaling pathways (i.e. MAPK), a process known as GPCR-dependent EGFR transactivation (Daub et al. 1996). This cross-talk mechanism plays a central role in the ability of GPCRs to mediate proliferative signaling and its deregulation can induce pathological states such as cancer. In fact, it has been reported for several cancer types like head and neck, lung, breast, ovarian, prostate and in cancer cell lines (Filardo et al. 2000; Fischer et al. 2003; Lautrette et al. 2005; Yasuda et al. 2007; Bhola and Grandis 2008).

GPCR-mediated EGFR transactivation is known to be a metalloproteinase-dependent process. It is initiated by agonist-mediated stimulation of a GPCR that leads to the activation of a metalloproteinase, resulting in EGFR ligand shedding and subsequent activation of EGFR-downstream pathways (Prenzel et al. 1999). A wide variety of GPCR agonists can trigger EGFR transactivation, including thrombin, angiotensin II (ANGII), endothelin-1, bradykinin, lysophosphatidic acid (LPA), carbachol, dopamine and hormones (Dorsam and Gutkind 2007; Liebman 2011). Although different ADAMs can regulate GPCR-mediated EGFR transactivation (Schäfer et al. 2004; Sanderson et al. 2006), ADAM17 seems to be the main protease involved, as it is the main protease involved in EGF ligand shedding (Ohtsu et al. 2006). Binding of an agonist to a GPCR induces a conformational change in the receptor and its associated heterotrimeric G proteins (α, β, γ), ultimately leading to the dissociation of the GTP bound α subunit from the βγ dimer (Choe et al. 2011). Both the α-GTP subunit and the βγ dimers may play a role in EGFR transactivation (Sugden and Clerk 1997). For transactivation mediated by the α subunit, the $G_{q/11}$ family members stimulate the hydrolysis of the membrane phospholipid phosphatidylinositol-4, 5, bisphosphate ($PtdInsP_2$) producing diacylglycerol (DG) and $InsP_3$, which regulate PKC and intracellular Ca^{2+} respectively, both known to play a role in ADAM17 activation. In this regard, it was demonstrated that ADAM17-dependent EGFR transactivation mediated by the ANGII receptors required G_q (Mifune et al. 2005; Ohtsu et al. 2008; Wang et al. 2009b). A similar mechanism and G_q requirement have been postulated for the P2Y6 receptor in response to γ irradiation (Tamaishi et al. 2011). In addition, the G_i subunit has been implicated in the ADAM17-dependent EGFR transactivation by LPA in squamous cell carcinoma and breast cancer cells (Gschwind et al. 2003).

Involvement of the $G_{\beta\gamma}$ subunits has also been proposed in EGFR transactivation. In this regard, $G_{\beta\gamma}$ subunits derived from pertussis toxin-sensitive $G_{i/o}$ proteins are necessary for α_{2A}AR-mediated transactivation of the EGFR (Luttrell et al. 1997; Pierce et al. 2001) and for EP2 receptor-mediated EGFR transactivation by prostaglandin E2 in colon cancer cells (Pai et al. 2002). Also, in sertoli cells, it has been demonstrated that G protein-coupled estrogen receptor 1 (GPER1)-mediated MAPK3/1 activation occurs via EGFR transactivation through G protein βγ subunits that promote Src kinase-mediated metalloproteinase-dependent release of EGFR ligands (Lucas et al. 2010). However, the metalloproteinases involved in these processes are not known.

Downstream of the G protein activation, several signaling components have been implicated as intracellular mediators in GPCR-initiated ADAM17-mediated transactivation. These intracellular signals include: an increase in the levels of calcium (Eguchi et al. 1998; Nagano et al. 2004; Mifune et al. 2005), reactive oxygen species (ROS; Zhang et al. 2001; Myers et al. 2009; Tai and Ascoli 2011) and PKC activation (Kodama et al. 2002; Rozengurt 2007). Src-like and PYK2 non-receptor tyrosine kinases have also been recognized as intermediates in the process (Luttrell et al. 1997; Andreev et al. 2001; Luttrell and Luttrell 2004).

Mechanisms of Activation of the Sheddase

While the signaling pathways leading to sheddase activation have been identified, the modification/changes in the sheddase molecule that trigger its activation are not clear. Ectodomain cleavage by ADAM17 is induced within minutes after treatment with stimuli, suggesting that it occurs through post-translational mechanisms. This is supported by the fact that rapid shedding is not inhibited by compounds targeting RNA or protein synthesis (Fan and Derynck 1999). Removal of the prodomain of ADAM17 was proposed as the mechanism for activation however, several reports demonstrated that this is not the case (Le Gall et al. 2010; Black et al. 1997; Black 2002). Since ADAM17 contains Tyr and Ser/Thr phosphorylation sites in its cytoplasmic domain, it was also proposed that intracellular signaling would lead to modification of the cytoplasmic tail of ADAM17 and that this effect would be necessary for its activation. This possibility is supported by the fact that overexpression of certain tyrosine phosphatases can inhibit ADAM17-mediated shedding (Zheng et al. 2002). In fact, a recent study indicates that p38α MAPK interacts with and phosphorylates ADAM17 on Thr735 and that this modification is necessary for ectodomain shedding (Xu and Derynck 2010). Similarly, Soond et al. (2005)

and Díaz-Rodríguez et al. (2002) demonstrated that phorbol esters promote ERK-mediated association and phosphorylation of ADAM17 at Thr735; however, this last study concluded that p38 MAPK was able to stimulate TrkA shedding in cells lacking ADAM17 activity, which is in conflict with the results by Xu and Derynck. Other phosphorylation mechanisms –i.e. association with c-Src and PDK1 and subsequent phosphorylation of ADAM17 (Zhang et al. 2006), or other phosphorylation sites, i.e. Ser819 and Ser791, have also been proposed (Fan et al. 2003). The role for the phosphorylation at these sites is not clear, adding to the controversial role of phosphorylation in ADAM17 activation. Perhaps the most striking result is that a truncated form of ADAM17 lacking the cytoplasmic domain and therefore all of the intracellular phosphorylation sites, is still able to promote ectodomain shedding of HB-EGF, TNFα, p75 TNFR and IL-1R-II (Reddy et al. 2000; Doedens et al. 2003; Le Gall et al. 2010; Maretzky et al. 2011a). It is possible that phosphorylation is required for proper trafficking as suggested by Soond et al. (2005) or for maturation or localization of the enzyme; however, more investigation is needed to understand the role of phosphorylation in ADAM17 activity. For an in depth review, see Murphy 2008.

In addition to phosphorylation, modifications that involve the ADAM17 cysteine-rich domain have been proposed to have regulatory activity (Li and Fan 2004). ADAM17 has two cysteine-sulfydryl (thioredoxin) motifs in the disintegrin/cysteine-rich domain that are necessary for the L-selectin cleavage activity of ADAM17. It has been proposed that ADAM17 activity might be regulated through reduction–oxidation modifications of those cysteine-sulfydryl groups in response to extracellular redox conditions, either via protein disulfide isomerase (PDI) or other thiol isomerase activities (Wang et al. 2009a). In support of this hypothesis, Willems et al. demonstrated that PDI inhibitors enhance PMA-stimulated ADAM17-mediated shedding of HB-EGF-AP in several cell types, also suggesting that this could be a general regulatory mechanism. Based on their studies using conformation specific antibodies, the authors conclude that PDI acts by isomerizing the disulfide bonds in the noncatalytic domains of ADAM17, causing a conformational shift in the enzyme structure and restricting access to the ADAM17 catalytic site. Therefore, conformational changes due to isomerization of the disulfide bonds are responsible for modulation in ADAM17 activity (Willems et al. 2010). It is worth mentioning that disulfide bonds in the ADAM17 are thought to be key elements in the determination of its tertiary structure (Takeda 2009). The idea that conformational changes are responsible for ADAM17 activation is attractive since recently, Le Gall et al. (2010) have proposed that activation involves a rapid and reversible exposure of the ADAM17 catalytic site.

The authors demonstrate that the tight-binding active site inhibitor DPC333 can quickly bind to ADAM17 in stimulated, but not quiescent cells and speculate that in unstimulated cells ADAM17 assumes a conformation in which its catalytic site is not accessible.

Besides direct modifications to the enzyme, ADAM17 activity can be regulated through protein interactions. Several proteins have been shown to interact with the cytoplasmic domain of ADAM17 including the protein tyrosine phosphatase (PTPH1; Zheng et al. 2002) and the synapse associated protein 97 (SAP97; Peiretti et al. 2003) which negatively regulate its activity, and Eve-1 which is required for HB-EGF shedding induced by PMA or ANG II (Tanaka et al. 2004). However, the relevance of these interactions is not clear since the cytoplasmic domain is not required for the shedding activity. Other proteins interact with different domains of ADAM17, including TIMP-3 that inhibits ADAM17 by binding to the catalytic center (Gooz 2010) and integrin αβ1 which binds to the disintegrin/cystein rich domain and has been proposed to target ADAM17 activity (Bax et al. 2004). In addition, although the site of interaction has not been explored, it has been demonstrated that Nardilysin (N-arginine dibasic convertase) regulates ADAM17 activity by mediating the interaction between ADAM17 and HB-EGF or neuregulin-1 (NRG1) and regulating the shedding event (Nishi et al. 2006). This suggests that Nardilysin may work to bring the enzyme and substrate to proximity.

Possible regulation of ADAM17 activity by localization/trafficking has also been suggested (see for example Hooper 2005; Tellier et al. 2006), but will not be discussed here.

THE SIGNIFICANCE OF ECTODOMAIN SHEDDING IN CANCER

Numerous studies suggest that sheddases play an important role in the pathogenesis of cancer. In fact, several MMPs and/or ADAMs are upregulated in human cancers and their expression levels seem to correlate with tumor stage (Tanaka et al. 2005; Zhou et al. 2006; Kenny and Bissell 2007; Kveiborg et al. 2011). The importance of ectodomain shedding in tumorigenesis is supported by the observation that inhibitors of metalloproteinases are able to delay cancer growth and block metastasis in different pre-clinical tumor models (Hidalgo and Eckhardt 2001; Bloomston et al. 2002). For example, specific targeting of ADAM17 with small molecule inhibitors or siRNA can revert the malignant phenotype of breast cancer cell lines (Kenny and Bissell 2007). Moreover, transgenic mice that overexpress wild-type transmembrane transforming growth factor alpha (TGFα) develop tumors, while expression of an uncleavable form of this protein does not (Sandgren et al. 1990).

Several mechanisms can be envisioned to explain how ectodomain shedding can contribute to the pathogenesis of cancer (Fig. 2): (i) shedding may lead to the production of a soluble protein that can either activate potentially oncogenic surface receptors or bind to ligands and prevent them from binding to their receptors inhibiting a tumor suppressor signaling cascade, (ii) shedding may lead to depletion of surface receptors with a potential tumor suppressor signaling activity and (iii) shedding produces an intramembrane protein fragment that can be further processed into an intracellular protein with an oncogenic function. These mechanisms are described in more detail below.

EGFR signaling, mainly through activation of MAP kinases, controls a wide variety of biological responses such as proliferation, differentiation, migration and apoptosis, which when deregulated can promote tumorigenesis (Fischer et al. 2003; Bublil and Yarden 2007). Since ectodomain shedding, particularly via ADAM17 and ADAM10, is responsible for EGFR (also called ErbB) activation by proteolytic processing of ErbB-ligand precursors (Kveiborg et al. 2011; Sahin et al. 2004; Hayashida et al. 2010), shedding comes to be an important regulator of tumorigenesis. This is particularly obvious in the case of growth factor induced shedding. As depicted in Figure 2A, EGF ligand stimulation of EGFR signaling can promote ADAM17 activation. This triggers increased release of EGF ligands by ectodomain shedding and subsequent EGFR signaling amplification. As a consequence, autocrine and paracrine growth stimulation of tumor cells occurs (Fig. 2A). This mechanism suggests a two-phase stimulation of EGFR signaling that has been demonstrated for the VEGF and PDGFβ receptors. Growth factor activation of these receptors results in an initial early metalloproteinase-independent stimulation of EKR1/2 and a later wave of ERK1/2 phosphorylation that depends on ADAM17 stimulation and the release of EGFR ligands via ectodomain shedding (Swendeman et al. 2008; Mendelson et al. 2010). Only in one instance, that is the stimulation of binding of fibroblast growth factor 7 (FGF7) to FGF receptor 2b (FGFR2b), activation of ERK1/2 has been shown to be strictly dependent on stimulation of ADAM17 (Maretzky et al. 2011a). These observations indicate that ectodomain shedding may provide a positive feedback mechanism to amplify growth factor/EFGR signaling by promoting cross-talk between pathways. It is expected that the increased ERK-MAPK activity which is observed in a third of all tumors (Dhillon et al. 2007) should also contribute to ectodomain shedding through ADAM17 activation.

Ectodomain shedding also contributes to tumorigenesis by modulating the activity of the TGFβ signaling pathway. TGFβ signals through binding to the TGFβ receptor II (TGFβR-II), which then binds and phosphorylates TGFβR-I, the kinase responsible for phosphorylation of Smad2 and 3. Activation of ADAM17 by the ERK-MAPK pathway promotes shedding of the TGFβ receptor I (TGFβR-I) and Vasorin, a TGFβ binding, type I transmembrane protein (Fig. 2B). Shedding of the TGFβR-I depletes the cells of a receptor necessary for TGFβ ligand signaling, therefore decreasing the antiproliferative activity of TGFβ (Liu et al. 2009), while soluble Vasorin binds to TGFβ and prevents it from binding to its receptor inhibiting TGFβ signaling (Ikeda et al. 2004; Malapeira et al. 2011). Therefore, activation of ADAM17 inhibits TGFβ signaling by two different mechanisms. Conversely, ectodomain shedding of TGFβR-III (betacellulin), a TGFβ co-receptor, seems to inhibit tumor cell invasion and angiogenesis (Bandyopadhyay et al. 2002a, b; Lei et al. 2002; Dong et al. 2007). This apparently contradictory result is probably the consequence of the dual role of TGFβ in cancer, as a tumor suppressor and as a promoter of invasion and metastasis (Meulmeester and Ten Dijke 2011).

Ectodomain shedding has a yet different role in the modulation of the vascular endothelial growth factor receptor (VEGFR) signaling in non-endothelial cells (Fig. 2C). In endothelial cells, VEGFR-2 is the main receptor implicated in vasculogenesis and angiogenesis in response to the ligand VEGF-A, while VEGFR-1 does not seem to have signaling capacity. However, in non-endothelial cells, specifically in cancer cells, VEGFR-1 activation is linked to cell proliferation (Vincent et al. 2005; Fragoso et al. 2006) and tumor growth in tumor mouse models (Wang et al. 2004; Fischer et al. 2007). In addition, soluble VEGFR-1 is increased in patients with leukemia (Hu et al. 2004; Aref et al. 2005; Medinger et al. 2010), suggesting an important role of the ectodomain, although it is not clear whether increased production of the ectodomain is due to alternative splicing, ectodomain shedding or both (Kendall and Thomas 1993; Rahimi et al. 2009). Using B-cell lymphoma cells, it has been shown that VEGF-A activation of VEGFR-1 promotes increased VEGFR-1 shedding by a mechanism involving PKC phosphorylation and subsequent ADAM17 activation. As a consequence of shedding: (i) the cell surface levels of VEGFR-1 decrease preventing binding to ligand and signaling, (ii) the shed VEGFR-1 modulates VEGF signaling possibly by binding to VEGF ligands or receptors and modifying their activity (Rahimi et al. 2009), and (iii) after cleavage the cytoplasmic part of VEGFR-1 is further processed (see below) and can mediate additional roles inside the cell.

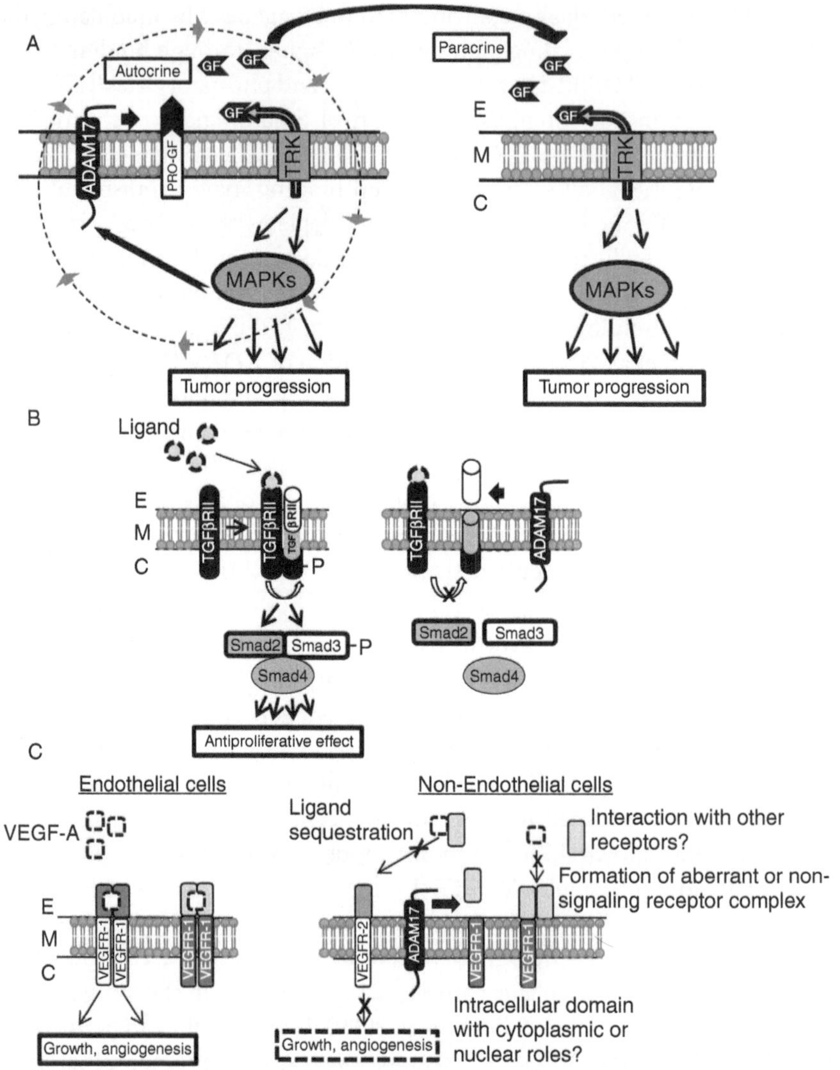

Fig. 2 Ectodomain Shedding and Cancer. A) Ectodomain shedding leads to the production of a soluble protein that can activate potentially oncogenic surface receptors. EGF ligand stimulation promotes ADAM17 activation which causes increased release of EGF ligands and subsequent EGFR signaling amplification. This results in autocrine and paracrine growth stimulation of tumor cells. B) Ectodomain shedding contributes to tumorigenesis by modulating the activity of the TGFβ signaling. Activation of ADAM17 by the ERK-MAPK pathway promotes shedding of the TGFβ receptor, depleting the cells of the receptor necessary for the antiproliferative activity of TGFβ. C) In non-endothelial cells, increased VEGFR-1 shedding by ADAM17 leads to a decrease in the cell surface levels of VEGFR-1 preventing signaling and the intracellular domain may be released and acquire additional roles. In addition, the shed ectodomain can modulate VEGF signaling by binding to VEGF ligands or receptors and modifying their activity.

Finally, regulated ectodomain shedding can contribute to cancer as it is the rate limiting step in the process of regulated intramembrane proteolysis (RIP). In RIP, the transmembrane protein undergoes ectodomain shedding followed by a second cleavage within the transmembrane domain resulting in the release of the intracellular domain into the cytosol. The proteolytic cleavage fragment can function as signaling molecule or travel to the nucleus where it affects gene expression and cellular responses (Bao et al. 2003; Lyu et al. 2008; Ma et al. 2008; Carpenter and Liao 2009; Allison et al. 2011; Higashiyama et al. 2011). For example, the VEGFR-1, just described, is regulated by RIP (Boulton et al. 2008). Other receptors, growth factors and cytokines utilize RIP to release a functional protein fragment with roles in numerous biological roles including growth or apoptosis, processes that are relevant to tumorigenesis. In fact, deregulation of RIP is found in diseases including cancer (Maetzel et al. 2009; Jorissen and De Strooper 2010; Lichtenthaler et al. 2011).

POTENTIAL BIOMARKERS GENERATED BY ECTODOMAIN SHEDDING (ECTOMARKERS)

ADAMs have been shown to be upregulated in several malignancies and several reports suggest that the level of circulating metalloproteinases may themselves be a marker for disease (see for example Ishikawa et al. 2004; Gack et al. 2005; Rausch et al. 2011).

In addition to their role as potential markers, metalloproteinases are responsible for the generation of soluble circulating ectodomains from different transmembrane proteins that can potentially function as prognostic/diagnostic biomarkers, ectomarkers. In the following section we will focus on some of these potential shedding-generated biomarkers and their value in cancer diagnosis/prognosis. See also Table 2 for additional information on these biomarkers.

Table 2 Ectomarkers present in body fluids of cancer patients.

Soluble Molecule	*Cancer*	*Correlation*	*References*
HER2	Breast	See text	See text.
	Colorectal	Metastasis Advanced stage/ metastases	Tsigris et al. 2002; Polykronidis et al. 2003.
	Ovarian		McKenzie et al. 1993.
	Lung		Osaki et al. 1995.
	Gastric		Kono et al. 2000.

Table 2 Contd. ...

Table 2 Contd. ...

CD44	Breast	See text	See text.
	Colorectal	High levels but no association	Masson et al. 1999; Guo et al. 1994.
	Gastric		Guo et al. 1994.
	Head and neck		Franzmann et al. 2007.
	Leukemia	Advanced disease	Samy et al. 2010.
E-cadherin	Gastric	See text	See text.
	Bladder		Banks et al. 1995; Shariat et al. 2005.
	Prostate	Metastases	Kuefer et al. 2003;
		Advance stage/ metastasis	Kuefer et al. 2005.
		Advanced stage	Weiss et al. 2011.
	Gastric		
	Hepatocellular		Katayama et al. 1994
	Lung	Metastasis	Katayama et al. 1994
	Melanoma		Charalabopoulos et al. 2006
	Esophageal		Shirahama et al. 1996
			Chung et al. 2011b
V-CAM/I-CAM	Colorectal	See text	See text
	Gastric	Metastasis	Maruo et al. 2002; Kaihara et al. 1998;
		Advanced stage/ metastatic	Velikova et al. 1997;
			Nakata et al. 2000.
	Breast		Köstler et al. 2001
	Lymhoma		Terol et al. 2 003.
	Cervical		Okamoto et al. 1999.
	Lung		Grothey et al. 1998.

Breast Cancer Ectomarkers

Her2

Human epidermal growth factor receptor 2 (HER2 or ErbB2), is a 185 kDa member of the EGFR (or ErbB) family of transmembrane tyrosine kinase receptors, which are expressed in various tissues of epithelial, mesenchymal and neuronal origin, as well as in human cancers (Hynes and Lane 2005). The ErbB receptors and their ligands are involved in many essential cellular processes like proliferation, differentiation, migration and invasion and therefore their therapeutic potential has been extensively investigated. There are four members in the ErbB family, EGFR (ErbB1), HER2 (ErbB2), HER3 (ErbB3) and HER4 (ErbB4), characterized by the presence of an extracellular ligand binding domain and an intracellular tyrosine kinase domain, which is lacking in HER3. ErbB receptor activation is initiated by ligand binding which triggers a conformational change that induces the formation of homo- or heterodimers, transphosphorylation of their intracellular domains and activation of downstream pathways (Citri and Yarden 2006). HER2 is the only ErbB receptor that does not have a known ligand but due to its open conformation in the absence of ligand, it is the preferred dimerization partner (Graus-Porta et al. 1997; Cho et al. 2003) and can form dimers with a strong kinase activity (Moasser 2007).

Activation of HER2 can also occur through ADAM10 mediated ectodomain shedding (Liu et al. 2006). Cleavage of HER2 results in the production of two fragments: a soluble fragment that contains the extracellular domain (HER2 ECD, 105-110 kDa) and is measurable in serum, and a truncated membrane-associated fragment (95 kDa, HER2p95; Christianson et al. 1998; Arribas et al. 2011) with enhanced kinase/signaling activity (Segatto et al. 1988) and therefore clinically relevant (Sáez et al. 2006). This implies that HER2 ectodomain shedding not only provides a serum marker but it can also provide an indication of the degree of receptor activation (Molina et al. 2001, 2002). Moreover, recently Ghedini et al. (2010) uncovered a functional role for the HER2 ECD. Using human ovarian carcinoma cells the authors demonstrated that soluble recombinant HER2 ECD inhibited the growth of HER2-overexpressing cells by potentially promoting HER2 ECD dimerization with HER2, HER1 and HER3 transmembrane receptors leading to impaired phosphorylation and activation of the receptors. Similarly, in mice, high levels of circulating recombinant HER2 ECD inhibited growth of HER2-expressing xenotransplanted cells but not HER2 negative cells (Ghedini et al. 2010).

The clinical relevance of HER2 ECD has been thoroughly discussed and the reader is referred to recent reviews for an in-depth coverage of the subject (Carney 2007; Leary et al. 2009; Leyland-Jones and Smith 2011; Tše et al. 2012). Here we will summarize some of the most relevant arguments that support its role as a biomarker. HER2 protein is expressed at low levels on the epithelia of the gastro-intestinal, respiratory, reproductive, and urinary tracts as well as in the skin, breast and placenta (Press et al. 1990), but it is overexpressed in 20 percent of breast, and in some ovarian and gastric cancers, and this is associated with poor clinical prognosis and patient survival (Slamon et al. 1987; Baselga and Swain 2009). Interestingly, high serum levels of HER2 ECD are correlated with a poor prognosis and decreased responsiveness to hormonal- and chemo-therapy (Yamauchi et al. 1997; Colomer et al. 2000). Moreover, several studies have reported that the observed increase in serum HER2 ECD levels in patients with metastatic breast cancer is not the result of the increase in tumor burden (Ali et al. 2002; Fehm et al. 2004). These studies suggest that HER2 ECD can function as a prognostic and/or diagnostic biomarker and in fact, the FDA has approved a serum HER2 test for the managing and monitoring of patients with metastatic breast cancer who have an initial serum HER2 ECD value above the normal cut off of 15 ng/ml (Carney et al. 2003). The following observations refer to the applicability of the HER2 ECD serum levels as a marker at different stages of breast cancer or during treatment of the disease: (i) in primary breast cancer, high serum HER2 ECD levels (≥15 ng/ml) have been observed for 7-9 percent of the patients (Isola et al. 1994; Saghatchian et al. 2004; Ludovini et al. 2008). Several reports have demonstrated a correlation between preoperative HER2 ECD and tumor size or nodal involvement (Krainer et al. 1997; Pallud et al. 2005), or a significant association with high histological grade, lymph node involvement, negative estrogen and progesterone receptors and low disease-free survival. Multivariate analysis suggested that serum HER2 was an independent prognostic factor only in patients with HER2-overexpressing tumors (Ludovini et al. 2008; Molina et al. 2010). In conclusion, these studies suggest that patients with positive HER2 status and high serum HER2 ECD levels have a worse outcome. (ii) In patients with metastatic disease a high variability in the levels of serum HER2 has been reported (Tśe et al. 2012; Leary et al. 2009; Carney 2007), although in general, there is a prevalence for high level serum HER2 ECD in these patients. As reported by Carney (2007), on average 48.6 percent of patients with metastatic disease demonstrate high levels of serum HER2 (>15 ng/ml). In a recent study including 158 women with newly diagnosed metastatic breast cancer, 22 percent had elevated serum HER2 ECD levels and this was associated with short survival (Bramwell et al. 2009). (iii) A possible correlation between serum and tissue HER2 levels although of high clinical relevance because if positive, serum HER2 would provide a non-invasive surrogate marker for HER2 tissue diagnosis, is still uncertain. While Witzel et al. (2010) and Ludovini et al. (2008) reported high concordance between serum HER ECD and tissue HER2 status, Quaranta et al. (2006) and Fontana et al. (1994) failed to demonstrate a positive correlation in studies using a larger number of patients. Interestingly several reports indicate that patients with primary HER2 negative tumors can develop HER2 positive metastatic tumors (Gancberg et al. 2002; Regitnig et al. 2004; Meng et al. 2004) and therefore high levels of serum HER2 ECD (Fehm et al. 2002). A high percentage of these patients responded to Trastuzumab therapy, a HER2 ECD targeted antibody. Since only candidates with HER2 positive tumors are candidates for Trastuzumab, high levels of circulating HER2 ECD can be a critical clue that the tumor has changed from a negative to a positive HER2 status at metastasis and therefore that the patient could benefit from Trastuzumab therapy. iv) The prognostic value of serum HER2 in response to treatments is also uncertain. For neoadjuvant treatment Witzel et al. (2010) reported that a 15 ng/ml HER2 ECD baseline together with a 20 percent of more decrease after treatment was associated with a complete pathological response in HER2 positive patients, shorter time to progression (TTP) and decreased overall survival. Several studies confirm the association between response to Trastuzumab therapy and a post-treatment decrease in serum HER2 ECD and therefore that changes in serum HER2 ECD may represent a valuable tool to predict pathological response to neoadjuvant therapy (Bethune-Volters et al. 2004; Köstler et al. 2004; Esteva et al. 2005; Tśe et al. 2005; Ali et al. 2008). There are very few trials for the predictive value of serum HER2 in response to treatment with Lapatinib. These studies indicate that although baseline does not predict response to Lapatinib (Cameron et al. 2008; Finn et al. 2009), a decrease of more than 20 percent in the levels 4-8 weeks after treatment initiation is associated with longer progression free survival (Gomez et al. 2008). v) Recurrence: increased serum HER2 ECD levels are associated with increased recurrence. Several studies have indicated that there is an increase in the serum levels of the HER2 ECD that occurs 4-9 months before clinical signs or clinical diagnosis and therefore, increasing serum HER2 ECD levels may be a powerful means to detect recurrence earlier (Isola et al. 1994; Fehm et al. 2002; Carney et al. 2003).

In conclusion, although the studies on serum concentrations of HER2 ECD as a biomarker for breast cancer are promising, numerous studies suggest that no single biomarker may have sufficient sensitivity and specificity to be used clinically, especially for early detection. Therefore, in addition to HER2 many other groups of circulating proteins have been examined as potential prognostic factors in the diagnosis of breast cancer. Below we summarize some of those studies.

CD44

CD44 is a type I transmembrane glycoprotein with a wide variety of structural heterogeneity resulting from protein polymorphisms and posttranslational modifications (Zöller 2011). The smallest CD44 isoform (CD44s) is expressed ubiquitously on the membrane of most vertebrate cells (Naor et al. 2008), with highest expression in hematopoietic cells. Expression of the other isoforms comprising variable exons (CD44v) is more restricted to some epithelial cells, or to embryonic development, lymphocyte maturation and activation, and to several types of carcinoma (Ruiz et al. 1995). CD44 functions as the major cellular adhesion molecule for hyaluronic acid (HA), a component of the extracellular matrix (ECM; Knudson and Knudson 1993, Lesley et al. 1993). In addition, CD44 can interact with various other ECM components including chondroitin sulfate, fibrinogen, fibronectin, collagen and laminin and can bind mucosal vascular addressin, osteopontin, serglycin/gp600, and the MHC class II invariant chain (Naor et al.1997). The interaction of CD44 with the ECM is through its ectodomain (CD44 ECD) while the intracellular domain associates with the actin cytoskeleton via binding to the ezrin, radixin and moesin proteins (Ponta et al. 2003). This arrangement makes CD44 a key molecule to direct cell-matrix interactions. Put together, the high heterogeneity in its structure and binding partners as well as its signaling capabilities are a reflection of the numerous physiological and pathological processes in which it is involved, such as lymphocyte homing and activation, wound healing, cell migration, cell growth, metastasis and tumorigenesis (reviewed in Aruffo et al. 1990; Lesley and Hyman 1998; Hertweck et al. 2011).

Cleavage of the CD44 ECD is triggered and regulated by multiple signaling pathways, including activation of PKC, Ca^{2+} influx, and activation of small Rho GTPases (reviewed in Nagano and Saya 2004), and is mediated by several metalloproteinases. Extracellular Ca^{2+} influx promotes CD44 ECD cleavage through the activation of ADAM10, while phorbol ester stimulation activates ADAM17 through the activation of PKC and small GTPase Rac, inducing CD44 cleavage (Nagano et al. 2004; Takamune et al. 2007). In melanoma cells, ADAM10 is responsible for basal shedding of CD44 (Anderegg et al. 2009). In addition to ADAM10 and ADAM17, the membrane type-1 metalloproteinase (MT-1 MMP) has also been shown to be able to cleave CD44 (Kajita et al. 2001). Cleavage by metalloproteinases releases the soluble CD44 ECD and the fragment that remains attached to the membrane is subsequently cleaved by presenillin-dependent γ-secretase, releasing the CD44 intracellular domain (CD44 ICD) into the cytoplasm. The CD44 ICD travels into the nucleus and activates gene transcription. Therefore the soluble ECD can potentially be a surrogate marker for level and/or activity of the CD44 ICD (Stamenkovic and Yu 2009).

While enhanced CD44 ECD cleavage has been reported in various human tumors, including breast, non-small cell lung, colon and ovarian carcinomas and gliomas (Okamoto et al. 2002), the physiological role of the soluble CD44 ECD is still unclear. Some reports suggest that it competes with surface CD44 for binding to HA preventing the activity of CD44 (Yu et al. 1997; Ahrens et al. 2001). Supporting this, some results indicate that *in vivo*, overexpression of soluble CD44 ECD suppresses growth and metastasis of injected tumor cells in mice (Bartolazzi et al. 1994; Sy et al. 1992; Peterson et al. 2000; Zawadzki et al. 1998; Yu et al. 1997). Ultimately these observations indicate that CD44 cleavage contributes to the regulation of surface CD44 interaction with HA which is required for the tumor cell migration and invasion.

Clinically, several studies have associated CD44 expression with tumor progression and metastasis, whereas others argue that it is associated with a favorable prognosis. These contradictory results are in part due to the structural variability of the CD44 molecule and may depend on isoform and ligand expression (Orian-Rousseau 2010). CD44 is upregulated in both *in situ* and invasive breast ductal carcinoma (Auvinen et al. 2005) and its increased expression correlates with overall patient survival (Diaz et al. 2005). Furthermore, CD44 is expressed at a higher level in tubular carcinoma, compared with invasive micropapillary carcinoma, a form of the disease with a high metastatic potential and poor clinical outcome (Gong et al. 2005). Stage differences have also been reported. In primary tumors of stage III breast cancer, overexpression of the CD44 protein has been associated with overall poor survival, while no correlation could be made when stage II tumors were analyzed (Bhatavdekar et al. 2000).

Aberrant expression of CD44 isoforms carrying the variable regions v5 and/or v6 (CD44v5 or CD44v6) has also been detected in breast cancer (Kaufmann et al. 1995; Sinn et al. 1995). Interestingly, the expression of these variants was shown to correlate with the levels of the corresponding soluble forms in the patient's serum. Moreover, when metastases were detected, soluble CD44v5 and v6 ECD serum levels were significantly increased with respect to patients with no metastases, and elevated CD44v6 ECD levels but not CD44v5 ECD were associated with metastases in liver or bone (Lackner et al. 1998), and with larger tumor size and increased risk for lymph node metastasis (Mayer et al. 2008). All together these observations suggest a potential role for the soluble CD44 as a prognostic marker. In support of this, a very recent study including 110 patients with stage I-IV breast cancer, reported that the serum level of CD44 ECD significantly correlates with tumor stage although it does not predict treatment response. Interestingly, these authors demonstrate that in patients with positive HER2 breast cancer the serum levels of CD44 ECD are a significant predictor of overall survival (Baek et al. 2011). Of note, a recent study suggests that CD44 expression is a good prognostic marker only for hormone receptor negative disease (Kim et al. 2011). The reasons for this apparent discordance are not known.

Gastric Cancer Ectomarkers

E-cadherin

Cadherins are calcium-dependent integral transmembrane single-chain glycoproteins that mediate intercellular cell–cell adhesion through homotypic interactions (Steinberg et al. 1999; Hulpiau and van Roy 2009). Classic cadherins are composed of five conserved extracellular domains repeats involved in cell binding, a single transmembrane segment, and an intracellular domain that interacts with the cytoskeleton through the catenin complex (Harrington and Syrigos 2000; Nelson 2008).

The epidermal specific cadherin, E-cadherin, is preferentially expressed on epithelial cells of a variety of tissues in different species and it has a crucial role in the maintenance of epithelial integrity and polarity and in many aspects of cell adhesion and epithelial development. In addition to its role in development, E-cadherin has anti-invasive and anti-metastatic properties (Vleminckx et al. 1991; Perl et al. 1998). In most cancers of epithelial origin, E-cadherin-mediated cell adhesion is lost concomitantly with the acquisition of an invasive and aggressive phenotype (Pokutta and Weis 2007). In fact, an inverse correlation exists between E-cadherin levels, tumor grade, and patient survival rates (Syrigos et al. 1995; Karayiannakis et al. 1998; Karatzas et al. 1999). Loss of E-cadherin has also been associated with progression to dysplasia and adenocarcinoma in precancerous conditions, such as in Barrett's esophagus (Bailey et al. 1998).

Several mechanisms have been described for E-cadherin inactivation in tumors, including mutations, epigenetic silencing, endocytosis, proteosomal degradation, increased expression of non-epithelial cadherins such as N-cadherin and proteolytic cleavage (van Roy and Berx 2008). *In vitro*, several metalloproteinases including MMP9, MMP7, MMP3, MT1-MMP and ADAM10 have been shown to cleave mature E-cadherin (De Wever et al. 2007) releasing the soluble E-cadherin ECD. This mechanism not only decreases the level of full length E-cadherin but the soluble E-cadherin fragments produced after shedding may have several functions: (i) as a paracrine or autocrine competitor of E-cadherin function, (ii) as an anchor for migration when bound to the ECM, or (iii) as an enhancer of metalloproteinase expression (Nawrocki-Raby et al. 2003; De Wever et al. 2007). Recently, it has been demonstrated that soluble E-cadherin prevents apoptosis via a mechanism that requires a functional interaction with cellular E-cadherin and the activation of the EGFR, PI3K and ERK1/2 signaling pathways (Inge et al. 2011).

Clinically, decreased membranous expression of E-cadherin molecules has been found not only in gastric cancer (Mayer et al. 1993; Chan 2006), but also in colon (Dorudi et al. 1995), pancreas (Pignatelli et al. 1994; Fei et al. 2010), esophagus (Jankowski et al. 1994; Takeno et al. 2004), liver (Endo et al. 2000), prostate (Umbas et al. 1994; Jaggi et al. 2005), ovarian (Faleiro-Rodrigues et al. 2004), bladder (Bringuier et al. 1993; Garcia del Muro et al. 2000), breast (Siitonen et al. 1996; Kowalski et al. 2003) and head and neck tumours (Muramatsu et al. 2008; Andrews et al. 1997).

With respect to soluble E-cadherin ECD, Katayama et al. (1994) reported that its levels were significantly elevated in the urine of 53 percent of patients suffering from various types of cancers including lung, liver, stomach, colon and rectal cancers, as compared with those in the age-matched healthy subjects. In gastric cancers, this data is supported by several other studies. For example, Gofuku et al. reported abnormally high levels of soluble E-cadherin ECD in 67 percent of patients with gastric cancer, although they could not find a significant correlation between the amount of soluble E-cadherin ECD and clinicopathological factors (Gofuku et al. 1998). Chan et al. (2001) also report elevated levels of serum E-cadherin ECD in patients with gastric cancer and demonstrated that elevated serum E-cadherin ECD can predict T4 invasion and palliative/conservative treatment. In a study using only pre-treatment data from patients, the same group demonstrated that serum E-cadherin can function as an independent prognostic marker. Level of serum E-cadherin ECD greater than 10,000 ng/ml could i) predict that 90 percent of patients would have a survival time of less than 3 years and ii) provide good sensitivity (59%) and specificity (75%) for predicting recurrence at 6 months postsurgery (Chan et al. 2003, 2005). This is of paramount importance in determining the course of treatment and for early detection of cancer recurrence. Contradictory results have been reported by Velikova et al. that failed to observe an increase in the serum levels of E-cadherin ECD between gastric patients and normal subjects (Velikova et al. 1997). Since it has been described that the serum levels of E-cadherin rise with increasing age in both gastric cancer patients and control subjects preventing the utility of soluble E-cadherin as a marker in older patients (Pedrazzani et al. 2008), it is possible that the Velikova study is biased because of the age of the subjects. Finally, it is worth mentioning that elevated levels of ADAM10, a metalloproteinase that cleaves E-cadherin, have been demonstrated in patients with gastric cancer and are significantly associated with the presence of lymph nodes and distant metastases (Wang et al. 2011).

Colorectal Cancer Ectomarkers

VCAM/ICAM

Intercellular adhesion molecules (ICAMs) and vascular adhesion molecules (VCAMs) are members of the immunoglobulin superfamily of adhesion molecules. These molecules, present in the surface of leukocytes and endothelial cells, serve as ligands for the integrins and play a critical role in cell adhesion and in facilitating leukocyte transmigration. Although some of these molecules –i.e. ICAM-1-- are constitutively expressed at low levels in resting cells, most of them are induced upon activation by proinflammatory cytokines such as IL-1 and TNF-α (Carlos and Harlan 1994). In cancer, these molecules play a very important role in the process of metastasis by regulating cellular interactions and signal transduction (Aplin et al. 1998; Konstantopoulos and Thomas 2009).

Ectodomain cleavage of VCAM-1 can be induced by PMA, and in stomach epithelial cells and fibroblast derived from genetically deficient mice, ADAM17 was identified as the protease responsible for the shedding (Garton et al. 2003). ADAM17 is also responsible of VCAM-1 induced shedding from murine endothelial cells in response to proinflammatory cytokines IL-1beta and TNFα (Singh et al. 2005), and for ICAM shedding in HEK293 cells upon PMA stimulation and in endothelial cells upon TNFα stimulation (Tsakadze et al. 2006). In addition, ectodomain cleavage of VCAM-1 has also been reported to be mediated by ADAM9 in COS-7 cells (Guaiquil et al. 2009) and by ADAM8 in human umbilical endothelial cells (Matsuno et al. 2006). Finally, MT-MMP1 has been reported to be responsible for ectodomain shedding of ICAM upon pervanadate induction (Essick et al. 2008).

Soluble ICAM ECD can compete with surface ICAM to bind lymphocyte function-associated antigen 1, preventing recognition and lysis of target cells by leukocytes. This could be an important mechanism for tumors to escape from immune surveillance. Shedding of ICAM by tumor cells may allow escape from cytotoxic T cells and natural killer cells, promoting metastasis (Becker et al. 1993; Banks et al. 1993).

Clinically, cell surface levels of VCAM-1 and ICAM-1 have been shown to be upregulated in patients with breast and/or liver cancer and elevated serum levels of soluble ICAM-1 and VCAM-1 ECD have been described in patients with several types of cancers including breast, gastrointestinal, liver, bladder, lymphoma and myeloma with respect to the healthy subjects (Banks et al. 1993; Kobayashi et al. 2007). In patients with colorectal cancer both ICAM-1 and VCAM-1 are expressed on endothelial cells of small blood vessels adjacent to colorectal cancer cell nests (Nelson et al. 1994), and their expression is upregulated by cytokines produced by the tumor. Soluble forms of both adhesion molecules have been detected in the supernatant of cytokine-activated endothelial cells and in the culture media of colonic carcinoma cell lines (Pigott et al. 1992; Dippold et al. 1993). More importantly, increased levels of soluble ICAM-1 and VCAM-1 have been detected in patients with colorectal tumors (Sánchez-Rovira et al. 1998). The potential for serum ICAM-1 and/or VCAM-1 ECD levels as a marker for colorectal cancer disease has been intensively studied. Reports by Alexiou et al. and Mantur et al. demonstrated a significant association between the serum levels of ICAM-1 and VCAM-1 ECDs with disease stage and the presence of both lymph node and distant metastases and the TNM classification stage, suggesting that serum concentrations of these molecules may reflect tumor progression and metastasis. The serum levels of these molecules were shown to decrease significantly after radical resection of the tumor indicating their origin in the tumor cells (Alexiou et al. 2001; Mantur et al. 2009). In addition, Alexiou et al. showed that elevated pre-operative ICAM-1 and VCAM-1 ECDs levels were significantly correlated with disease survival. The prognostic value of the soluble ICAM-1 ECD was supported by a study conducted by Kang et al. Patients with lymph node and liver invasions and poorly differentiated colorectal cancer had significantly higher serum concentrations of soluble ICAM-1 ECD (Kang et al. 2005). More recently Toiyama et al. demonstrated that the prognosis of colorectal cancer patients with an elevated ICAM-1 ECD level was significantly worse than that of patients with a normal ICAM-1 ECD level. Interestingly, serum ICAM-1 ECD levels were shown to be a very strong prognostic marker for patients with stage II grade of the disease (Toiyama et al. 2008). The prognostic value of the soluble VCAM-1 ECD in rectal and colorectal cancer patients has been supported in recent studies by Kemik et al. (2011) and Okugawa et al. (2009). The authors conclude that the preoperative level of soluble VCAM-1 reflects disease progression. Rectal cancer patients had significantly higher levels of serum soluble VCAM-1 ECD when compared with healthy controls and those levels were significantly correlated with tumor size, the presence of nodal and distant metastases and poor prognosis. Distant metastasis and elevated serum VCAM-1 ECD level were independent risk factors predicting poor prognosis (Okugawa et al. 2009; Kemik et al. 2011).

THE POTENTIAL ROLE OF TMEFF2 AS AN ECTOMARKER

Changes in the levels of a many ectomarkers can be observed in multiple cancer types (Table 2), demonstrating a lack of specificity. In the last section of this chapter, we discuss the potential to use the ectodomain of the transmembrane protein with an epidermal growth factor and two follistatin motifs 2 (TMEFF2), as a tissue specific biomarker. The rationale being that TMEFF2 is a protein with limited distribution to brain and prostate and therefore highly specific for these tissue types.

TMEFF2 is an evolutionarily conserved protein expressed in the embryo (Uchida et al. 1999; Heanue and Pachnis 2006) and selectively in the adult brain and prostate (Glynne-Jones et al. 2001; Gery et al. 2002; Afar et al. 2004). It belongs to the TMEFF family of proteins, comprised of only two members TMEFF1 and TMEFF2. Both of them are single-pass type I transmembrane proteins characterized by the presence of an EGF-like domain and two follistatin modules in the extracellular portion of the protein. The cytoplasmic domain of these proteins contains a putative G-protein activating motif, suggesting a receptor function (see below; Eib and Martens 1996; Uchida et al. 1999). While TMEFF1 and TMEFF2 have a similar structural organization and both are expressed in the embryo, they differ grossly in their adult tissue distribution. TMEFF2 is restricted to brain and prostate while TMEFF1 is more widely distributed, although it is predominantly expressed in the brain (Eib et al. 2000; Gery et al. 2003). In brain, they also exhibit different patterns of expression; while TMEFF1 is expressed in the neocortex, CA3, CA1 and subiculum regions of the hippocampus, locus coeruleus, and dentate cerebellar nucleus, TMEFF2 is highly expressed in the medial habenular, CA2, CA3 and dentate gyrus region of the hippocampus, corpus callosum, cerebellar cortex and cranial nerve nuclei. In spite of these differences, their high level of sequence similarity suggests that TMEFF1 and TMEFF2 could be playing a similar role in different tissues/developmental stages (Kanemoto et al. 2001).

The features in the TMEFF proteins and their expression in the embryo suggest a role in development however, little is known about their role in this process. Using Xenopus embryos as a model system, it has been demonstrated that TMEFF1 can inhibit nodal and BMP signaling. TMEFF1 inhibits nodal signaling by binding to its co-receptor Cripto (Harms and Chang 2003). The extracellular portion of the protein and its anchorage to the membrane are necessary for this activity suggesting that TMEFF1 blocks nodal signaling at the ligand or receptor level. TMEFF1 can also block BMP signaling and this activity is dependent only on the presence of the C-terminal domain (Chang et al. 2003). In addition, a role for TMEFF1 in development is indicated by the observation that it is upregulated in the blastema during amphibian limb regeneration (Morais da Silva et al. 2001). Although the possible role of TMEFF2 in development has not been investigated, Kanemoto et al. (2001) described that while TMEFF2 knockout mice were born normally, 85 percent exhibited retarded growth and died at weaning apparently from an impaired ability to feed, suggesting possible neurological defects.

A role for TMEFF1 in cancer was proposed by Gery et al. (2003) based on the observation that its expression is downregulated in 96 percent of brain tumors samples relative to normal brain tissue. Furthermore, overexpressing TMEFF1 in the glioblastoma cell line U118 had a growth inhibitory effect on the cells, consistent with a role as a tumor suppressor gene (Gery et al. 2003).

TMEFF2 may also play a role in brain disease as it is present in Alzheimer's plaques (Siegel et al. 2006). Studies of TMEFF2 in cancer suggest dual oncogenic roles. In human samples, TMEFF2 is up-regulated in a significant fraction of primary and metastatic prostate tumors (Glynne-Jones et al. 2001; Afar et al. 2004) and in leiomyomas (Wu et al. 2002), suggesting an oncogenic/growth promoting role, while it is hypermethylated and therefore downregulated in gliomas and several other cancers suggesting a tumor suppressor role (Suzuki et al. 2005; Brücher et al. 2006; Hellwinkel et al. 2008; Ivanauskas et al. 2008; Tsunoda et al. 2009; Gonzalo et al. 2010; Lin et al. 2011, and references therein). Data obtained from *in vitro* or mouse xenograft studies also suggest dual functions. While ectopic expression or addition of the purified extracellular domain (ECD) of TMEFF2 promotes neuronal cell survival (Horie et al. 2000) and cell growth (Ali and Knaüper 2007; Chen et al. 2011), ectopic expression of full length TMEFF2 demonstrates *in vitro* anti-proliferative effects (Gery et al. 2002; Elahi et al. 2008; Chen et al. 2011) and suppresses tumor growth *in vivo* in nude mouse xenografts (Elahi et al. 2008). Based on these results, we believe that the varied biology of TMEFF2 in cancer likely arises from the complexity of the molecule with the extracellular domain and the full length protein having diametrically opposed functions. This is supported by the fact that TMEFF2 can be cleaved from the membrane by ADAM17 to release a functional TMEFF2 ECD that can promote growth (Ali and Knaüper 2007). ADAM17-dependent cleavage of TMEFF2 is followed by γ-secretase cleavage within the transmembrane domain and can be induced by inflammatory cytokines like TNFα; (Lin et al. 2003; Ali and Knaüper 2007; Chen et al. 2011) which are characteristic of the tumor microenvironment. Therefore, regulated TMEFF2 cleavage can modulate the function of TMEFF2, as it plays a role during normal development or during tumor establishment and progression.

Little is known about the molecular mechanisms involved in TMEFF2 activity although the TMEFF2 ectodomain has been shown to induce tyrosine phosphorylation of ErbB4 (Uchida et al. 1999) or an increase in ErbB1-dependent ERK1/2 phosphorylation (Ali and Knaüper 2007). Using microarray analysis Elahi et al. (2008) demonstrated that overexpression of full length TMEFF2 induces upregulation of STAT1 in a human colorectal carcinoma cell line, an effect that the authors proposed to be mediated by the EGF-like motif in TMEFF2. In addition, Lin et al. (2011) have shown that TMEFF2 interacts with platelet derived growth factor AA (PDGF-AA) and regulates PDGF signaling. Importantly, we have recently demonstrated that full length TMEFF2 interacts with sarcosine dehydrogenase (SARDH), the enzyme responsible for conversion of sarcosine into glycine (Chen et al. 2011). The relevance of these findings to cancer is illustrated by the fact that metabolomic analysis of prostate cancer tissue, serum and urine samples has identified sarcosine as a metabolite highly increased during prostate cancer progression to metastasis and therefore as a potential marker for prostate cancer. Moreover, SARDH and other enzymes involved in sarcosine metabolism can also act as regulators of cell invasion and therefore are potential therapeutic targets for prostate cancer (Sreekumar et al. 2009). Our results indicate that TMEFF2 can affect sarcosine metabolism and decrease cellular sacosine levels through its interaction with SARDH resulting in decreased cellular motility and invasion. Our results provide a mechanistic link between metabolomic prostate cancer profiling identifying sarcosine as a tumor promoter, and a regulatory pathway involving TMEFF2 functioning as a tumor suppressor.

Clinically, several studies have suggested that changes in TMEFF2 methylation pattern could be developed as a marker for detection and/or progression of cancer lesions in lung (Hanabata et al. 2004; Chung et al. 2011a; Selamat et al. 2011), breast (Park et al. 2011), bladder (Hellwinkel et al. 2008; Costa et al. 2010), colorectal (Young et al. 2001; Sato et al. 2002; Ebert et al. 2005; Huang et al. 2007; Lofton-Day et al. 2008; Gonzalo et al. 2010; He et al.

2010), cholangiocarcinoma (Kim et al. 2009b), esophageal (Smith et al. 2008; Zhao et al. 2008), Barret's esophagus (Schulmann et al. 2005; Smith et al. 2008; Jin et al. 2009), gastric (Shibata et al. 2002; Geddert et al. 2004; Ivanauskas et al. 2008) and small bowel (Brücher et al. 2006). As mentioned above, TMEFF2 seems to have a dual role in cancer. In this respect, it is upregulated in primary and metastatic prostate tumors, although no study has so far established a significant correlation between TMEFF2 overexpression and prostate cancer that could readily suggest its utility as a marker. Using primary tumor tissue from 74 prostate cancer patients and 347 samples of nonmalignant adult tissues, Afar et al. (2004) reported increased TMEFF2 expression in the malignant samples compared to the tissue from the control subjects. However, no correlation with Gleason score or stage of the disease was found. On the contrary, another study based only on quantitative RT-PCR suggest that levels of TMEFF2 expression does not allow differentiation of malignant prostate tissue from benign prostatic hyperplasia (Antunes et al. 2009)

Therapeutically, since TMEFF2 is a membrane protein with restricted expression to brain and prostate and overexpressed in prostate cancer and metastases, it represents a good target for antibody-directed therapy for prostate cancer. In fact, using human prostate tumors xenografts in mice, two reports have been published that successfully validate TMEFF2 as a target for radioimmunotherapy (Zhao et al. 2005) and for auristatin-conjugated antibody therapy for the treatment of this disease (Afar et al. 2004).

Can the TMEFF2 ectodomain be used as a diagnostic/prognostic marker? So far, a strong correlation between TMEFF2 expression and prostate cancer has not been demonstrated. In addition, the fact that TMEFF2 is overexpressed in prostate cancer is difficult to reconcile with its proposed tumor suppressor activity. Our model described below attempts to resolve this apparent contradiction while providing a role for the TMEFF2 ectodomain as a diagnostic and/or prognostic marker for prostate cancer. Based on the literature and our results, we propose that in the early stages of the disease TMEFF2 is overexpressed to signal as a tumor suppressor. The inflammation and signaling transformations that occur in the tumor during more advanced stages of the disease lead to TMEFF2 ectodomain shedding and accumulation which, as described above, has a growth promoting activity. If this is the case, once shed, the TMEFF2 ECD could be detected in serum or urine of prostate cancer patients and considering that TMEFF2 has a very restricted expression, the TMEFF2 ECD could prove to be a very attractive biomarker. Confirmation of this hypothesis awaits further evaluation.

CONCLUDING REMARKS

ADAMs are zinc-dependent transmembrane metalloproteinases that cleave the extracellular domains of membrane bound growth factors, cytokines and receptors among them, the EGFR ligands which are currently the targets for anti-tumor therapies. Since ADAMs are also responsible for shedding of other factors involved in angiogenesis and metastasis, targeting ADAMs may provide a therapeutic strategy on their own or a complement to existing anti-EGFR therapies. ADAMs have been shown to be upregulated in several types of cancer and based on the fact that ADAMs are themselves cleaved, or that they can encode soluble forms by alternative splicing, it is conceivable that circulating metalloproteinases may themselves be a marker for disease. In addition to their role as potential therapeutics and markers, metalloproteinases are responsible for the generation of soluble circulating ectodomains from different transmembrane proteins that can potentially function as prognostic/diagnostic biomarkers, ectomarkers.

Emerging soluble biomarkers detectable in body fluids–serum, urine, saliva—have earned considerable interest as potential contributors to improving detection and response to cancer therapies. Soluble biomarkers represent a great advantage over tissue biomarkers since they are not invasive a biopsy is not necessary, and they allow determining variations in real time. However, to date only a few randomized trials have included circulating biomarkers, and unfortunately, none have yet identified a strong one. The discovery of soluble cancer prognostic and predictive biomarkers generated by shedding remains complicated, with lack of sensitivity and specificity being some of the most difficult challenges. In fact, as shown in Table 2, changes in the levels of an ectomarker can be observed in different cancer types indicating lack of specificity (this is not a comprehensive list and only a few examples were included). Therefore, in addition to reliable methods to allow sensitive detection of circulating proteins, there is a need to identify ectomarkers that offer specificity. In that sense, proteins with a restricted expression, like TMEFF2, could show promise.

ACKNOWLEDGEMENTS

We are deeply grateful to Dr. AS Asch for the critical reading of this chapter and helpful discussions during its conception. We apologize to those authors whose work has not been cited due to lack of space.

REFERENCES

Afar, D.E., V. Bhaskar, E. Ibsen, D. Breinberg, S.M. Henshall, J.G. Kench, M. Drobnjak, R. Powers, M. Wong, F. Evangelista, C. O'Hara, D. Powers, R.B. DuBridge, I. Caras, R. Winter, T. Anderson, N. Solvason, P.D. Stricker, C. Cordon-Cardo, H.I. Scher, J.J. Grygiel, R.L. Sutherland, R. Murray, V. Ramakrishnan, and D.A. Law. 2004. Preclinical validation of anti-TMEFF2-auristatin E-conjugated antibodies in the treatment of prostate cancer. Mol. Cancer. Ther. 3: 921-932.

Ahrens, T., J.P. Sleeman, C.M. Schempp, N. Howells, M. Hofmann, H. Ponta, P. Herrlich, and J.C. Simon. 2001. Soluble CD44 inhibits melanoma tumor growth by blocking cell surface CD44 binding to hyaluronic acid. Oncogene. 20: 3399-3408.

Alexiou, D., A.J. Karayiannakis, K.N. Syrigos, A. Zbar, A. Kremmyda, I. Bramis, and C. Tsigris. 2001. Serum levels of E-selectin, ICAM-1 and VCAM-1 in colorectal cancer patients: correlations with clinicopathological features, patient survival and tumour surgery. Eur. J. Cancer. 37: 2392-2397.

Ali, N. and V. Knaüper. 2007. Phorbol ester-induced shedding of the prostate cancer marker transmembrane protein with epidermal growth factor and two follistatin motifs 2 is mediated by the disintegrin and metalloproteinase-17. J. Biol. Chem. 282: 37378-37388.

Ali, S.M., W.P. Carney, F.J. Esteva, M. Fornier, L. Harris, W.J. Kostler, J.P. Lotz, D. Luftner, M.F. Pichon, A. Lipton, and Serum HER-2/neu Study Group. 2008. Serum HER-2/neu and relative resistance to trastuzumab-based therapy in patients with metastatic breast cancer. Cancer. 113: 1294-1301.

Ali, S.M., K. Leitzel, V.M. Chinchilli, L. Engle, L. Demers, H.A. Harvey, W. Carney, J.W. Allard, and A. Lipton. 2002. Relationship of serum HER-2/neu and serum CA 15-3 in patients with metastatic breast cancer. Clin. Chem. 48: 1314-1320.

Allison, J.G., P.M. Das, J. Ma, F.M. Inglis, and F.E. Jones. 2011. The ERBB4 intracellular domain (4ICD) regulates NRG1-induced gene expression in hippocampal neurons. Neurosci. Res. 70: 155-163.

Anderegg, U., T. Eichenberg, T. Parthaune, C. Haiduk, A. Saalbach, L. Milkova, A. Ludwig, J. Grosche, M. Averbeck, C. Gebhardt, V. Voelcker, J.P. Sleeman, and J.C. Simon. 2009. ADAM10 is the constitutive functional sheddase of CD44 in human melanoma cells. J. Invest. Dermatol. 129: 1471-1482.

Anders, A., S. Gilbert, W. Garten, R. Postina, and F. Fahrenholz. 2001. Regulation of the alpha-secretase ADAM10 by its prodomain and proprotein convertases. FASEB J. 15: 1837-1839.

Andreev, J., M.L. Galisteo, O. Kranenburg, S.K. Logan, E.S. Chiu, M. Okigaki, L.A. Cary, W.H. Moolenaar, and J. Schlessinger. 2001. Src and Pyk2 mediate G-protein-coupled receptor activation of epidermal growth factor receptor (EGFR) but are not required for coupling to the mitogen-activated protein (MAP) kinase signaling cascade. J. Biol. Chem. 276: 20130-20135.

Andrews, N.A., A.S. Jones, T.R. Helliwell, and A.R. Kinsella. 1997. Expression of the E-cadherin-catenin cell adhesion complex in primary squamous cell carcinomas of the head and neck and their nodal metastases. Br. J. Cancer. 75: 1474-1480.

Antunes, A.A., K.R. Leite, S.T. Reis, J.M. Sousa-Canavez, L.H. Camara-Lopes, M.F. Dall'oglio, and M. Srougi. 2009. GREB1 tissue expression is associated with organ-confined prostate cancer. Urol. Oncol.

Aplin, A.E., A. Howe, S.K. Alahari, and R.L. Juliano. 1998. Signal transduction and signal modulation by cell adhesion receptors: the role of integrins, cadherins, immunoglobulin-cell adhesion molecules, and selectins. Pharmacol. Rev. 50: 197-263.

Aref, S., M. El Sherbiny, T. Goda, M. Fouda, H. Al Askalany, and D. Abdalla. 2005. Soluble VEGF/sFLt1 ratio is an independent predictor of AML patient out come. Hematology. 10: 131-134.

Arima, T., H. Enokida, H. Kubo, I. Kagara, R. Matsuda, K. Toki, H. Nishimura, T. Chiyomaru, S. Tatarano, T. Idesako, K. Nishiyama, and M. Nakagawa. 2007. Nuclear translocation of ADAM-10 contributes to the pathogenesis and progression of human prostate cancer. Cancer. Sci. 98: 1720-1726.

Armanious, H., P. Gelebart, M. Anand, A. Belch, and R. Lai. 2011. Constitutive activation of metalloproteinase ADAM10 in mantle cell lymphoma promotes cell growth and activates the TNFalpha/NFkappaB pathway. Blood. 117: 6237-6246.

Arribas, J., J. Baselga, K. Pedersen, and J.L. Parra-Palau. 2011. p95HER2 and breast cancer. Cancer Res. 71: 1515-1519.

Arribas, J. and C. Esselens. 2009. ADAM17 as a therapeutic target in multiple diseases. Curr. Pharm. Des. 15: 2319-2335.

Aruffo, A., I. Stamenkovic, M. Melnick, C.B. Underhill, and B. Seed. 1990. CD44 is the principal cell surface receptor for hyaluronate. Cell. 61: 1303-1313.

Auvinen, P., R. Tammi, M. Tammi, R. Johansson, and V.M. Kosma. 2005. Expression of CD 44 s, CD 44 v 3 and CD 44 v 6 in benign and malignant breast lesions: correlation and colocalization with hyaluronan. Histopathology. 47: 420-428.

Baek, J.M., Q. Jin, J. Ensor, D.R. Boulbes, and F.J. Esteva. 2011. Serum CD44 levels and overall survival in patients with HER2-positive breast cancer. Breast Cancer Res. Treat. 130: 1029-1036.

Bailey, T., L. Biddlestone, N. Shepherd, H. Barr, P. Warner, and J. Jankowski. 1998. Altered cadherin and catenin complexes in the Barrett's esophagus-dysplasia-adenocarcinoma sequence: correlation with disease progression and dedifferentiation. Am. J. Pathol. 152: 135-144.

Bandyopadhyay, A., F. Lopez-Casillas, S.N. Malik, J.L. Montiel, V. Mendoza, J. Yang, and L.Z. Sun. 2002a. Antitumor activity of a recombinant soluble betaglycan in human breast cancer xenograft. Cancer Res. 62: 4690-4695.

Bandyopadhyay, A., Y. Zhu, S.N. Malik, J. Kreisberg, M.G. Brattain, E.A. Sprague, J. Luo, F. Lopez-Casillas, and L.Z. Sun. 2002b. Extracellular domain of TGFbeta type III receptor inhibits angiogenesis and tumor growth in human cancer cells. Oncogene. 21: 3541-3551.

Banks, R.E., A.J. Gearing, I.K. Hemingway, D.R. Norfolk, T.J. Perren, and P.J. Selby. 1993. Circulating intercellular adhesion molecule-1 (ICAM-1), E-selectin and vascular cell adhesion molecule-1 (VCAM-1) in human malignancies. Br. J. Cancer. 68: 122-124.

Bao, J., D. Wolpowitz, L.W. Role, and D.A. Talmage. 2003. Back signaling by the Nrg-1 intracellular domain. J. Cell Biol. 161: 1133-1141.

Bartolazzi, A., R. Peach, A. Aruffo, and I. Stamenkovic. 1994. Interaction between CD44 and hyaluronate is directly implicated in the regulation of tumor development. J. Exp. Med. 180: 53-66.

Baselga, J. and S.M. Swain. 2009. Novel anticancer targets: revisiting ERBB2 and discovering ERBB3. Nat. Rev. Cancer. 9: 463-475.

Bax, D.V., A.J. Messent, J. Tart, M. van Hoang, J. Kott, R.A. Maciewicz, and M.J. Humphries. 2004. Integrin alpha5beta1 and ADAM-17 interact in vitro and co-localize in migrating HeLa cells. J. Biol. Chem. 279: 22377-22386.

Becherer, J.D. and C.P. Blobel. 2003. Biochemical properties and functions of membrane-anchored metalloprotease-disintegrin proteins (ADAMs). Curr. Top. Dev. Biol. 54: 101-123.

Becker, J.C., C. Termeer, R.E. Schmidt, and E.B. Brocker. 1993. Soluble intercellular adhesion molecule-1 inhibits MHC-restricted specific T cell/tumor interaction. J. Immunol. 151: 7224-7232.

Bethune-Volters, A., M. Labroquere, S. Gueprattte, K. Hacene, R. Neumann, W. Carney, and M.F. Pichon. 2004. Longitudinal changes in serum HER-2/neu oncoprotein levels in trastuzumab-treated metastatic breast cancer patients. Anticancer Res. 24: 1083-1089.

Bezuhly, M., R. Cullen, C.T. Esmon, S.F. Morris, K.A. West, B. Johnston, and R.S. Liwski. 2009. Role of activated protein C and its receptor in inhibition of tumor metastasis. Blood. 113: 3371-3374.

Bhatavdekar, J.M., D.D. Patel, N.G. Shah, H.H. Vora, T.P. Suthar, P.R. Chikhlikar, N. Ghosh, and T.I. Trivedi. 2000. Prognostic significance of immunohistochemically localized biomarkers in stage II and stage III breast cancer: a multivariate analysis. Ann. Surg. Oncol. 7: 305-311.

Bhola, N.E. and J.R. Grandis. 2008. Crosstalk between G-protein-coupled receptors and epidermal growth factor receptor in cancer. Front. Biosci. 13: 1857-1865.

Black, R.A. 2002. Tumor necrosis factor-alpha converting enzyme. Int. J. Biochem. Cell Biol. 34: 1-5.

Black, R.A., C.T. Rauch, C.J. Kozlosky, J.J. Peschon, J.L. Slack, M.F. Wolfson, B.J. Castner, K.L. Stocking, P. Reddy, S. Srinivasan, N. Nelson, N. Boiani, K.A. Schooley, M. Gerhart, R. Davis, J.N. Fitzner, R.S. Johnson, R.J. Paxton, C.J. March, and D.P. Cerretti. 1997. A metalloproteinase disintegrin that releases tumour-necrosis factor-alpha from cells. Nature. 385: 729-733.

Blanchot-Jossic, F., A. Jarry, D. Masson, K. Bach-Ngohou, J. Paineau, M.G. Denis, C.L. Laboisse, and J.F. Mosnier. 2005. Up-regulated expression of ADAM17 in human colon carcinoma: co-expression with EGFR in neoplastic and endothelial cells. J. Pathol. 207: 156-163.

Blobel, C.P. 2005. ADAMs: key components in EGFR signalling and development. Nat. Rev. Mol. Cell Biol. 6: 32-43.

Blobel, C.P. and J.M. White. 1992. Structure, function and evolutionary relationship of proteins containing a disintegrin domain. Curr. Opin. Cell Biol. 4: 760-765.

Bloomston, M., E.E. Zervos, and A.S. Rosemurgy 2nd. 2002. Matrix metalloproteinases and their role in pancreatic cancer: a review of preclinical studies and clinical trials. Ann. Surg. Oncol. 9: 668-674.

Bode, W., F.X. Gomis-Ruth, and W. Stockler. 1993. Astacins, serralysins, snake venom and matrix metalloproteinases exhibit identical zinc-binding environments (HEXXHXXGXXH and Met-turn) and topologies and should be grouped into a common family, the 'metzincins'. FEBS Lett. 331: 134-140.

Borrell-Pages, M., F. Rojo, J. Albanell, J. Baselga, and J. Arribas. 2003. TACE is required for the activation of the EGFR by TGF-alpha in tumors. EMBO J. 22: 1114-1124.

Boulton, M.E., J. Cai, M.B. Grant, and Y. Zhang. 2008. Gamma-secretase regulates VEGFR-1 signalling in vascular endothelium and RPE. Adv. Exp. Med. Biol. 613: 313-319.

Bozkulak, E.C. and G. Weinmaster. 2009. Selective use of ADAM10 and ADAM17 in activation of Notch1 signaling. Mol. Cell. Biol. 29: 5679-5695.

Bramwell, V.H., G.S. Doig, A.B. Tuck, S.M. Wilson, K.S. Tonkin, A. Tomiak, F. Perera, T.A. Vandenberg, and A.F. Chambers. 2009. Changes over time of extracellular domain of HER2 (ECD/HER2) serum levels have prognostic value in metastatic breast cancer. Breast Cancer Res. Treat. 114: 503-511.

Bringuier, P.P., R. Umbas, H.E. Schaafsma, H.F. Karthaus, F.M. Debruyne, and J.A. Schalken. 1993. Decreased E-cadherin immunoreactivity correlates with poor survival in patients with bladder tumors. Cancer Res. 53: 3241-3245.

Brou, C., F. Logeat, N. Gupta, C. Bessia, O. LeBail, J.R. Doedens, A. Cumano, P. Roux, R.A. Black, and A. Israel. 2000. A novel proteolytic cleavage involved in Notch signaling: the role of the disintegrin-metalloprotease TACE. Mol. Cell. 5: 207-216.

Brücher, B.L., H. Geddert, C. Langner, H. Hofler, U. Fink, J.R. Siewert, and M. Sarbia. 2006. Hypermethylation of hMLH1, HPP1, p14(ARF), p16(INK4A) and APC in primary adenocarcinomas of the small bowel. Int. J. Cancer. 119: 1298-1302.

Bublil, E.M. and Y. Yarden. 2007. The EGF receptor family: spearheading a merger of signaling and therapeutics. Curr. Opin. Cell Biol. 19: 124-134.

Cameron, D., M. Casey, M. Press, D. Lindquist, T. Pienkowski, C.G. Romieu, S. Chan, A. Jagiello-Gruszfeld, B. Kaufman, J. Crown, A. Chan, M. Campone, P. Viens, N. Davidson, V. Gorbounova, J.I. Raats, D. Skarlos, B. Newstat, D. Roychowdhury, P. Paoletti, C. Oliva, S. Rubin, S. Stein, and C.E. Geyer. 2008. A phase III randomized comparison of lapatinib plus capecitabine versus capecitabine alone in women with advanced breast cancer that has progressed on trastuzumab: updated efficacy and biomarker analyses. Breast Cancer Res. Treat. 112: 533-543.

Carl-McGrath, S., U. Lendeckel, M. Ebert, A. Roessner, and C. Rocken. 2005. The disintegrin-metalloproteinases ADAM9, ADAM12, and ADAM15 are upregulated in gastric cancer. Int. J. Oncol. 26: 17-24.

Carlos, T.M. and J.M. Harlan. 1994. Leukocyte-endothelial adhesion molecules. Blood. 84: 2068-2101.

Carney, W.P. 2007. Circulating oncoproteins HER2/neu, EGFR and CAIX (MN) as novel cancer biomarkers. Expert Rev. Mol. Diagn. 7: 309-319.

Carney, W.P., R. Neumann, A. Lipton, K. Leitzel, S. Ali, and C.P. Price. 2003. Potential clinical utility of serum HER-2/neu oncoprotein concentrations in patients with breast cancer. Clin. Chem. 49: 1579-1598.

Carpenter, G. and H.J. Liao. 2009. Trafficking of receptor tyrosine kinases to the nucleus. Exp. Cell Res. 315: 1556-1566.

Chan, A.O. 2006. E-cadherin in gastric cancer. World J. Gastroenterol. 12: 199-203.

Chan, A.O., K.M. Chu, S.K. Lam, K.L. Cheung, S. Law, K.F. Kwok, W.M. Wong, M.F. Yuen, and B.C. Wong. 2005. Early prediction of tumor recurrence after curative resection of gastric carcinoma by measuring soluble E-cadherin. Cancer. 104: 740-746.

Chan, A.O., K.M. Chu, S.K. Lam, B.C. Wong, K.F. Kwok, S. Law, S. Ko, W.M. Hui, Y.H. Yueng, and J. Wong. 2003. Soluble E-cadherin is an independent pretherapeutic factor for long-term survival in gastric cancer. J. Clin. Oncol. 21: 2288-2293.

Chan, A.O., S.K. Lam, K.M. Chu, C.M. Lam, E. Kwok, S.Y. Leung, S.T. Yuen, S.Y. Law, W.M. Hui, K.C. Lai, C.Y. Wong, H.C. Hu, C.L. Lai, and J. Wong. 2001. Soluble E-cadherin is a valid prognostic marker in gastric carcinoma. Gut. 48: 808-811.

Chang, C., B.J. Eggen, D.C. Weinstein, and A.H. Brivanlou. 2003. Regulation of nodal and BMP signaling by tomoregulin-1 (X7365) through novel mechanisms. Dev. Biol. 255: 1-11.

Charalabopoulos, K., A. Gogali, Y. Dalavaga, G. Daskalopoulos, M. Vassiliou, G. Bablekos, A. Karakosta, and S. Constantopoulos. 2006. The clinical significance of soluble E-cadherin in nonsmall cell lung cancer. Exp. Oncol. 28: 83-85.

Chen, H., A.L. Pyluck, M. Janik, and N.S. Sampson. 1998. Peptides corresponding to the epidermal growth factor-like domain of mouse fertilin: synthesis and biological activity. Biopolymers. 47: 299-307.

Chen, X., R. Overcash, T. Green, D. Hoffman, A.S. Asch, and M.J. Ruiz-Echevarria. 2011. The tumor suppressor activity of the transmembrane protein with epidermal growth factor and two follistatin motifs 2 (TMEFF2) correlates with its ability to modulate sarcosine levels. J. Biol. Chem. 286: 16091-16100.

Cho, H.S., K. Mason, K.X. Ramyar, A.M. Stanley, S.B. Gabelli, D.W. Denney Jr, and D.J. Leahy. 2003. Structure of the extracellular region of HER2 alone and in complex with the Herceptin Fab. Nature. 421: 756-760.

Choe, H.W., J.H. Park, Y.J. Kim, and O.P. Ernst. 2011. Transmembrane signaling by GPCRs: insight from rhodopsin and opsin structures. Neuropharmacology. 60: 52-57.

Christianson, T.A., J.K. Doherty, Y.J. Lin, E.E. Ramsey, R. Holmes, E.J. Keenan, and G.M. Clinton. 1998. NH2-terminally truncated HER-2/neu protein: relationship with shedding of the extracellular domain and with prognostic factors in breast cancer. Cancer Res. 58: 5123-5129.

Chung, J.H., H.J. Lee, B.H. Kim, N.Y. Cho, and G.H. Kang. 2011. DNA methylation profile during multistage progression of pulmonary adenocarcinomas. Virchows Arch. 459: 201-211.

Chung, Y., S. Law, D.L. Kwong, and J.M. Luk. 2011. Serum soluble E-cadherin is a potential prognostic marker in esophageal squamous cell carcinoma. Dis. Esophagus. 24: 49-55.

Citri, A. and Y. Yarden. 2006. EGF-ERBB signalling: towards the systems level. Nat. Rev. Mol. Cell Biol. 7: 505-516.

Colomer, R., S. Montero, A. Lluch, B. Ojeda, A. Barnadas, A. Casado, B. Massuti, H. Cortes-Funes, and B. Lloveras. 2000. Circulating HER2 extracellular domain and resistance to chemotherapy in advanced breast cancer. Clin. Cancer Res. 6: 2356-2362.

Cominetti, M.R., A.C. Martin, J.U. Ribeiro, I. Djaafri, F. Fauvel-Lafeve, M. Crepin, and H.S. Selistre-de-Araujo. 2009. Inhibition of platelets and tumor cell adhesion by the disintegrin domain of human ADAM9 to collagen I under dynamic flow conditions. Biochimie. 91: 1045-1052.

Costa, V.L., R. Henrique, S.A. Danielsen, S. Duarte-Pereira, M. Eknaes, R.I. Skotheim, A. Rodrigues, J.S. Magalhaes, J. Oliveira, R.A. Lothe, M.R. Teixeira, C. Jeronimo, and G.E. Lind. 2010. Three epigenetic biomarkers, GDF15, TMEFF2, and VIM, accurately predict bladder cancer from DNA-based analyses of urine samples. Clin. Cancer Res. 16: 5842-5851.

Crawford, H.C., P.J. Dempsey, G. Brown, L. Adam, and M.L. Moss. 2009. ADAM10 as a therapeutic target for cancer and inflammation. Curr. Pharm. Des. 15: 2288-2299.

Dang, M., K. Dubbin, A. D'Aiello, M. Hartmann, H. Lodish, and A. Herrlich. 2011. Epidermal growth factor (EGF) ligand release by substrate-specific a disintegrin and metalloproteases (ADAMs) involves different protein kinase C (PKC) isoenzymes depending on the stimulus. J. Biol. Chem. 286: 17704-17713.

Daub, H., F.U. Weiss, C. Wallasch, and A. Ullrich. 1996. Role of transactivation of the EGF receptor in signalling by G-protein-coupled receptors. Nature. 379: 557-560.

De Wever, O., L. Derycke, A. Hendrix, G. De Meerleer, F. Godeau, H. Depypere, and M. Bracke. 2007. Soluble cadherins as cancer biomarkers. Clin. Exp. Metastasis. 24: 685-697.

Dhillon, A.S., S. Hagan, O. Rath, and W. Kolch. 2007. MAP kinase signalling pathways in cancer. Oncogene. 26: 3279-3290.

Diaz, L.K., X. Zhou, E.T. Wright, M. Cristofanilli, T. Smith, Y. Yang, N. Sneige, A. Sahin, and M.Z. Gilcrease. 2005. CD44 expression is associated with increased survival in node-negative invasive breast carcinoma. Clin. Cancer Res. 11: 3309-3314.

Díaz-Rodríguez, E., J.C. Montero, A. Esparis-Ogando, L. Yuste, and A. Pandiella. 2002. Extracellular signal-regulated kinase phosphorylates tumor necrosis factor alpha-converting enzyme at threonine 735: a potential role in regulated shedding. Mol. Biol. Cell. 13: 2031-2044.

Dippold, W., B. Wittig, W. Schwaeble, W. Mayet, and K.H. Meyer zum Buschenfelde. 1993. Expression of intercellular adhesion molecule 1 (ICAM-1, CD54) in colonic epithelial cells. Gut. 34: 1593-1597.

Doedens, J.R., R.M. Mahimkar, and R.A. Black. 2003. TACE/ADAM-17 enzymatic activity is increased in response to cellular stimulation. Biochem. Biophys. Res. Commun. 308: 331-338.

Dong, M., T. How, K.C. Kirkbride, K.J. Gordon, J.D. Lee, N. Hempel, P. Kelly, B.J. Moeller, J.R. Marks, and G.C. Blobe. 2007. The type III TGF-beta receptor suppresses breast cancer progression. J. Clin. Invest. 117: 206-217.

Dorsam, R.T. and J.S. Gutkind. 2007. G-protein-coupled receptors and cancer. Nat. Rev. Cancer. 7: 79-94.

Dorudi, S., A.M. Hanby, R. Poulsom, J. Northover, and I.R. Hart. 1995. Level of expression of E-cadherin mRNA in colorectal cancer correlates with clinical outcome. Br. J. Cancer. 71: 614-616.

Duffy, M.J., D.J. Lynn, A.T. Lloyd, and C.M. O'Shea. 2003. The ADAMs family of proteins: from basic studies to potential clinical applications. Thromb. Haemost. 89: 622-631.

Duffy, M.J., E. McKiernan, N. O'Donovan, and P.M. McGowan. 2009. Role of ADAMs in cancer formation and progression. Clin. Cancer Res. 15: 1140-1144.

Ebert, M.P., S.H. Mooney, L. Tonnes-Priddy, J. Lograsso, J. Hoffmann, J. Chen, C. Rocken, H.U. Schulz, P. Malfertheiner, and C. Lofton-Day. 2005. Hypermethylation of the TPEF/HPP1 gene in primary and metastatic colorectal cancers. Neoplasia. 7: 771-778.

Ebi, M., H. Kataoka, T. Shimura, E. Kubota, Y. Hirata, T. Mizushima, T. Mizoshita, M. Tanaka, M. Mabuchi, H. Tsukamoto, S. Tanida, T. Kamiya, S. Higashiyama, and T. Joh. 2010. TGFbeta induces proHB-EGF shedding and EGFR transactivation through ADAM activation in gastric cancer cells. Biochem. Biophys. Res. Commun. 402: 449-454.

Edwards, D.R., M.M. Handsley, and C.J. Pennington. 2008. The ADAM metalloproteinases. Mol. Aspects Med. 29: 258-289.

Eguchi, S., K. Numaguchi, H. Iwasaki, T. Matsumoto, T. Yamakawa, H. Utsunomiya, E.D. Motley, H. Kawakatsu, K.M. Owada, Y. Hirata, F. Marumo, and T. Inagami. 1998. Calcium-dependent epidermal growth factor receptor transactivation mediates the angiotensin II-induced mitogen-activated protein kinase activation in vascular smooth muscle cells. J. Biol. Chem. 273: 8890-8896.

Eib, D.W., T.M. Holling, A. Zwijsen, N. Dewulf, E. de Groot, A.J. van den Eijnden-van Raaij, D. Huylebroeck, and G.J. Martens. 2000. Expression of the follistatin/EGF-containing transmembrane protein M7365 (tomoregulin-1) during mouse development. Mech. Dev. 97: 167-171.

Eib, D.W. and G.J. Martens. 1996. A novel transmembrane protein with epidermal growth factor and follistatin domains expressed in the hypothalamo-hypophysial axis of Xenopus laevis. J. Neurochem. 67: 1047-1055.

Elahi, A., L. Zhang, T.J. Yeatman, S. Gery, S. Sebti, and D. Shibata. 2008. HPP1-mediated tumor suppression requires activation of STAT1 pathways. Int. J. Cancer. 122: 1567-1572.

Endo, K., T. Ueda, J. Ueyama, T. Ohta, and T. Terada. 2000. Immunoreactive E-cadherin, alpha-catenin, beta-catenin, and gamma-catenin proteins in hepatocellular carcinoma: relationships with tumor grade, clinicopathologic parameters, and patients' survival. Hum. Pathol. 31: 558-565.

Esper, R.M. and J.A. Loeb. 2009. Neurotrophins induce neuregulin release through protein kinase Cdelta activation. J. Biol. Chem. 284: 26251-26260.

Essick, E., S. Sithu, W. Dean, and S. D'Souza. 2008. Pervanadate-induced shedding of the intercellular adhesion molecule (ICAM)-1 ectodomain is mediated by membrane type-1 matrix metalloproteinase (MT1-MMP). Mol. Cell. Biochem. 314: 151-159.

Esteva, F.J., C.D. Cheli, H. Fritsche, M. Fornier, D. Slamon, R.P. Thiel, D. Luftner, and F. Ghani. 2005. Clinical utility of serum HER2/neu in monitoring and prediction of progression-free survival in metastatic breast cancer patients treated with trastuzumab-based therapies. Breast Cancer Res. 7: R436-43.

Eto, K., C. Huet, T. Tarui, S. Kupriyanov, H.Z. Liu, W. Puzon-McLaughlin, X.P. Zhang, D. Sheppard, E. Engvall, and Y. Takada. 2002. Functional classification of ADAMs based on a conserved motif for binding to integrin alpha 9beta 1: implications for sperm-egg binding and other cell interactions. J. Biol. Chem. 277: 17804-17810.

Faleiro-Rodrigues, C., I. Macedo-Pinto, D. Pereira, and C.S. Lopes. 2004. Prognostic value of E-cadherin immunoexpression in patients with primary ovarian carcinomas. Ann. Oncol. 15: 1535-1542.

Fan, H. and R. Derynck. 1999. Ectodomain shedding of TGF-alpha and other transmembrane proteins is induced by receptor tyrosine kinase activation and MAP kinase signaling cascades. EMBO J. 18: 6962-6972.

Fan, H., C.W. Turck, and R. Derynck. 2003. Characterization of growth factor-induced serine phosphorylation of tumor necrosis factor-alpha converting enzyme and of an alternatively translated polypeptide. J. Biol. Chem. 278: 18617-18627.

Fehm, T., G. Gebauer, and W. Jager. 2002. Clinical utility of serial serum c-erbB-2 determinations in the follow-up of breast cancer patients. Breast Cancer Res. Treat. 75: 97-106.

Fehm, T., W. Jager, S. Kramer, C. Sohn, E. Solomayer, D. Wallwiener, and G. Gebauer. 2004. Prognostic significance of serum HER2 and CA 15-3 at the time of diagnosis of metastatic breast cancer. Anticancer Res. 24: 1987-1992.

Fei, Y., X. Liu, F. Wang, W. Wang, and S. Liu. 2010. E-cadherin expression in normal and abnormal tissue specimens from patients with pancreatic carcinoma. Lab. Medicine. 41: 473-477.

Filardo, E.J., J.A. Quinn, K.I. Bland, and A.R. Frackelton Jr. 2000. Estrogen-induced activation of Erk-1 and Erk-2 requires the G protein-coupled receptor homolog, GPR30, and occurs via trans-activation of the epidermal growth factor receptor through release of HB-EGF. Mol. Endocrinol. 14: 1649-1660.

Finn, R.S., R. Gagnon, A. Di Leo, M.F. Press, M. Arbushites, and M. Koehler. 2009. Prognostic and predictive value of HER2 extracellular domain in metastatic breast cancer treated with lapatinib and paclitaxel in a randomized phase III study. J. Clin. Oncol. 27: 5552-5558.

Fischer, C., B. Jonckx, M. Mazzone, S. Zacchigna, S. Loges, L. Pattarini, E. Chorianopoulos, L. Liesenborghs, M. Koch, M. De Mol, M. Autiero, S. Wyns, S. Plaisance, L. Moons, N. van Rooijen, M. Giacca, J.M. Stassen, M. Dewerchin, D. Collen, and P. Carmeliet. 2007. Anti-PlGF inhibits growth of VEGF(R)-inhibitor-resistant tumors without affecting healthy vessels. Cell. 131: 463-475.

Fischer, O.M., S. Hart, A. Gschwind, N. Prenzel, and A. Ullrich. 2004. Oxidative and osmotic stress signaling in tumor cells is mediated by ADAM proteases and heparin-binding epidermal growth factor. Mol. Cell. Biol. 24: 5172-5183.

Fischer, O.M., S. Hart, A. Gschwind, and A. Ullrich. 2003. EGFR signal transactivation in cancer cells. Biochem. Soc. Trans. 31: 1203-1208.

Fogel, M., P. Gutwein, S. Mechtersheimer, S. Riedle, A. Stoeck, A. Smirnov, L. Edler, A. Ben-Arie, M. Huszar, and P. Altevogt. 2003. L1 expression as a predictor of progression and survival in patients with uterine and ovarian carcinomas. Lancet. 362: 869-875.

Fong, K.P., C. Barry, A.N. Tran, E.A. Traxler, K.M. Wannemacher, H.Y. Tang, K.D. Speicher, I.A. Blair, D.W. Speicher, T. Grosser, and L.F. Brass. 2011. Deciphering the human platelet sheddome. Blood. 117: e15-26.

Fontana, X., P. Ferrari, M. Namer, R. Peysson, C. Salanon, and F. Bussiere. 1994. C-erb-B2 gene amplification and serum level of c-erb-B2 oncoprotein at primary breast cancer diagnosis. Anticancer Res. 14: 2099-2104.

Fragoso, R., T. Pereira, Y. Wu, Z. Zhu, J. Cabecadas, and S. Dias. 2006. VEGFR-1 (FLT-1) activation modulates acute lymphoblastic leukemia localization and survival within the bone marrow, determining the onset of extramedullary disease. Blood. 107: 1608-1616.

Franzmann, E.J., E.P. Reategui, F. Pedroso, F.G. Pernas, B.M. Karakullukcu, K.L. Carraway, K. Hamilton, R. Singal, and W.J. Goodwin. 2007. Soluble CD44 is a potential marker for the early detection of head and neck cancer. Cancer Epidemiol. Biomarkers Prev. 16: 1348-1355.

Fritzsche, F.R., M. Jung, A. Tolle, P. Wild, A. Hartmann, K. Wassermann, A. Rabien, M. Lein, M. Dietel, C. Pilarsky, D. Calvano, R. Grutzmann, K. Jung, and G. Kristiansen. 2008. ADAM9 expression is a significant and independent prognostic marker of PSA relapse in prostate cancer. Eur. Urol. 54: 1097-1106.

Fritzsche, F.R., M. Jung, C. Xu, A. Rabien, H. Schicktanz, C. Stephan, M. Dietel, K. Jung, and G. Kristiansen. 2006. ADAM8 expression in prostate cancer is associated with parameters of unfavorable prognosis. Virchows Arch. 449: 628-636.

Fritzsche, F.R., K. Wassermann, M. Jung, A. Tolle, I. Kristiansen, M. Lein, M. Johannsen, M. Dietel, K. Jung, and G. Kristiansen. 2008. ADAM9 is highly expressed in renal cell cancer and is associated with tumour progression. BMC Cancer. 8: 179.

Frohlich, C., R. Albrechtsen, L. Dyrskjot, L. Rudkjaer, T.F. Orntoft, and U.M. Wewer. 2006. Molecular profiling of ADAM12 in human bladder cancer. Clin. Cancer Res. 12: 7359-7368.

Fry, J.L. and A. Toker. 2010. Secreted and membrane-bound isoforms of protease ADAM9 have opposing effects on breast cancer cell migration. Cancer Res. 70: 8187-8198.

Gack, S., A. Marme, F. Marme, G. Wrobel, B. Vonderstrass, G. Bastert, P. Lichter, P. Angel, and M. Schorpp-Kistner. 2005. Preeclampsia: increased expression of soluble ADAM 12. J. Mol. Med. (Berl). 83: 887-896.

Gaida, M.M., N. Haag, F. Gunther, D.F. Tschaharganeh, P. Schirmacher, H. Friess, N.A. Giese, J. Schmidt, and M.N. Wente. 2010. Expression of A disintegrin and metalloprotease 10 in pancreatic carcinoma. Int. J. Mol. Med. 26: 281-288.

Gancberg, D., A. Di Leo, F. Cardoso, G. Rouas, M. Pedrocchi, M. Paesmans, A. Verhest, C. Bernard-Marty, M.J. Piccart, and D. Larsimont. 2002. Comparison of HER-2 status between primary breast cancer and corresponding distant metastatic sites. Ann. Oncol. 13: 1036-1043.

Garcia del Muro, X., A. Torregrosa, J. Munoz, X. Castellsague, E. Condom, F. Vigues, A. Arance, A. Fabra, and J.R. Germa. 2000. Prognostic value of the expression of E-cadherin and beta-catenin in bladder cancer. Eur. J. Cancer. 36: 357-362.

Garton, K.J., P.J. Gough, J. Philalay, P.T. Wille, C.P. Blobel, R.H. Whitehead, P.J. Dempsey, and E.W. Raines. 2003. Stimulated shedding of vascular cell adhesion molecule 1 (VCAM-1) is mediated by tumor necrosis factor-alpha-converting enzyme (ADAM 17). J. Biol. Chem. 278: 37459-37464.

Gaultier, A., H. Cousin, T. Darribere, and D. Alfandari. 2002. ADAM13 disintegrin and cysteine-rich domains bind to the second heparin-binding domain of fibronectin. J. Biol. Chem. 277: 23336-23344.

Gavert, N., M. Sheffer, S. Raveh, S. Spaderna, M. Shtutman, T. Brabletz, F. Barany, P. Paty, D. Notterman, E. Domany, and A. Ben-Ze'ev. 2007. Expression of L1-CAM and ADAM10 in human colon cancer cells induces metastasis. Cancer Res. 67: 7703-7712.

Gechtman, Z., J.L. Alonso, G. Raab, D.E. Ingber, and M. Klagsbrun. 1999. The shedding of membrane-anchored heparin-binding epidermal-like growth factor is regulated by the Raf/mitogen-activated protein kinase cascade and by cell adhesion and spreading. J. Biol. Chem. 274: 28828-28835.

Geddert, H., S. Kiel, E. Iskender, A.R. Florl, T. Krieg, S. Vossen, H.E. Gabbert, and M. Sarbia. 2004. Correlation of hMLH1 and HPP1 hypermethylation in gastric, but not in esophageal and cardiac adenocarcinoma. Int. J. Cancer. 110: 208-211.

Gery, S., C.L. Sawyers, D.B. Agus, J.W. Said, and H.P. Koeffler. 2002. TMEFF2 is an androgen-regulated gene exhibiting antiproliferative effects in prostate cancer cells. Oncogene. 21: 4739-4746.

Gery, S., D. Yin, D. Xie, K.L. Black, and H.P. Koeffler. 2003. TMEFF1 and brain tumors. Oncogene. 22: 2723-2727.

Ghedini, G.C., V. Ciravolo, M. Tortoreto, S. Giuffre, F. Bianchi, M. Campiglio, M. Mortarino, M. Figini, A. Coliva, M.L. Carcangiu, M. Zambetti, T. Piazza, S. Ferrini, S. Menard, E. Tagliabue, and S.M. Pupa. 2010. Shed HER2 extracellular domain in HER2-mediated tumor growth and in trastuzumab susceptibility. J. Cell. Physiol. 225: 256-265.

Gilpin, B.J., F. Loechel, M.G. Mattei, E. Engvall, R. Albrechtsen, and U.M. Wewer. 1998. A novel, secreted form of human ADAM 12 (meltrin alpha) provokes myogenesis in vivo. J. Biol. Chem. 273: 157-166.

Glynne-Jones, E., M.E. Harper, L.T. Seery, R. James, I. Anglin, H.E. Morgan, K.M. Taylor, J.M. Gee, and R.I. Nicholson. 2001. TENB2, a proteoglycan identified in prostate cancer that is associated with disease progression and androgen independence. Int. J. Cancer. 94: 178-184.

Gofuku, J., H. Shiozaki, Y. Doki, M. Inoue, M. Hirao, N. Fukuchi, and M. Monden. 1998. Characterization of soluble E-cadherin as a disease marker in gastric cancer patients. Br. J. Cancer. 78: 1095-1101.

Gomez, H.L., D.C. Doval, M.A. Chavez, P.C. Ang, Z. Aziz, S. Nag, C. Ng, S.X. Franco, L.W. Chow, M.C. Arbushites, M.A. Casey, M.S. Berger, S.H. Stein, and G.W. Sledge.

2008. Efficacy and safety of lapatinib as first-line therapy for ErbB2-amplified locally advanced or metastatic breast cancer. J. Clin. Oncol. 26: 2999-3005.

Gong, Y., X. Sun, L. Huo, E.L. Wiley, and M.S. Rao. 2005. Expression of cell adhesion molecules, CD44s and E-cadherin, and microvessel density in invasive micropapillary carcinoma of the breast. Histopathology. 46: 24-30.

Gonzalo, V., J.J. Lozano, J. Munoz, F. Balaguer, M. Pellise, C. Rodriguez de Miguel, M. Andreu, R. Jover, X. Llor, M.D. Giraldez, T. Ocana, A. Serradesanferm, V. Alonso-Espinaco, M. Jimeno, M. Cuatrecasas, O. Sendino, S. Castellvi-Bel, A. Castells, and Gastrointestinal Oncology Group of the Spanish Gastroenterological Association. 2010. Aberrant gene promoter methylation associated with sporadic multiple colorectal cancer. PLoS One. 5: e8777.

Gooz, M. 2010. ADAM-17: the enzyme that does it all. Crit. Rev. Biochem. Mol. Biol. 45: 146-169.

Graus-Porta, D., R.R. Beerli, J.M. Daly, and N.E. Hynes. 1997. ErbB-2, the preferred heterodimerization partner of all ErbB receptors, is a mediator of lateral signaling. EMBO J. 16: 1647-1655.

Grothey, A., P. Heistermann, S. Philippou, and R. Voigtmann. 1998. Serum levels of soluble intercellular adhesion molecule-1 (ICAM-1, CD54) in patients with non-small-cell lung cancer: correlation with histological expression of ICAM-1 and tumour stage. Br. J. Cancer. 77: 801-807.

Grutzmann, R., J. Luttges, B. Sipos, O. Ammerpohl, F. Dobrowolski, I. Alldinger, S. Kersting, D. Ockert, R. Koch, H. Kalthoff, H.K. Schackert, H.D. Saeger, G. Kloppel, and C. Pilarsky. 2004. ADAM9 expression in pancreatic cancer is associated with tumour type and is a prognostic factor in ductal adenocarcinoma. Br. J. Cancer. 90: 1053-1058.

Gschwind, A., S. Hart, O.M. Fischer, and A. Ullrich. 2003. TACE cleavage of proamphiregulin regulates GPCR-induced proliferation and motility of cancer cells. EMBO J. 22: 2411-2421.

Guaiquil, V., S. Swendeman, T. Yoshida, S. Chavala, P.A. Campochiaro, and C.P. Blobel. 2009. ADAM9 is involved in pathological retinal neovascularization. Mol. Cell. Biol. 29: 2694-2703.

Guo, Y.J., G. Liu, X. Wang, D. Jin, M. Wu, J. Ma, and M.S. Sy. 1994. Potential use of soluble CD44 in serum as indicator of tumor burden and metastasis in patients with gastric or colon cancer. Cancer Res. 54: 422-426.

Gutwein, P., S. Mechtersheimer, S. Riedle, A. Stoeck, D. Gast, S. Joumaa, H. Zentgraf, M. Fogel, and D.P. Altevogt. 2003. ADAM10-mediated cleavage of L1 adhesion molecule at the cell surface and in released membrane vesicles. FASEB J. 17: 292-294.

Hanabata, T., K. Tsukuda, S. Toyooka, M. Yano, M. Aoe, I. Nagahiro, Y. Sano, H. Date, and N. Shimizu. 2004. DNA methylation of multiple genes and clinicopathological relationship of non-small cell lung cancers. Oncol. Rep. 12: 177-180.

Harms, P.W. and C. Chang. 2003. Tomoregulin-1 (TMEFF1) inhibits nodal signaling through direct binding to the nodal coreceptor Cripto. Genes Dev. 17: 2624-2629.

Harrington, K.J. and K.N. Syrigos. 2000. The role of E-cadherin-catenin complex: more than an intercellular glue? Ann. Surg. Oncol. 7: 783-788.

Hayashida, K., A.H. Bartlett, Y. Chen, and P.W. Park. 2010. Molecular and cellular mechanisms of ectodomain shedding. Anat. Rec. (Hoboken). 293: 925-937.

He, Q., H.Y. Chen, E.Q. Bai, Y.X. Luo, R.J. Fu, Y.S. He, J. Jiang, and H.Q. Wang. 2010. Development of a multiplex MethyLight assay for the detection of multigene methylation in human colorectal cancer. Cancer Genet. Cytogenet. 202: 1-10.

Heanue, T.A. and V. Pachnis. 2006. Expression profiling the developing mammalian enteric nervous system identifies marker and candidate Hirschsprung disease genes. Proc. Natl. Acad. Sci. U. S. A. 103: 6919-6924.

Hellwinkel, O.J., M. Kedia, H. Isbarn, L. Budaus, and M.G. Friedrich. 2008. Methylation of the TPEF- and PAX6-promoters is increased in early bladder cancer and in normal mucosa adjacent to pTa tumours. BJU Int. 101: 753-757.

Hernandez, I., J.L. Moreno, C. Zandueta, L. Montuenga, and F. Lecanda. 2010. Novel alternatively spliced ADAM8 isoforms contribute to the aggressive bone metastatic phenotype of lung cancer. Oncogene. 29: 3758-3769.

Herrlich, A., E. Klinman, J. Fu, C. Sadegh, and H. Lodish. 2008. Ectodomain cleavage of the EGF ligands HB-EGF, neuregulin1-beta, and TGF-alpha is specifically triggered by different stimuli and involves different PKC isoenzymes. FASEB J. 22: 4281-4295.

Hertweck, M.K., F. Erdfelder, and K.A. Kreuzer. 2011. CD44 in hematological neoplasias. Ann. Hematol. 90: 493-508.

Hidalgo, M. and S.G. Eckhardt. 2001. Development of matrix metalloproteinase inhibitors in cancer therapy. J. Natl. Cancer Inst. 93: 178-193.

Higashiyama, S., D. Nanba, H. Nakayama, H. Inoue, and S. Fukuda. 2011. Ectodomain shedding and remnant peptide signalling of EGFRs and their ligands. J. Biochem. 150: 15-22.

Hooper, N.M. 2005. Roles of proteolysis and lipid rafts in the processing of the amyloid precursor protein and prion protein. Biochem. Soc. Trans. 33: 335-338.

Horie, M., Y. Mitsumoto, H. Kyushiki, N. Kanemoto, A. Watanabe, Y. Taniguchi, N. Nishino, T. Okamoto, M. Kondo, T. Mori, K. Noguchi, Y. Nakamura, E. Takahashi, and A. Tanigami. 2000. Identification and characterization of TMEFF2, a novel survival factor for hippocampal and mesencephalic neurons. Genomics. 67: 146-152.

Howard, L., R.A. Maciewicz, and C.P. Blobel. 2000. Cloning and characterization of ADAM28: evidence for autocatalytic pro-domain removal and for cell surface localization of mature ADAM28. Biochem. J. 348 Pt 1: 21-27.

Hu, Q., A.L. Dey, Y. Yang, Y. Shen, I.B. Jilani, E.H. Estey, H.M. Kantarjian, F.J. Giles, and M. Albitar. 2004. Soluble vascular endothelial growth factor receptor 1, and not receptor 2, is an independent prognostic factor in acute myeloid leukemia and myelodysplastic syndromes. Cancer. 100: 1884-1891.

Huang, Z.H., L.H. Li, F. Yang, and J.F. Wang. 2007. Detection of aberrant methylation in fecal DNA as a molecular screening tool for colorectal cancer and precancerous lesions. World J. Gastroenterol. 13: 950-954.

Hulpiau, P. and F. van Roy. 2009. Molecular evolution of the cadherin superfamily. Int. J. Biochem. Cell Biol. 41: 349-369.

Hynes, N.E. and H.A. Lane. 2005. ERBB receptors and cancer: the complexity of targeted inhibitors. Nat. Rev. Cancer. 5: 341-354.

Iba, K., R. Albrechtsen, B. Gilpin, C. Frohlich, F. Loechel, A. Zolkiewska, K. Ishiguro, T. Kojima, W. Liu, J.K. Langford, R.D. Sanderson, C. Brakebusch, R. Fassler, and U.M. Wewer. 2000. The cysteine-rich domain of human ADAM 12 supports cell adhesion through syndecans and triggers signaling events that lead to beta1 integrin-dependent cell spreading. J. Cell Biol. 149: 1143-1156.

Iba, K., R. Albrechtsen, B.J. Gilpin, F. Loechel, and U.M. Wewer. 1999. Cysteine-rich domain of human ADAM 12 (meltrin alpha) supports tumor cell adhesion. Am. J. Pathol. 154: 1489-1501.

Ikeda, Y., Y. Imai, H. Kumagai, T. Nosaka, Y. Morikawa, T. Hisaoka, I. Manabe, K. Maemura, T. Nakaoka, T. Imamura, K. Miyazono, I. Komuro, R. Nagai, and T. Kitamura. 2004. Vasorin, a transforming growth factor beta-binding protein expressed in vascular smooth muscle cells, modulates the arterial response to injury in vivo. Proc. Natl. Acad. Sci. U. S. A. 101: 10732-10737.

Inge, L. J., S.P. Barwe, J. D'Ambrosio, J. Gopal, K. Lu, S. Ryazantsev, S.A. Rajasekaran, and A.K. Rajasekaran. 2011. Soluble E-cadherin promotes cell survival by activating epidermal growth factor receptor. Exp. Cell Res. 317: 838-848.

Ishikawa, N., Y. Daigo, W. Yasui, K. Inai, H. Nishimura, E. Tsuchiya, N. Kohno, and Y. Nakamura. 2004. ADAM8 as a novel serological and histochemical marker for lung cancer. Clin. Cancer Res. 10: 8363-8370.

Isola, J.J., K. Holli, H. Oksa, Y. Teramoto, and O.P. Kallioniemi. 1994. Elevated erbB-2 oncoprotein levels in preoperative and follow-up serum samples define an aggressive disease course in patients with breast cancer. Cancer. 73: 652-658.

Ivanauskas, A., J. Hoffmann, L.V. Jonaitis, R. Markelis, E. Juozaityte, L. Kupcinskas, C. Lofton-Day, C. Rocken, and P. Malfertheiner. 2008. Distinct TPEF/HPP1 gene methylation patterns in gastric cancer indicate a field effect in gastric carcinogenesis. Dig. Liver Dis. 40: 920-926.

Jaggi, M., S.L. Johansson, J.J. Baker, L.M. Smith, A. Galich, and K.C. Balaji. 2005. Aberrant expression of E-cadherin and beta-catenin in human prostate cancer. Urol. Oncol. 23: 402-406.

Janes, P.W., N. Saha, W.A. Barton, M.V. Kolev, S.H. Wimmer-Kleikamp, E. Nievergall, C.P. Blobel, J.P. Himanen, M. Lackmann, and D.B. Nikolov. 2005. Adam meets Eph: an ADAM substrate recognition module acts as a molecular switch for ephrin cleavage in trans. Cell. 123: 291-304.

Jankowski, J., P. Newham, O. Kandemir, S. Hirano, M. Takeichi, and M. Pignatelli. 1994. Differential expression of e-cadherin in normal, metaplastic and dysplastic esophageal mucosa - a putative biomarker. Int. J. Oncol. 4: 441-448.

Jin, Z., Y. Cheng, W. Gu, Y. Zheng, F. Sato, Y. Mori, A.V. Olaru, B.C. Paun, J. Yang, T. Kan, T. Ito, J.P. Hamilton, F.M. Selaru, R. Agarwal, S. David, J.M. Abraham, H.C. Wolfsen, M.B. Wallace, N.J. Shaheen, K. Washington, J. Wang, M.I. Canto, A. Bhattacharyya, M.A. Nelson, P.D. Wagner, Y. Romero, K.K. Wang, Z. Feng, R.E. Sampliner, and S.J. Meltzer. 2009. A multicenter, double-blinded validation study of methylation biomarkers for progression prediction in Barrett's esophagus. Cancer Res. 69: 4112-4115.

Jorissen, E. and B. De Strooper. 2010. Gamma-secretase and the intramembrane proteolysis of Notch. Curr. Top. Dev. Biol. 92: 201-230.

Kaihara, A., H. Iwagaki, A. Gouchi, A. Hizuta, H. Isozaki, N. Takakura, and N. Tanaka. 1998. Soluble intercellular adhesion molecule-1 and natural killer cell activity in gastric cancer patients. Res. Commun. Mol. Pathol. Pharmacol. 100: 283-300.

Kajita, M., Y. Itoh, T. Chiba, H. Mori, A. Okada, H. Kinoh, and M. Seiki. 2001. Membrane-type 1 matrix metalloproteinase cleaves CD44 and promotes cell migration. J. Cell Biol. 153: 893-904.

Kanemoto, N., M. Horie, K. Omori, N. Nishino, M. Kondo, K. Noguchi, and A. Tanigami. 2001. Expression of TMEFF1 mRNA in the mouse central nervous system: precise examination and comparative studies of TMEFF1 and TMEFF2. Brain Res. Mol. Brain Res. 86: 48-55.

Kang, X., F. Wang, J.D. Xie, J. Cao, and P.Z. Xian. 2005. Clinical evaluation of serum concentrations of intercellular adhesion molecule-1 in patients with colorectal cancer. World J. Gastroenterol. 11: 4250-4253.

Karadag, A., M. Zhou, and P.I. Croucher. 2006. ADAM-9 (MDC-9/meltrin-gamma), a member of the a disintegrin and metalloproteinase family, regulates myeloma-cell-induced interleukin-6 production in osteoblasts by direct interaction with the alpha(v)beta5 integrin. Blood. 107: 3271-3278.

Karan, D., F.C. Lin, M. Bryan, J. Ringel, N. Moniaux, M.F. Lin, and S.K. Batra. 2003. Expression of ADAMs (a disintegrin and metalloproteases) and TIMP-3 (tissue inhibitor of metalloproteinase-3) in human prostatic adenocarcinomas. Int. J. Oncol. 23: 1365-1371.

Karatzas, G., A.J. Karayiannakis, K.N. Syrigos, E. Chatzigianni, S. Papanikolaou, F. Riza, and D. Papanikolaou. 1999. E-cadherin expression correlates with tumor differentiation in colorectal cancer. Hepatogastroenterology. 46: 232-235.

Karayiannakis, A.J., K.N. Syrigos, E. Chatzigianni, S. Papanikolaou, D. Alexiou, N. Kalahanis, T. Rosenberg, and E. Bastounis. 1998. Aberrant E-cadherin expression associated with loss of differentiation and advanced stage in human pancreatic cancer. Anticancer Res. 18: 4177-4180.

Katagiri, T., Y. Harada, M. Emi, and Y. Nakamura. 1995. Human metalloprotease/disintegrin-like (MDC) gene: exon-intron organization and alternative splicing. Cytogenet. Cell Genet. 68: 39-44.

Kataoka, H. 2009. EGFR ligands and their signaling scissors, ADAMs, as new molecular targets for anticancer treatments. J. Dermatol. Sci. 56: 148-153.

Katayama, M., S. Hirai, M. Yasumoto, K. Nishikawa, S. Nagata, M. Otsuka, K. Kamihagi, and I. Kato. 1994. Soluble fragments of e-cadherin cell-adhesion molecule increase in urinary-excretion of cancer-patients, potentially indicating its shedding from epithelial tumor-cells. Int. J. Oncol. 5: 1049-1057.

Kaufmann, M., K.H. Heider, H.P. Sinn, G. von Minckwitz, H. Ponta, and P. Herrlich. 1995. CD44 variant exon epitopes in primary breast cancer and length of survival. Lancet. 345: 615-619.

Kemik, O., A.S. Kemik, S. Purisa, I. Hasirci, A.C. Dulger, M. Adas, and S. Tuzun. 2011. Serum level of soluble vascular adhesion molecule 1 in patients with rectal cancer. Eur. J. Gen. Med. 8: 105-109.

Kenchappa, R.S., C. Tep, Z. Korade, S. Urra, F.C. Bronfman, S.O. Yoon, and B.D. Carter. 2010. p75 neurotrophin receptor-mediated apoptosis in sympathetic neurons involves a biphasic activation of JNK and up-regulation of tumor necrosis factor-alpha-converting enzyme/ADAM17. J. Biol. Chem. 285: 20358-20368.

Kendall, R.L. and K.A. Thomas. 1993. Inhibition of vascular endothelial cell growth factor activity by an endogenously encoded soluble receptor. Proc. Natl. Acad. Sci. U.S.A. 90: 10705-10709.

Kenny, P.A. and M.J. Bissell. 2007. Targeting TACE-dependent EGFR ligand shedding in breast cancer. J. Clin. Invest. 117: 337-345.

Keshet, Y. and R. Seger. 2010. The MAP kinase signaling cascades: a system of hundreds of components regulates a diverse array of physiological functions. Methods Mol. Biol. 661: 3-38.

Killock, D.J. and A. Ivetic. 2010. The cytoplasmic domains of TNFalpha-converting enzyme (TACE/ADAM17) and L-selectin are regulated differently by p38 MAPK and PKC to promote ectodomain shedding. Biochem. J. 428: 293-304.

Kim, B.H., N.Y. Cho, S.H. Shin, H.J. Kwon, J.J. Jang, and G.H. Kang. 2009. CpG island hypermethylation and repetitive DNA hypomethylation in premalignant lesion of extrahepatic cholangiocarcinoma. Virchows Arch. 455: 343-351.

Kim, H.J., M.J. Kim, S.H. Ahn, B.H. Son, S.B. Kim, J.H. Ahn, W.C. Noh, and G. Gong. 2011. Different prognostic significance of CD24 and CD44 expression in breast cancer according to hormone receptor status. Breast. 20: 78-85.

Kim, K.E., H. Song, C. Hahm, S.Y. Yoon, S. Park, H.R. Lee, D.Y. Hur, T. Kim, C.H. Kim, S.I. Bang, J.W. Bang, H. Park, and D.H. Cho. 2009. Expression of ADAM33 is a novel regulatory mechanism in IL-18-secreted process in gastric cancer. J. Immunol. 182: 3548-3555.

Knudson, C.B. and W. Knudson. 1993. Hyaluronan-binding proteins in development, tissue homeostasis, and disease. FASEB J. 7: 1233-1241.

Ko, S.Y., S.C. Lin, Y.K. Wong, C.J. Liu, K.W. Chang, and T.Y. Liu. 2007. Increase of disintergin metalloprotease 10 (ADAM10) expression in oral squamous cell carcinoma. Cancer Lett. 245: 33-43.

Kobayashi, H., K.C. Boelte, and P.C. Lin. 2007. Endothelial cell adhesion molecules and cancer progression. Curr. Med. Chem. 14: 377-386.

Kodama, H., K. Fukuda, T. Takahashi, M. Sano, T. Kato, S. Tahara, D. Hakuno, T. Sato, T. Manabe, F. Konishi, and S. Ogawa. 2002. Role of EGF Receptor and Pyk2 in endothelin-1-induced ERK activation in rat cardiomyocytes. J. Mol. Cell. Cardiol. 34: 139-150.

Kodama, T., E. Ikeda, A. Okada, T. Ohtsuka, M. Shimoda, T. Shiomi, K. Yoshida, M. Nakada, E. Ohuchi, and Y. Okada. 2004. ADAM12 is selectively overexpressed in human glioblastomas and is associated with glioblastoma cell proliferation and shedding of heparin-binding epidermal growth factor. Am. J. Pathol. 165: 1743-1753.

Kohga, K., T. Takehara, T. Tatsumi, T. Miyagi, H. Ishida, K. Ohkawa, T. Kanto, N. Hiramatsu, and N. Hayashi. 2009. Anticancer chemotherapy inhibits MHC class I-related chain a ectodomain shedding by downregulating ADAM10 expression in hepatocellular carcinoma. Cancer Res. 69: 8050-8057.

Kono, K., H. Naganuma, T. Sekikawa, H. Amemiya, A. Takahashi, H. Iizuka, and Y. Matsumoto. 2000. Serum level of HER-2/neu in patients with gastric cancer: correlation with HER-2/neu overexpression in gastric carcinoma tissue. Tumour Biol. 21: 139-144.

Konstantopoulos, K. and S.N. Thomas. 2009. Cancer cells in transit: the vascular interactions of tumor cells. Annu. Rev. Biomed. Eng. 11: 177-202.

Köstler, W.J., G.G. Steger, A. Soleiman, B. Schwab, C.F. Singer, S. Tomek, T. Brodowicz, M. Krainer, C. Wiltschke, R. Horvat, R. Jakesz, and C.C. Zielinski. 2004. Monitoring of serum Her-2/neu predicts histopathological response to neoadjuvant trastuzumab-based therapy for breast cancer. Anticancer Res. 24: 1127-1130.

Köstler, W.J., S. Tomek, T. Brodowicz, A.C. Budinsky, M. Flamm, M. Hejna, M. Krainer, C. Wiltschke, and C.C. Zielinski. 2001. Soluble ICAM-1 in breast cancer: clinical significance and biological implications. Cancer Immunol. Immunother. 50: 483-490.

Kowalski, P.J., M.A. Rubin, and C.G. Kleer. 2003. E-cadherin expression in primary carcinomas of the breast and its distant metastases. Breast Cancer Res. 5: R217-22.

Krainer, M., T. Brodowicz, R. Zeillinger, C. Wiltschke, C. Scholten, M. Seifert, E. Kubista, and C.C. Zielinski. 1997. Tissue expression and serum levels of HER-2/neu in patients with breast cancer. Oncology. 54: 475-481.

Kubota, Y., P. O'Grady, H. Saito, and M. Takekawa. 2011. Oncogenic Ras abrogates MEK SUMOylation that suppresses the ERK pathway and cell transformation. Nat. Cell Biol. 13: 282-291.

Kuefer, R., K.C. Day, C.G. Kleer, M.S. Sabel, M.D. Hofer, S. Varambally, C.S. Zorn, A.M. Chinnaiyan, M.A. Rubin, and M.L. Day. 2006. ADAM15 disintegrin is associated with aggressive prostate and breast cancer disease. Neoplasia. 8: 319-329.

Kuefer, R., M.D. Hofer, J.E. Gschwend, K.J. Pienta, M.G. Sanda, A.M. Chinnaiyan, M.A. Rubin, and M.L. Day. 2003. The role of an 80 kDa fragment of E-cadherin in the metastatic progression of prostate cancer. Clin. Cancer Res. 9: 6447-6452.

Kuefer, R., M.D. Hofer, C.S. Zorn, O. Engel, B.G. Volkmer, M.A. Juarez-Brito, M. Eggel, J.E. Gschwend, M.A. Rubin, and M.L. Day. 2005. Assessment of a fragment of e-cadherin as a serum biomarker with predictive value for prostate cancer. Br. J. Cancer. 92: 2018-2023.

Kuroda, H., S. Mochizuki, M. Shimoda, M. Chijiiwa, K. Kamiya, Y. Izumi, M. Watanabe, H. Horinouchi, M. Kawamura, K. Kobayashi, and Y. Okada. 2010. ADAM28 is a serological and histochemical marker for non-small-cell lung cancers. Int. J. Cancer. 127: 1844-1856.

Kveiborg, M., C. Frohlich, R. Albrechtsen, V. Tischler, N. Dietrich, P. Holck, P. Kronqvist, F. Rank, A.M. Mercurio, and U.M. Wewer. 2005. A role for ADAM12 in breast tumor progression and stromal cell apoptosis. Cancer Res. 65: 4754-4761.

Kveiborg, M., R. Instrell, C. Rowlands, M. Howell, and P.J. Parker. 2011. PKCalpha and PKCdelta regulate ADAM17-mediated ectodomain shedding of heparin binding-EGF through separate pathways. PLoS One. 6: e17168.

Lackner, C., R. Moser, T. Bauernhofer, M. Wilders-Truschnig, H. Samonigg, A. Berghold, and K. Zatloukal. 1998. Soluble CD44 v5 and v6 in serum of patients with breast cancer. Correlation with expression of CD44 v5 and v6 variants in primary tumors and location of distant metastasis. Breast Cancer Res. Treat. 47: 29-40.

Lautrette, A., S. Li, R. Alili, S.W. Sunnarborg, M. Burtin, D.C. Lee, G. Friedlander, and F. Terzi. 2005. Angiotensin II and EGF receptor cross-talk in chronic kidney diseases: a new therapeutic approach. Nat. Med. 11: 867-874.

Le Gall, S.M., T. Maretzky, P.D. Issuree, X.D. Niu, K. Reiss, P. Saftig, R. Khokha, D. Lundell, and C.P. Blobel. 2010. ADAM17 is regulated by a rapid and reversible mechanism that controls access to its catalytic site. J. Cell. Sci. 123: 3913-3922.

Le Pabic, H., D. Bonnier, U.M. Wewer, A. Coutand, O. Musso, G. Baffet, B. Clement, and N. Theret. 2003. ADAM12 in human liver cancers: TGF-beta-regulated expression in stellate cells is associated with matrix remodeling. Hepatology. 37: 1056-1066.

Leary, A.F., W.M. Hanna, M.J. van de Vijver, F. Penault-Llorca, J. Ruschoff, R.Y. Osamura, M. Bilous, and M. Dowsett. 2009. Value and limitations of measuring

HER-2 extracellular domain in the serum of breast cancer patients. J. Clin. Oncol. 27: 1694-1705.

Lee, S.B., A. Schramme, K. Doberstein, R. Dummer, M.S. Abdel-Bakky, S. Keller, P. Altevogt, S.T. Oh, J. Reichrath, D. Oxmann, J. Pfeilschifter, D. Mihic-Probst, and P. Gutwein. 2010. ADAM10 is upregulated in melanoma metastasis compared with primary melanoma. J. Invest. Dermatol. 130: 763-773.

Lei, X., A. Bandyopadhyay, T. Le, and L. Sun. 2002. Autocrine TGFbeta supports growth and survival of human breast cancer MDA-MB-231 cells. Oncogene. 21: 7514-7523.

Lemjabbar, H., D. Li, M. Gallup, S. Sidhu, E. Drori, and C. Basbaum. 2003. Tobacco smoke-induced lung cell proliferation mediated by tumor necrosis factor alpha-converting enzyme and amphiregulin. J. Biol. Chem. 278: 26202-26207.

Lendeckel, U., J. Kohl, M. Arndt, S. Carl-McGrath, H. Donat, and C. Rocken. 2005. Increased expression of ADAM family members in human breast cancer and breast cancer cell lines. J. Cancer Res. Clin. Oncol. 131: 41-48.

Lesley, J. and R. Hyman. 1998. CD44 structure and function. Front. Biosci. 3: d616-30.

Lesley, J., R. Hyman, and P.W. Kincade. 1993. CD44 and its interaction with extracellular matrix. Adv. Immunol. 54: 271-335.

Leyland-Jones, B. and B.R. Smith. 2011. Serum HER2 testing in patients with HER2-positive breast cancer: the death knell tolls. Lancet Oncol. 12: 286-295.

Li, X. and H. Fan. 2004. Loss of ectodomain shedding due to mutations in the metalloprotease and cysteine-rich/disintegrin domains of the tumor necrosis factor-alpha converting enzyme (TACE). J. Biol. Chem. 279: 27365-27375.

Lichtenthaler, S.F., C. Haass, and H. Steiner. 2011. Regulated intramembrane proteolysis--lessons from amyloid precursor protein processing. J. Neurochem. 117: 779-796.

Liebmann, C. 2011. EGF receptor activation by GPCRs: an universal pathway reveals different versions. Mol. Cell. Endocrinol. 331: 222-231.

Lin, H., K. Wada, M. Yonezawa, K. Shinoki, T. Akamatsu, T. Tsukui, and C. Sakamoto. 2003. Tomoregulin ectodomain shedding by proinflammatory cytokines. Life Sci. 73: 1617-1627.

Lin, K., J.R. Taylor Jr, T.D. Wu, J. Gutierrez, J.M. Elliott, J.M. Vernes, H. Koeppen, H.S. Phillips, F.J. de Sauvage, and Y.G. Meng. 2011. TMEFF2 is a PDGF-AA binding protein with methylation-associated gene silencing in multiple cancer types including glioma. PLoS One. 6: e18608.

Lin, P., X. Sun, T. Feng, H. Zou, Y. Jiang, Z. Liu, D. Zhao, and X. Yu. 2012. ADAM17 regulates prostate cancer cell proliferation through mediating cell cycle progression by EGFR/PI3K/AKT pathway. Mol. Cell. Biochem. 359: 235-243.

Liu, C., P. Xu, S. Lamouille, J. Xu, and R. Derynck. 2009. TACE-mediated ectodomain shedding of the type I TGF-beta receptor downregulates TGF-beta signaling. Mol. Cell. 35: 26-36.

Liu, P.C., X. Liu, Y. Li, M. Covington, R. Wynn, R. Huber, M. Hillman, G. Yang, D. Ellis, C. Marando, K. Katiyar, J. Bradley, K. Abremski, M. Stow, M. Rupar, J. Zhuo, Y.L. Li, Q. Lin, D. Burns, M. Xu, C. Zhang, D.Q. Qian, C. He, V. Sharief, L. Weng, C. Agrios, E. Shi, B. Metcalf, R. Newton, S. Friedman, W. Yao, P. Scherle, G. Hollis, and T.C. Burn. 2006. Identification of ADAM10 as a major source of HER2 ectodomain sheddase activity in HER2 overexpressing breast cancer cells. Cancer. Biol. Ther. 5: 657-664.

Lofton-Day, C., F. Model, T. Devos, R. Tetzner, J. Distler, M. Schuster, X. Song, R. Lesche, V. Liebenberg, M. Ebert, B. Molnar, R. Grutzmann, C. Pilarsky, and A. Sledziewski. 2008. DNA methylation biomarkers for blood-based colorectal cancer screening. Clin. Chem. 54: 414-423.

Lu, X., D. Lu, M. Scully, and V. Kakkar. 2008. ADAM proteins - therapeutic potential in cancer. Curr. Cancer. Drug Targets. 8: 720-732.

Lucas, N. and M.L. Day. 2009. The role of the disintegrin metalloproteinase ADAM15 in prostate cancer progression. J. Cell. Biochem. 106: 967-974.

Lucas, T.F., C. Royer, E.R. Siu, M.F. Lazari, and C.S. Porto. 2010. Expression and signaling of G protein-coupled estrogen receptor 1 (GPER) in rat sertoli cells. Biol. Reprod. 83: 307-317.

Ludovini, V., S. Gori, M. Colozza, L. Pistola, E. Rulli, I. Floriani, E. Pacifico, F.R. Tofanetti, A. Sidoni, C. Basurto, A. Rulli, and L. Crino. 2008. Evaluation of serum HER2 extracellular domain in early breast cancer patients: correlation with clinicopathological parameters and survival. Ann. Oncol. 19: 883-890.

Luetteke, N.C., T.H. Qiu, R.L. Peiffer, P. Oliver, O. Smithies, and D.C. Lee. 1993. TGF alpha deficiency results in hair follicle and eye abnormalities in targeted and waved-1 mice. Cell. 73: 263-278.

Lum, L., M.S. Reid, and C.P. Blobel. 1998. Intracellular maturation of the mouse metalloprotease disintegrin MDC15. J. Biol. Chem. 273: 26236-26247.

Luttrell, D.K. and L.M. Luttrell. 2004. Not so strange bedfellows: G-protein-coupled receptors and Src family kinases. Oncogene. 23: 7969-7978.

Luttrell, L.M., G.J. Della Rocca, T. van Biesen, D.K. Luttrell, and R.J. Lefkowitz. 1997. Gbetagamma subunits mediate Src-dependent phosphorylation of the epidermal growth factor receptor. A scaffold for G protein-coupled receptor-mediated Ras activation. J. Biol. Chem. 272: 4637-4644.

Lyu, J., V. Yamamoto, and W. Lu. 2008. Cleavage of the Wnt receptor Ryk regulates neuronal differentiation during cortical neurogenesis. Dev. Cell. 15: 773-780.

Ma, Q.H., T. Futagawa, W.L. Yang, X.D. Jiang, L. Zeng, Y. Takeda, R.X. Xu, D. Bagnard, M. Schachner, A.J. Furley, D. Karagogeos, K. Watanabe, G.S. Dawe, and Z.C.

Xiao. 2008. A TAG1-APP signalling pathway through Fe65 negatively modulates neurogenesis. Nat. Cell Biol. 10: 283-294.

Maetzel, D., S. Denzel, B. Mack, M. Canis, P. Went, M. Benk, C. Kieu, P. Papior, P.A. Baeuerle, M. Munz, and O. Gires. 2009. Nuclear signalling by tumour-associated antigen EpCAM. Nat. Cell Biol. 11: 162-171.

Malapeira, J., C. Esselens, J.J. Bech-Serra, F. Canals, and J. Arribas. 2011. ADAM17 (TACE) regulates TGFbeta signaling through the cleavage of vasorin. Oncogene. 30: 1912-1922.

Mantur, M., J. Snarska, O. Koper, J. Dzieciol, A. Plonski, and D. Lemancewicz. 2009. Serum sICAM, sVCAM and sE-selectin levels in colorectal cancer patients. Folia Histochem. Cytobiol. 47: 621-625.

Maretzky, T., A. Evers, W. Zhou, S.L. Swendeman, P.M. Wong, S. Rafii, K. Reiss, and C.P. Blobel. 2011a. Migration of growth factor-stimulated epithelial and endothelial cells depends on EGFR transactivation by ADAM17. Nat. Commun. 2: 229.

Maretzky, T., W. Zhou, X.Y. Huang, and C.P. Blobel. 2011b. A transforming Src mutant increases the bioavailability of EGFR ligands via stimulation of the cell-surface metalloproteinase ADAM17. Oncogene. 30: 611-618.

Maretzky, T., M. Schulte, A. Ludwig, S. Rose-John, C. Blobel, D. Hartmann, P. Altevogt, P. Saftig, and K. Reiss. 2005. L1 is sequentially processed by two differently activated metalloproteases and presenilin/gamma-secretase and regulates neural cell adhesion, cell migration, and neurite outgrowth. Mol. Cell. Biol. 25: 9040-9053.

Maruo, Y., A. Gochi, A. Kaihara, H. Shimamura, T. Yamada, N. Tanaka, and K. Orita. 2002. ICAM-1 expression and the soluble ICAM-1 level for evaluating the metastatic potential of gastric cancer. Int. J. Cancer. 100: 486-490.

Masson, D., M.G. Denis, M. Denis, D. Blanchard, M.J. Loirat, E. Cassagnau, and P. Lustenberger. 1999. Soluble CD44: quantification and molecular repartition in plasma of patients with colorectal cancer. Br. J. Cancer. 80: 1995-2000.

Matsuno, O., E. Miyazaki, S. Nureki, T. Ueno, T. Kumamoto, and Y. Higuchi. 2006. Role of ADAM8 in experimental asthma. Immunol. Lett. 102: 67-73.

Matsuura, S., Y. Oda, H. Matono, T. Izumi, H. Yamamoto, S. Tamiya, Y. Iwamoto, and M. Tsuneyoshi. 2010. Overexpression of A disintegrin and metalloproteinase 28 is correlated with high histologic grade in conventional chondrosarcoma. Hum. Pathol. 41: 343-351.

Mayer, B., J.P. Johnson, F. Leitl, K.W. Jauch, M.M. Heiss, F.W. Schildberg, W. Birchmeier, and I. Funke. 1993. E-cadherin expression in primary and metastatic gastric cancer: down-regulation correlates with cellular dedifferentiation and glandular disintegration. Cancer Res. 53: 1690-1695.

Mayer, S., A. zur Hausen, D.O. Watermann, S. Stamm, M. Jager, G. Gitsch, and E. Stickeler. 2008. Increased soluble CD44 concentrations are associated with larger tumor size and lymph node metastasis in breast cancer patients. J. Cancer Res. Clin. Oncol. 134: 1229-1235.

Mazzocca, A., R. Coppari, R. De Franco, J.Y. Cho, T.A. Libermann, M. Pinzani, and A. Toker. 2005. A secreted form of ADAM9 promotes carcinoma invasion through tumor-stromal interactions. Cancer Res. 65: 4728-4738.

McGowan, P.M., E. McKiernan, F. Bolster, B.M. Ryan, A.D. Hill, E.W. McDermott, D. Evoy, N. O'Higgins, J. Crown, and M.J. Duffy. 2008. ADAM-17 predicts adverse outcome in patients with breast cancer. Ann. Oncol. 19: 1075-1081.

McKenzie, S.J., K.A. DeSombre, B.S. Bast, D.R. Hollis, R.S. Whitaker, A. Berchuck, C.M. Boyer, and R.C. Bast Jr. 1993. Serum levels of HER-2 neu (C-erbB-2) correlate with overexpression of p185neu in human ovarian cancer. Cancer. 71: 3942-3946.

Mechtersheimer, S., P. Gutwein, N. Agmon-Levin, A. Stoeck, M. Oleszewski, S. Riedle, R. Postina, F. Fahrenholz, M. Fogel, V. Lemmon, and P. Altevogt. 2001. Ectodomain shedding of L1 adhesion molecule promotes cell migration by autocrine binding to integrins. J. Cell Biol. 155: 661-673.

Medinger, M., N. Fischer, and A. Tzankov. 2010. Vascular endothelial growth factor-related pathways in hemato-lymphoid malignancies. J. Oncol. 2010: 729725.

Mendelson, K., S. Swendeman, P. Saftig, and C.P. Blobel. 2010. Stimulation of platelet-derived growth factor receptor beta (PDGFRbeta) activates ADAM17 and promotes metalloproteinase-dependent cross-talk between the PDGFRbeta and epidermal growth factor receptor (EGFR) signaling pathways. J. Biol. Chem. 285: 25024-25032.

Meng, S., D. Tripathy, S. Shete, R. Ashfaq, B. Haley, S. Perkins, P. Beitsch, A. Khan, D. Euhus, C. Osborne, E. Frenkel, S. Hoover, M. Leitch, E. Clifford, E. Vitetta, L. Morrison, D. Herlyn, L.W. Terstappen, T. Fleming, T. Fehm, T. Tucker, N. Lane, J. Wang, and J. Uhr. 2004. HER-2 gene amplification can be acquired as breast cancer progresses. Proc. Natl. Acad. Sci. U. S. A. 101: 9393-9398.

Menschikowski, M., A. Hagelgans, G. Eisenhofer, and G. Siegert. 2009. Regulation of endothelial protein C receptor shedding by cytokines is mediated through differential activation of MAP kinase signaling pathways. Exp. Cell Res. 315: 2673-2682.

Meulmeester, E. and P. Ten Dijke. 2011. The dynamic roles of TGF-beta in cancer. J. Pathol. 223: 205-218.

Mifune, M., H. Ohtsu, H. Suzuki, H. Nakashima, E. Brailoiu, N.J. Dun, G.D. Frank, T. Inagami, S. Higashiyama, W.G. Thomas, A.D. Eckhart, P.J. Dempsey, and S. Eguchi. 2005. G protein coupling and second messenger generation are indispensable for metalloprotease-dependent, heparin-binding epidermal growth factor shedding through angiotensin II type-1 receptor. J. Biol. Chem. 280: 26592-26599.

Milla, M.E., M.A. Leesnitzer, M.L. Moss, W.C. Clay, H.L. Carter, A.B. Miller, J.L. Su, M.H. Lambert, D.H. Willard, D.M. Sheeley, T.A. Kost, W. Burkhart, M. Moyer, R.K. Blackburn, G.L. Pahel, J.L. Mitchell, C.R. Hoffman, and J.D. Becherer. 1999. Specific sequence elements are required for the expression of functional tumor necrosis factor-alpha-converting enzyme (TACE). J. Biol. Chem. 274: 30563-30570.

Mine, N., R. Iwamoto, and E. Mekada. 2005. HB-EGF promotes epithelial cell migration in eyelid development. Development. 132: 4317-4326.

Mitsui, Y., S. Mochizuki, T. Kodama, M. Shimoda, T. Ohtsuka, T. Shiomi, M. Chijiiwa, T. Ikeda, M. Kitajima, and Y. Okada. 2006. ADAM28 is overexpressed in human breast carcinomas: implications for carcinoma cell proliferation through cleavage of insulin-like growth factor binding protein-3. Cancer Res. 66: 9913-9920.

Moasser, M.M. 2007. The oncogene HER2: its signaling and transforming functions and its role in human cancer pathogenesis. Oncogene. 26: 6469-6487.

Mochizuki, S. and Y. Okada. 2009. ADAM28 as a target for human cancers. Curr. Pharm. Des. 15: 2349-2358.

Mochizuki, S. and Y. Okada. 2007. ADAMs in cancer cell proliferation and progression. Cancer. Sci. 98: 621-628.

Mochizuki, S., R. Tanaka, M. Shimoda, J. Onuma, Y. Fujii, H. Jinno, and Y. Okada. 2010. Connective tissue growth factor is a substrate of ADAM28. Biochem. Biophys. Res. Commun. 402: 651-657.

Molina, M.A., J. Codony-Servat, J. Albanell, F. Rojo, J. Arribas, and J. Baselga. 2001. Trastuzumab (herceptin), a humanized anti-Her2 receptor monoclonal antibody, inhibits basal and activated Her2 ectodomain cleavage in breast cancer cells. Cancer Res. 61: 4744-4749.

Molina, M.A., R. Saez, E.E. Ramsey, M.J. Garcia-Barchino, F. Rojo, A.J. Evans, J. Albanell, E.J. Keenan, A. Lluch, J. Garcia-Conde, J. Baselga, and G.M. Clinton. 2002. NH(2)-terminal truncated HER-2 protein but not full-length receptor is associated with nodal metastasis in human breast cancer. Clin. Cancer Res. 8: 347-353.

Molina, R., J.M. Auge, J.M. Escudero, X. Filella, G. Zanon, J. Pahisa, B. Farrus, M. Munoz, and M. Velasco. 2010. Evaluation of tumor markers (HER-2/neu oncoprotein, CEA, and CA 15.3) in patients with locoregional breast cancer: prognostic value. Tumour Biol. 31: 171-180.

Mongaret, C., J. Alexandre, A. Thomas-Schoemann, E. Bermudez, C. Chereau, C. Nicco, F. Goldwasser, B. Weill, F. Batteux, and F. Lemare. 2011. Tumor invasion induced by oxidative stress is dependent on membrane ADAM 9 protein and its secreted form. Int. J. Cancer. 129: 791-798.

Montero, J.C., L. Yuste, E. Diaz-Rodriguez, A. Esparis-Ogando, and A. Pandiella. 2002. Mitogen-activated protein kinase-dependent and -independent routes control shedding of transmembrane growth factors through multiple secretases. Biochem. J. 363: 211-221.

Morais da Silva, S., P.B. Gates, D.W. Eib, G.J. Martens, and J.P. Brockes. 2001. The expression pattern of tomoregulin-1 in urodele limb regeneration and mouse limb development. Mech. Dev. 104: 125-128.

Moss, M.L., A. Stoeck, W. Yan, and P.J. Dempsey. 2008. ADAM10 as a target for anti-cancer therapy. Curr. Pharm. Biotechnol. 9: 2-8.

Muramatsu, H., T. Akimoto, K. Maebayashi, M. Kita, and N. Mitsuhashi. 2008. Prognostic significance of dysadherin and E-cadherin expression in patients with head and neck cancer treated by radiation therapy. Anticancer Res. 28: 3859-3864.

Murphy, G. 2008. The ADAMs: signalling scissors in the tumour microenvironment. Nat. Rev. Cancer. 8: 929-941.

Myers, T.J., L.H. Brennaman, M. Stevenson, S. Higashiyama, W.E. Russell, D.C. Lee, and S.W. Sunnarborg. 2009. Mitochondrial reactive oxygen species mediate GPCR-induced TACE/ADAM17-dependent transforming growth factor-alpha shedding. Mol. Biol. Cell. 20: 5236-5249.

Nagano, O., D. Murakami, D. Hartmann, B. De Strooper, P. Saftig, T. Iwatsubo, M. Nakajima, M. Shinohara, and H. Saya. 2004. Cell-matrix interaction via CD44 is independently regulated by different metalloproteinases activated in response to extracellular Ca(2+) influx and PKC activation. J. Cell Biol. 165: 893-902.

Nagano, O. and H. Saya. 2004. Mechanism and biological significance of CD44 cleavage. Cancer. Sci. 95: 930-935.

Najy, A.J., K.C. Day, and M.L. Day. 2008. ADAM15 supports prostate cancer metastasis by modulating tumor cell-endothelial cell interaction. Cancer Res. 68: 1092-1099.

Najy, A.J., K.C. Day, and M.L. Day. 2008. The ectodomain shedding of E-cadherin by ADAM15 supports ErbB receptor activation. J. Biol. Chem. 283: 18393-18401.

Nakata, B., T. Hori, T. Sunami, Y. Ogawa, M. Yashiro, K. Maeda, T. Sawada, Y. Kato, T. Ishikawa, and K. Hirakawa. 2000. Clinical significance of serum soluble intercellular adhesion molecule 1 in gastric cancer. Clin. Cancer Res. 6: 1175-1179.

Naor, D., R.V. Sionov, and D. Ish-Shalom. 1997. CD44: structure, function, and association with the malignant process. Adv. Cancer Res. 71: 241-319.

Naor, D., S.B. Wallach-Dayan, M.A. Zahalka, and R.V. Sionov. 2008. Involvement of CD44, a molecule with a thousand faces, in cancer dissemination. Semin. Cancer Biol. 18: 260-267.

Narita, D., A. Anghel, E. Seclaman, R. Ilina, N. Cireap, and S. Ursoniu. 2010. Molecular profiling of ADAM12 gene in breast cancers. Rom. J. Morphol. Embryol. 51: 669-676.

Nath, D., P.M. Slocombe, P.E. Stephens, A. Warn, G.R. Hutchinson, K.M. Yamada, A.J. Docherty, and G. Murphy. 1999. Interaction of metargidin (ADAM-15) with alphavbeta3 and alpha5beta1 integrins on different haemopoietic cells. J. Cell. Sci. 112 (Pt 4): 579-587.

Nath, D., P.M. Slocombe, A. Webster, P.E. Stephens, A.J. Docherty, and G. Murphy. 2000. Meltrin gamma(ADAM-9) mediates cellular adhesion through alpha(6)beta(1) integrin, leading to a marked induction of fibroblast cell motility. J. Cell. Sci. 113 (Pt 12): 2319-2328.

Nawrocki-Raby, B., C. Gilles, M. Polette, E. Bruyneel, J.Y. Laronze, N. Bonnet, J.M. Foidart, M. Mareel, and P. Birembaut. 2003. Upregulation of MMPs by soluble E-cadherin in human lung tumor cells. Int. J. Cancer. 105: 790-795.

Nelson, H., P.S. Ramsey, J.H. Donohue, and L.E. Wold. 1994. Cell adhesion molecule expression within the microvasculature of human colorectal malignancies. Clin. Immunol. Immunopathol. 72: 129-136.

Nelson, W.J. 2008. Regulation of cell-cell adhesion by the cadherin-catenin complex. Biochem. Soc. Trans. 36: 149-155.

Niewiarowski, S., M.A. McLane, M. Kloczewiak, and G.J. Stewart. 1994. Disintegrins and other naturally occurring antagonists of platelet fibrinogen receptors. Semin. Hematol. 31: 289-300.

Nishi, E., Y. Hiraoka, K. Yoshida, K. Okawa, and T. Kita. 2006. Nardilysin enhances ectodomain shedding of heparin-binding epidermal growth factor-like growth factor through activation of tumor necrosis factor-alpha-converting enzyme. J. Biol. Chem. 281: 31164-31172.

Ohtsu, H., P.J. Dempsey, and S. Eguchi. 2006. ADAMs as mediators of EGF receptor transactivation by G protein-coupled receptors. Am. J. Physiol. Cell. Physiol. 291: C1-10.

Ohtsu, H., S. Higuchi, H. Shirai, K. Eguchi, H. Suzuki, A. Hinoki, E. Brailoiu, A.D. Eckhart, G.D. Frank, and S. Eguchi. 2008. Central role of Gq in the hypertrophic signal transduction of angiotensin II in vascular smooth muscle cells. Endocrinology. 149: 3569-3575.

Ohtsuka, T., T. Shiomi, M. Shimoda, T. Kodama, A. Amour, G. Murphy, E. Ohuchi, K. Kobayashi, and Y. Okada. 2006. ADAM28 is overexpressed in human non-small cell lung carcinomas and correlates with cell proliferation and lymph node metastasis. Int. J. Cancer. 118: 263-273.

Okamoto, I., H. Tsuiki, L.C. Kenyon, A.K. Godwin, D.R. Emlet, M. Holgado-Madruga, I.S. Lanham, C.J. Joynes, K.T. Vo, A. Guha, M. Matsumoto, Y. Ushio, H. Saya, and A.J. Wong. 2002. Proteolytic cleavage of the CD44 adhesion molecule in multiple human tumors. Am. J. Pathol. 160: 441-447.

Okamoto, Y., T. Tsurunaga, and M. Ueki. 1999. Serum soluble ICAM-1 levels in the patients with cervical cancer. Acta Obstet. Gynecol. Scand. 78: 60-65.

Okugawa, Y., C. Miki, Y. Toiyama, Y. Koike, Y. Inoue, and M. Kusunoki. 2009. Serum level of soluble vascular cell adhesion molecule 1 is a valuable prognostic marker in colorectal carcinoma. Dis. Colon Rectum. 52: 1330-1336.

Orian-Rousseau, V. 2010. CD44, a therapeutic target for metastasising tumours. Eur. J. Cancer. 46: 1271-1277.

Osaki, T., T. Mitsudomi, T. Oyama, R. Nakanishi, and K. Yasumoto. 1995. Serum level and tissue expression of c-erbB-2 protein in lung adenocarcinoma. Chest. 108: 157-162.

O'Shea, C., N. McKie, Y. Buggy, C. Duggan, A.D. Hill, E. McDermott, N. O'Higgins, and M.J. Duffy. 2003. Expression of ADAM-9 mRNA and protein in human breast cancer. Int. J. Cancer. 105: 754-761.

Pai, R., B. Soreghan, I.L. Szabo, M. Pavelka, D. Baatar, and A.S. Tarnawski. 2002. Prostaglandin E2 transactivates EGF receptor: a novel mechanism for promoting colon cancer growth and gastrointestinal hypertrophy. Nat. Med. 8: 289-293.

Pallud, C., J.M. Guinebretiere, S. Guepratte, K. Hacene, R. Neumann, W. Carney, and M.F. Pichon. 2005. Tissue expression and serum levels of the oncoprotein HER-2/neu in 157 primary breast tumours. Anticancer Res. 25: 1433-1440.

Pandiella, A., M.W. Bosenberg, E.J. Huang, P. Besmer, and J. Massague. 1992. Cleavage of membrane-anchored growth factors involves distinct protease activities regulated through common mechanisms. J. Biol. Chem. 267: 24028-24033.

Park, S.Y., H.J. Kwon, H.E. Lee, H.S. Ryu, S.W. Kim, J.H. Kim, I.A. Kim, N. Jung, N.Y. Cho, and G.H. Kang. 2011. Promoter CpG island hypermethylation during breast cancer progression. Virchows Arch. 458: 73-84.

Parr-Sturgess, C.A., D.J. Rushton, and E.T. Parkin. 2010. Ectodomain shedding of the Notch ligand Jagged1 is mediated by ADAM17, but is not a lipid-raft-associated event. Biochem. J. 432: 283-294.

Pedrazzani, C., S. Caruso, G. Corso, D. Marrelli, A. Neri, A. Berardi, and F. Roviello. 2008. Influence of age on soluble E-cadherin serum levels prevents its utility as a disease marker in gastric cancer patients. Scand. J. Gastroenterol. 43: 765-766.

Peduto, L., V.E. Reuter, A. Sehara-Fujisawa, D.R. Shaffer, H.I. Scher, and C.P. Blobel. 2006. ADAM12 is highly expressed in carcinoma-associated stroma and is required for mouse prostate tumor progression. Oncogene. 25: 5462-5466.

Peduto, L., V.E. Reuter, D.R. Shaffer, H.I. Scher, and C.P. Blobel. 2005. Critical function for ADAM9 in mouse prostate cancer. Cancer Res. 65: 9312-9319.

Peiretti, F., P. Deprez-Beauclair, B. Bonardo, H. Aubert, I. Juhan-Vague, and G. Nalbone. 2003. Identification of SAP97 as an intracellular binding partner of TACE. J. Cell. Sci. 116: 1949-1957.

Perl, A.K., P. Wilgenbus, U. Dahl, H. Semb, and G. Christofori. 1998. A causal role for E-cadherin in the transition from adenoma to carcinoma. Nature. 392: 190-193.

Peschon, J.J., J.L. Slack, P. Reddy, K.L. Stocking, S.W. Sunnarborg, D.C. Lee, W.E. Russell, B.J. Castner, R.S. Johnson, J.N. Fitzner, R.W. Boyce, N. Nelson, C.J. Kozlosky, M.F. Wolfson, C.T. Rauch, D.P. Cerretti, R.J. Paxton, C.J. March, and R.A. Black. 1998. An essential role for ectodomain shedding in mammalian development. Science. 282: 1281-1284.

Peterson, R.M., Q. Yu, I. Stamenkovic, and B.P. Toole. 2000. Perturbation of hyaluronan interactions by soluble CD44 inhibits growth of murine mammary carcinoma cells in ascites. Am. J. Pathol. 156: 2159-2167.

Pierce, K.L., A. Tohgo, S. Ahn, M.E. Field, L.M. Luttrell, and R.J. Lefkowitz. 2001. Epidermal growth factor (EGF) receptor-dependent ERK activation by

G protein-coupled receptors: a co-culture system for identifying intermediates upstream and downstream of heparin-binding EGF shedding. J. Biol. Chem. 276: 23155-23160.

Pignatelli, M., T.W. Ansari, P. Gunter, D. Liu, S. Hirano, M. Takeichi, G. Kloppel, and N.R. Lemoine. 1994. Loss of membranous E-cadherin expression in pancreatic cancer: correlation with lymph node metastasis, high grade, and advanced stage. J. Pathol. 174: 243-248.

Pigott, R., L.P. Dillon, I.H. Hemingway, and A.J. Gearing. 1992. Soluble forms of E-selectin, ICAM-1 and VCAM-1 are present in the supernatants of cytokine activated cultured endothelial cells. Biochem. Biophys. Res. Commun. 187: 584-589.

Pokutta, S. and W.I. Weis. 2007. Structure and mechanism of cadherins and catenins in cell-cell contacts. Annu. Rev. Cell Dev. Biol. 23: 237-261.

Polychronidis, A.C., A.K. Tsaroucha, S.K. Anagnostoulis, E.E. Efstathiou, P.G. Georgiadis and C.E. Simopoulos. 2003. Serum levels of c-erbb-2 correlate with advanced stage and liver metastasis in colorectal cancer. Folia. Med. (Plovdiv) 45: 12-16.

Ponta, H., L. Sherman, and P.A. Herrlich. 2003. CD44: from adhesion molecules to signalling regulators. Nat. Rev. Mol. Cell Biol. 4: 33-45.

Prenzel, N., E. Zwick, H. Daub, M. Leserer, R. Abraham, C. Wallasch, and A. Ullrich. 1999. EGF receptor transactivation by G-protein-coupled receptors requires metalloproteinase cleavage of proHB-EGF. Nature. 402: 884-888.

Press, M.F., C. Cordon-Cardo, and D.J. Slamon. 1990. Expression of the HER-2/neu proto-oncogene in normal human adult and fetal tissues. Oncogene. 5: 953-962.

Pruessmeyer, J. and A. Ludwig. 2009. The good, the bad and the ugly substrates for ADAM10 and ADAM17 in brain pathology, inflammation and cancer. Semin. Cell Dev. Biol. 20: 164-174.

Qu, D., Y. Wang, N.L. Esmon, and C.T. Esmon. 2007. Regulated endothelial protein C receptor shedding is mediated by tumor necrosis factor-alpha converting enzyme/ADAM17. J. Thromb. Haemost. 5: 395-402.

Quaranta, M., A. Daniele, M. Coviello, A. Savonarola, I. Abbate, M.T. Venneri, A. Paradiso, B. Stea, A. Zito, A. Labriola, and F. Schittulli. 2006. c-erbB-2 protein level in tissue and sera of breast cancer patients: a possibly useful clinical correlation. Tumori. 92: 311-317.

Rahimi, N., T.E. Golde, and R.D. Meyer. 2009. Identification of ligand-induced proteolytic cleavage and ectodomain shedding of VEGFR-1/FLT1 in leukemic cancer cells. Cancer Res. 69: 2607-2614.

Rausch, M.E., L. Beer, M.D. Sammel, P. Takacs, K. Chung, A. Shaunik, D. Speicher, and K.T. Barnhart. 2011. A disintegrin and metalloprotease protein-12 as a novel marker for the diagnosis of ectopic pregnancy. Fertil. Steril. 95: 1373-1378.

Reddy, P., J.L. Slack, R. Davis, D.P. Cerretti, C.J. Kozlosky, R.A. Blanton, D. Shows, J.J. Peschon, and R.A. Black. 2000. Functional analysis of the domain structure of tumor necrosis factor-alpha converting enzyme. J. Biol. Chem. 275: 14608-14614.

Regitnig, P., W. Schippinger, M. Lindbauer, H. Samonigg, and S.F. Lax. 2004. Change of HER-2/neu status in a subset of distant metastases from breast carcinomas. J. Pathol. 203: 918-926.

Rizoli, S.B., O.D. Rotstein, and A. Kapus. 1999. Cell volume-dependent regulation of L-selectin shedding in neutrophils. A role for p38 mitogen-activated protein kinase. J. Biol. Chem. 274: 22072-22080.

Roberts, C.M., P.H. Tani, L.C. Bridges, Z. Laszik, and R.D. Bowditch. 1999. MDC-L, a novel metalloprotease disintegrin cysteine-rich protein family member expressed by human lymphocytes. J. Biol. Chem. 274: 29251-29259.

Rocks, N., G. Paulissen, F. Quesada Calvo, M. Polette, M. Gueders, C. Munaut, J.M. Foidart, A. Noel, P. Birembaut, and D. Cataldo. 2006. Expression of a disintegrin and metalloprotease (ADAM and ADAMTS) enzymes in human non-small-cell lung carcinomas (NSCLC). Br. J. Cancer. 94: 724-730.

Roemer, A., L. Schwettmann, M. Jung, J. Roigas, G. Kristiansen, D. Schnorr, S.A. Loening, K. Jung, and R. Lichtinghagen. 2004. Increased mRNA expression of ADAMs in renal cell carcinoma and their association with clinical outcome. Oncol. Rep. 11: 529-536.

Roemer, A., L. Schwettmann, M. Jung, C. Stephan, J. Roigas, G. Kristiansen, S.A. Loening, R. Lichtinghagen, and K. Jung. 2004. The membrane proteases adams and hepsin are differentially expressed in renal cell carcinoma. Are they potential tumor markers? J. Urol. 172: 2162-2166.

Rosso, O., T. Piazza, I. Bongarzone, A. Rossello, D. Mezzanzanica, S. Canevari, A.M. Orengo, A. Puppo, S. Ferrini, and M. Fabbi. 2007. The ALCAM shedding by the metalloprotease ADAM17/TACE is involved in motility of ovarian carcinoma cells. Mol. Cancer. Res. 5: 1246-1253.

Roy, R., S. Rodig, D. Bielenberg, D. Zurakowski, and M.A. Moses. 2011. ADAM12 transmembrane and secreted isoforms promote breast tumor growth: a distinct role for ADAM12-S protein in tumor metastasis. J. Biol. Chem. 286: 20758-20768.

Roy, R., U.M. Wewer, D. Zurakowski, S.E. Pories, and M.A. Moses. 2004. ADAM 12 cleaves extracellular matrix proteins and correlates with cancer status and stage. J. Biol. Chem. 279: 51323-51330.

Rozengurt, E. 2007. Mitogenic signaling pathways induced by G protein-coupled receptors. J. Cell. Physiol. 213: 589-602.

Ruiz, P., C. Schwarzler, and U. Gunthert. 1995. CD44 isoforms during differentiation and development. Bioessays. 17: 17-24.

Sáez, R., M.A. Molina, E.E. Ramsey, F. Rojo, E.J. Keenan, J. Albanell, A. Lluch, J. Garcia-Conde, J. Baselga, and G.M. Clinton. 2006. p95HER-2 predicts worse outcome in patients with HER-2-positive breast cancer. Clin. Cancer Res. 12: 424-431.

Saftig, P. and K. Reiss. 2011. The "A Disintegrin And Metalloproteases" ADAM10 and ADAM17: novel drug targets with therapeutic potential? Eur. J. Cell Biol. 90: 527-535.

Saghatchian, M., S. Guepratte, K. Hacene, R. Neumann, J.L. Floiras, and M.F. Pichon. 2004. Serum HER-2 extracellular domain: relationship with clinicobiological presentation and prognostic value before and after primary treatment in 701 breast cancer patients. Int. J. Biol. Markers. 19: 14-22.

Sahasrabudhe, S.R., M.A. Spruyt, H.A. Muenkel, A.J. Blume, M.P. Vitek, and J.S. Jacobsen. 1992. Release of amino-terminal fragments from amyloid precursor protein reporter and mutated derivatives in cultured cells. J. Biol. Chem. 267: 25602-25608.

Sahin, U. and C.P. Blobel. 2007. Ectodomain shedding of the EGF-receptor ligand epigen is mediated by ADAM17. FEBS Lett. 581: 41-44.

Sahin, U., G. Weskamp, K. Kelly, H.M. Zhou, S. Higashiyama, J. Peschon, D. Hartmann, P. Saftig, and C.P. Blobel. 2004. Distinct roles for ADAM10 and ADAM17 in ectodomain shedding of six EGFR ligands. J. Cell Biol. 164: 769-779.

Samy, N., M.D. Abd El-Maksoud, T.E. Mousa, H.A. El-Mezayen, and M. Shaalan. 2010. Potential role of serum level of soluble CD44 and IFN-gamma in B-cell chronic lymphocytic leukemia. Med. Oncol. 28(1): 471-475.

Sánchez-Rovira, P., E. Jimenez, J. Carracedo, I.C. Barneto, R. Ramirez, and E. Aranda. 1998. Serum levels of intercellular adhesion molecule 1 (ICAM-1) in patients with colorectal cancer: inhibitory effect on cytotoxicity. Eur. J. Cancer. 34: 394-398.

Sanderson, M.P., C.A. Abbott, H. Tada, M. Seno, P.J. Dempsey, and A.J. Dunbar. 2006. Hydrogen peroxide and endothelin-1 are novel activators of betacellulin ectodomain shedding. J. Cell. Biochem. 99: 609-623.

Sandgren, E.P., N.C. Luetteke, R.D. Palmiter, R.L. Brinster, and D.C. Lee. 1990. Overexpression of TGF alpha in transgenic mice: induction of epithelial hyperplasia, pancreatic metaplasia, and carcinoma of the breast. Cell. 61: 1121-1135.

Sato, F., D. Shibata, N. Harpaz, Y. Xu, J. Yin, Y. Mori, S. Wang, A. Olaru, E. Deacu, F.M. Selaru, M.C. Kimos, P. Hytiroglou, J. Young, B. Leggett, A.F. Gazdar, S. Toyooka, J.M. Abraham, and S.J. Meltzer. 2002. Aberrant methylation of the HPP1 gene in ulcerative colitis-associated colorectal carcinoma. Cancer Res. 62: 6820-6822.

Schäfer, B., B. Marg, A. Gschwind, and A. Ullrich. 2004. Distinct ADAM metalloproteinases regulate G protein-coupled receptor-induced cell proliferation and survival. J. Biol. Chem. 279: 47929-47938.

Scheller, J., A. Chalaris, C. Garbers, and S. Rose-John. 2011. ADAM17: a molecular switch to control inflammation and tissue regeneration. Trends Immunol. 32: 380-387.

Schlomann, U., D. Wildeboer, A. Webster, O. Antropova, D. Zeuschner, C.G. Knight, A.J. Docherty, M. Lambert, L. Skelton, H. Jockusch, and J.W. Bartsch. 2002.

The metalloprotease disintegrin ADAM8. Processing by autocatalysis is required for proteolytic activity and cell adhesion. J. Biol. Chem. 277: 48210-48219.

Schulmann, K., A. Sterian, A. Berki, J. Yin, F. Sato, Y. Xu, A. Olaru, S. Wang, Y. Mori, E. Deacu, J. Hamilton, T. Kan, M.J. Krasna, D.G. Beer, M.S. Pepe, J.M. Abraham, Z. Feng, W. Schmiegel, B.D. Greenwald, and S.J. Meltzer. 2005. Inactivation of p16, RUNX3, and HPP1 occurs early in Barrett's-associated neoplastic progression and predicts progression risk. Oncogene. 24: 4138-4148.

Schutz, A., W. Hartig, M. Wobus, J. Grosche, C. Wittekind, and G. Aust. 2005. Expression of ADAM15 in lung carcinomas. Virchows Arch. 446: 421-429.

Seals, D.F. and S.A. Courtneidge. 2003. The ADAMs family of metalloproteases: multidomain proteins with multiple functions. Genes Dev. 17: 7-30.

Segatto, O., C.R. King, J.H. Pierce, P.P. Di Fiore, and S.A. Aaronson. 1988. Different structural alterations upregulate in vitro tyrosine kinase activity and transforming potency of the erbB-2 gene. Mol. Cell. Biol. 8: 5570-5574.

Selamat, S.A., J.S. Galler, A.D. Joshi, M.N. Fyfe, M. Campan, K.D. Siegmund, K.M. Kerr, and I.A. Laird-Offringa. 2011. DNA methylation changes in atypical adenomatous hyperplasia, adenocarcinoma in situ, and lung adenocarcinoma. PLoS One. 6: e21443.

Shariat, S.F., K. Matsumoto, R. Casella, W. Jian, and S.P. Lerner. 2005. Urinary levels of soluble e-cadherin in the detection of transitional cell carcinoma of the urinary bladder. Eur. Urol. 48: 69-76.

Shibata, D.M., F. Sato, Y. Mori, K. Perry, J. Yin, S. Wang, Y. Xu, A. Olaru, F. Selaru, K. Spring, J. Young, J.M. Abraham, and S.J. Meltzer. 2002. Hypermethylation of HPP1 is associated with hMLH1 hypermethylation in gastric adenocarcinomas. Cancer Res. 62: 5637-5640.

Shirahama, S., F. Furukawa, H. Wakita, and M. Takigawa. 1996. E- and P-cadherin expression in tumor tissues and soluble E-cadherin levels in sera of patients with skin cancer. J. Dermatol. Sci. 13: 30-36.

Sibilia, M. and E.F. Wagner. 1995. Strain-dependent epithelial defects in mice lacking the EGF receptor. Science. 269: 234-238.

Siegel, D.A., P. Davies, K. Dobrenis, and M. Huang. 2006. Tomoregulin-2 is found extensively in plaques in Alzheimer's disease brain. J. Neurochem. 98: 34-44.

Siitonen, S.M., J.T. Kononen, H.J. Helin, I.S. Rantala, K.A. Holli, and J.J. Isola. 1996. Reduced E-cadherin expression is associated with invasiveness and unfavorable prognosis in breast cancer. Am. J. Clin. Pathol. 105: 394-402.

Singh, R.J., J.C. Mason, E.A. Lidington, D.R. Edwards, R.K. Nuttall, R. Khokha, V. Knauper, G. Murphy, and J. Gavrilovic. 2005. Cytokine stimulated vascular cell adhesion molecule-1 (VCAM-1) ectodomain release is regulated by TIMP-3. Cardiovasc. Res. 67: 39-49.

Sinn, H.P., K.H. Heider, P. Skroch-Angel, G. von Minckwitz, M. Kaufmann, P. Herrlich, and H. Ponta. 1995. Human mammary carcinomas express homologues of rat metastasis-associated variants of CD44. Breast Cancer Res. Treat. 36: 307-313.

Sisodia, S.S. 1992. Secretion of the beta-amyloid precursor protein. Ann. N. Y. Acad. Sci. 674: 53-57.

Six, E., D. Ndiaye, Y. Laabi, C. Brou, N. Gupta-Rossi, A. Israel, and F. Logeat. 2003. The Notch ligand Delta1 is sequentially cleaved by an ADAM protease and gamma-secretase. Proc. Natl. Acad. Sci. U. S. A. 100: 7638-7643.

Slamon, D.J., G.M. Clark, S.G. Wong, W.J. Levin, A. Ullrich, and W.L. McGuire. 1987. Human breast cancer: correlation of relapse and survival with amplification of the HER-2/neu oncogene. Science. 235: 177-182.

Smalley, D.M. and K. Ley. 2005. L-selectin: mechanisms and physiological significance of ectodomain cleavage. J. Cell. Mol. Med. 9: 255-266.

Smith, E., N.J. De Young, S.J. Pavey, N.K. Hayward, D.J. Nancarrow, D.C. Whiteman, B.M. Smithers, A.R. Ruszkiewicz, A.D. Clouston, D.C. Gotley, P.G. Devitt, G.G. Jamieson, and P.A. Drew. 2008. Similarity of aberrant DNA methylation in Barrett's esophagus and esophageal adenocarcinoma. Mol. Cancer. 7: 75.

Smith, K.M., A. Gaultier, H. Cousin, D. Alfandari, J.M. White, and D.W. DeSimone. 2002. The cysteine-rich domain regulates ADAM protease function in vivo. J. Cell Biol. 159: 893-902.

Soond, S.M., B. Everson, D.W. Riches, and G. Murphy. 2005. ERK-mediated phosphorylation of Thr735 in TNFalpha-converting enzyme and its potential role in TACE protein trafficking. J. Cell. Sci. 118: 2371-2380.

Sreekumar, A., L.M. Poisson, T.M. Rajendiran, A.P. Khan, Q. Cao, J. Yu, B. Laxman, R. Mehra, R.J. Lonigro, Y. Li, M.K. Nyati, A. Ahsan, S. Kalyana-Sundaram, B. Han, X. Cao, J. Byun, G.S. Omenn, D. Ghosh, S. Pennathur, D.C. Alexander, A. Berger, J.R. Shuster, J.T. Wei, S. Varambally, C. Beecher, and A.M. Chinnaiyan. 2009. Metabolomic profiles delineate potential role for sarcosine in prostate cancer progression. Nature. 457: 910-914.

Stamenkovic, I. and Q. Yu. 2009. Shedding light on proteolytic cleavage of CD44: the responsible sheddase and functional significance of shedding. J. Invest. Dermatol. 129: 1321-1324.

Steinberg, M.S. and P.M. McNutt. 1999. Cadherins and their connections: adhesion junctions have broader functions. Curr. Opin. Cell Biol. 11: 554-560.

Sugden, P.H. and A. Clerk. 1997. Regulation of the ERK subgroup of MAP kinase cascades through G protein-coupled receptors. Cell. Signal. 9: 337-351.

Suzuki, M., H. Shigematsu, D.S. Shames, N. Sunaga, T. Takahashi, N. Shivapurkar, T. Iizasa, E.P. Frenkel, J.D. Minna, T. Fujisawa, and A.F. Gazdar. 2005. DNA methylation-associated inactivation of TGFbeta-related genes DRM/Gremlin, RUNX3, and HPP1 in human cancers. Br. J. Cancer. 93: 1029-1037.

Swendeman, S., K. Mendelson, G. Weskamp, K. Horiuchi, U. Deutsch, P. Scherle, A. Hooper, S. Rafii, and C.P. Blobel. 2008. VEGF-A stimulates ADAM17-dependent shedding of VEGFR2 and crosstalk between VEGFR2 and ERK signaling. Circ. Res. 103: 916-918.

Sy, M.S., Y.J. Guo, and I. Stamenkovic. 1992. Inhibition of tumor growth in vivo with a soluble CD44-immunoglobulin fusion protein. J. Exp. Med. 176: 623-627.

Syrigos, K.N., T. Krausz, J. Waxman, H. Pandha, G. Rowlinson-Busza, J. Verne, A.A. Epenetos, and M. Pignatelli. 1995. E-cadherin expression in bladder cancer using formalin-fixed, paraffin-embedded tissues: correlation with histopathological grade, tumour stage and survival. Int. J. Cancer. 64: 367-370.

Tai, P. and M. Ascoli. 2011. Reactive oxygen species (ROS) play a critical role in the cAMP-induced activation of Ras and the phosphorylation of ERK1/2 in Leydig cells. Mol. Endocrinol. 25: 885-893.

Takamune, Y., T. Ikebe, O. Nagano, H. Nakayama, K. Ota, T. Obayashi, H. Saya, and M. Shinohara. 2007. ADAM-17 associated with CD44 cleavage and metastasis in oral squamous cell carcinoma. Virchows Arch. 450: 169-177.

Takeda, S. 2009. Three-dimensional domain architecture of the ADAM family proteinases. Semin. Cell Dev. Biol. 20: 146-152.

Takeno, S., T. Noguchi, S. Fumoto, Y. Kimura, T. Shibata, and K. Kawahara. 2004. E-cadherin expression in patients with esophageal squamous cell carcinoma: promoter hypermethylation, Snail overexpression, and clinicopathologic implications. Am. J. Clin. Pathol. 122: 78-84.

Takenobu, H., A. Yamazaki, M. Hirata, T. Umata, and E. Mekada. 2003. The stress- and inflammatory cytokine-induced ectodomain shedding of heparin-binding epidermal growth factor-like growth factor is mediated by p38 MAPK, distinct from the 12-O-tetradecanoylphorbol-13-acetate- and lysophosphatidic acid-induced signaling cascades. J. Biol. Chem. 278: 17255-17262.

Tamaishi, N., M. Tsukimoto, A. Kitami, and S. Kojima. 2011. P2Y6 receptors and ADAM17 mediate low-dose gamma-ray-induced focus formation (activation) of EGF receptor. Radiat. Res. 175: 193-200.

Tanaka, M., D. Nanba, S. Mori, F. Shiba, H. Ishiguro, K. Yoshino, N. Matsuura, and S. Higashiyama. 2004. ADAM binding protein Eve-1 is required for ectodomain shedding of epidermal growth factor receptor ligands. J. Biol. Chem. 279: 41950-41959.

Tanaka, Y., S. Miyamoto, S.O. Suzuki, E. Oki, H. Yagi, K. Sonoda, A. Yamazaki, H. Mizushima, Y. Maehara, E. Mekada, and H. Nakano. 2005. Clinical significance of heparin-binding epidermal growth factor-like growth factor and a disintegrin and metalloprotease 17 expression in human ovarian cancer. Clin. Cancer Res. 11: 4783-4792.

Tannapfel, A., K. Anhalt, P. Hausermann, F. Sommerer, M. Benicke, D. Uhlmann, H. Witzigmann, J. Hauss, and C. Wittekind. 2003. Identification of novel proteins associated with hepatocellular carcinomas using protein microarrays. J. Pathol. 201: 238-249.

Tao, K., N. Qian, Y. Tang, Z. Ti, W. Song, D. Cao, and K. Dou. 2010. Increased expression of a disintegrin and metalloprotease-9 in hepatocellular carcinoma: implications for tumor progression and prognosis. Jpn. J. Clin. Oncol. 40: 645-651.

Tellier, E., M. Canault, L. Rebsomen, B. Bonardo, I. Juhan-Vague, G. Nalbone, and F. Peiretti. 2006. The shedding activity of ADAM17 is sequestered in lipid rafts. Exp. Cell Res. 312: 3969-3980.

Terol, M.J., M. Tormo, J.A. Martinez-Climent, I. Marugan, I. Benet, A. Ferrandez, A. Teruel, R. Ferrer, and J. Garcia-Conde. 2003. Soluble intercellular adhesion molecule-1 (s-ICAM-1/s-CD54) in diffuse large B-cell lymphoma: association with clinical characteristics and outcome. Ann. Oncol. 14: 467-474.

Thabard, W., M. Collette, R. Bataille, and M. Amiot. 2001. Protein kinase C delta and eta isoenzymes control the shedding of the interleukin 6 receptor alpha in myeloma cells. Biochem. J. 358: 193-200.

Toiyama, Y., C. Miki, Y. Inoue, Y. Okugawa, Y. Koike, T. Yokoe, K. Tanaka, and M. Kusunoki. 2008. Soluble intercellular adhesion molecule-1 as a prognostic marker for stage II colorectal cancer patients. Ann. Surg. Oncol. 15: 1617-1624.

Tsakadze, N.L., S.D. Sithu, U. Sen, W.R. English, G. Murphy, and S.E. D'Souza. 2006. Tumor necrosis factor-alpha-converting enzyme (TACE/ADAM-17) mediates the ectodomain cleavage of intercellular adhesion molecule-1 (ICAM-1). J. Biol. Chem. 281: 3157-3164.

Tśe, C., D. Brault, J. Gligorov, M. Antoine, R. Neumann, J.P. Lotz, and J. Capeau. 2005. Evaluation of the quantitative analytical methods real-time PCR for HER-2 gene quantification and ELISA of serum HER-2 protein and comparison with fluorescence in situ hybridization and immunohistochemistry for determining HER-2 status in breast cancer patients. Clin. Chem. 51: 1093-1101.

Tśe, C., A.S. Gauchez, W. Jacot, and P.J. Lamy. 2012. HER2 shedding and serum HER2 extracellular domain: Biology and clinical utility in breast cancer. Cancer Treat. Rev. 38(2): 133-142.

Tsigris, C., A.J. Karayiannakis, A. Zbar, K.N. Syrigos, N. Baibas, T. Diamantis, and D. Alexiou. 2002. Clinical significance of serum and urinary c-erbB-2 levels in colorectal cancer. Cancer Lett. 184: 215-222.

Tsunoda, S., E. Smith, N.J. De Young, X. Wang, Z.Q. Tian, J.F. Liu, G.G. Jamieson, and P.A. Drew. 2009. Methylation of CLDN6, FBN2, RBP1, RBP4, TFPI2, and TMEFF2 in esophageal squamous cell carcinoma. Oncol. Rep. 21: 1067-1073.

Uchida, T., K. Wada, T. Akamatsu, M. Yonezawa, H. Noguchi, A. Mizoguchi, M. Kasuga, and C. Sakamoto. 1999. A novel epidermal growth factor-like molecule containing two follistatin modules stimulates tyrosine phosphorylation of erbB-4 in MKN28 gastric cancer cells. Biochem. Biophys. Res. Commun. 266: 593-602.

Umbas, R., W.B. Isaacs, P.P. Bringuier, H.E. Schaafsma, H.F. Karthaus, G.O. Oosterhof, F.M. Debruyne, and J.A. Schalken. 1994. Decreased E-cadherin expression is associated with poor prognosis in patients with prostate cancer. Cancer Res. 54: 3929-3933.

Valkovskaya, N., H. Kayed, K. Felix, D. Hartmann, N.A. Giese, S.P. Osinsky, H. Friess, and J. Kleeff. 2007. ADAM8 expression is associated with increased invasiveness and reduced patient survival in pancreatic cancer. J. Cell. Mol. Med. 11: 1162-1174.

van Roy, F. and G. Berx. 2008. The cell-cell adhesion molecule E-cadherin. Cell Mol. Life Sci. 65: 3756-3788.

Van Schaeybroeck, S., J.N. Kyula, A. Fenton, C.S. Fenning, T. Sasazuki, S. Shirasawa, D.B. Longley, and P.G. Johnston. 2011. Oncogenic Kras promotes chemotherapy-induced growth factor shedding via ADAM17. Cancer Res. 71: 1071-1080.

van Tetering, G., P. van Diest, I. Verlaan, E. van der Wall, R. Kopan, and M. Vooijs. 2009. Metalloprotease ADAM10 is required for Notchl site 2 cleavage. J. Biol. Chem. 284: 31018-31027.

Van Wart, H.E. and H. Birkedal-Hansen. 1990. The cysteine switch: a principle of regulation of metalloproteinase activity with potential applicability to the entire matrix metalloproteinase gene family. Proc. Natl. Acad. Sci. U. S. A. 87: 5578-5582.

Velikova, G., R.E. Banks, A. Gearing, I. Hemingway, M.A. Forbes, S.R. Preston, M. Jones, J. Wyatt, K. Miller, U. Ward, J. Al-Maskatti, S.M. Singh, N.S. Ambrose, J.N. Primrose, and P.J. Selby. 1997. Circulating soluble adhesion molecules E-cadherin, E-selectin, intercellular adhesion molecule-1 (ICAM-1) and vascular cell adhesion molecule-1 (VCAM-1) in patients with gastric cancer. Br. J. Cancer. 76: 1398-1404.

Vincent, L., D.K. Jin, M.A. Karajannis, K. Shido, A.T. Hooper, W.K. Rashbaum, B. Pytowski, Y. Wu, D.J. Hicklin, Z. Zhu, P. Bohlen, R. Niesvizky, and S. Rafii. 2005. Fetal stromal-dependent paracrine and intracrine vascular endothelial growth factor-a/vascular endothelial growth factor receptor-1 signaling promotes proliferation and motility of human primary myeloma cells. Cancer Res. 65: 3185-3192.

Vleminckx, K., L. Vakaet Jr, M. Mareel, W. Fiers, and F. van Roy. 1991. Genetic manipulation of E-cadherin expression by epithelial tumor cells reveals an invasion suppressor role. Cell. 66: 107-119.

Waldhauer, I., D. Goehlsdorf, F. Gieseke, T. Weinschenk, M. Wittenbrink, A. Ludwig, S. Stevanovic, H.G. Rammensee, and A. Steinle. 2008. Tumor-associated MICA is shed by ADAM proteases. Cancer Res. 68: 6368-6376.

Wang, E.S., J. Teruya-Feldstein, Y. Wu, Z. Zhu, D.J. Hicklin, and M.A. Moore. 2004. Targeting autocrine and paracrine VEGF receptor pathways inhibits human lymphoma xenografts in vivo. Blood. 104: 2893-2902.

Wang, X., T. Oka, F.L. Chow, S.B. Cooper, J. Odenbach, G.D. Lopaschuk, Z. Kassiri, and C. Fernandez-Patron. 2009b. Tumor necrosis factor-alpha-converting enzyme is a key regulator of agonist-induced cardiac hypertrophy and fibrosis. Hypertension. 54: 575-582.

Wang, Y., A.H. Herrera, Y. Li, K.K. Belani, and B. Walcheck. 2009a. Regulation of mature ADAM17 by redox agents for L-selectin shedding. J. Immunol. 182: 2449-2457.

Wang, Y., A.C. Zhang, Z. Ni, A. Herrera, and B. Walcheck. 2010. ADAM17 activity and other mechanisms of soluble L-selectin production during death receptor-induced leukocyte apoptosis. J. Immunol. 184: 4447-4454.

Wang, Y.Y., Z.Y. Ye, L. Li, Z.S. Zhao, Q.S. Shao, and H.Q. Tao. 2011. ADAM 10 is associated with gastric cancer progression and prognosis of patients. J. Surg. Oncol. 103: 116-123.

Weiss, J.V., S. Klein-Scory, S. Kubler, A. Reinacher-Schick, I. Stricker, W. Schmiegel, and I. Schwarte-Waldhoff. 2011. Soluble E-cadherin as a serum biomarker candidate: elevated levels in patients with late-stage colorectal carcinoma and FAP. Int. J. Cancer. 128: 1384-1392.

Weskamp, G., J. Schlondorff, L. Lum, J.D. Becherer, T.W. Kim, P. Saftig, D. Hartmann, G. Murphy, and C.P. Blobel. 2004. Evidence for a critical role of the tumor necrosis factor alpha convertase (TACE) in ectodomain shedding of the p75 neurotrophin receptor (p75NTR). J. Biol. Chem. 279: 4241-4249.

Wheeler, D.L., K.J. Ness, T.D. Oberley, and A.K. Verma. 2003. Protein kinase Cepsilon is linked to 12-O-tetradecanoylphorbol-13-acetate-induced tumor necrosis factor-alpha ectodomain shedding and the development of metastatic squamous cell carcinoma in protein kinase Cepsilon transgenic mice. Cancer Res. 63: 6547-6555.

White, J.M. 2003. ADAMs: modulators of cell-cell and cell-matrix interactions. Curr. Opin. Cell Biol. 15: 598-606.

Wildeboer, D., S. Naus, Q.X. Amy Sang, J.W. Bartsch, and A. Pagenstecher. 2006. Metalloproteinase disintegrins ADAM8 and ADAM19 are highly regulated in human primary brain tumors and their expression levels and activities are associated with invasiveness. J. Neuropathol. Exp. Neurol. 65: 516-527.

Willems, S.H., C.J. Tape, P.L. Stanley, N.A. Taylor, I.G. Mills, D.E. Neal, J. McCafferty, and G. Murphy. 2010. Thiol isomerases negatively regulate the cellular shedding activity of ADAM17. Biochem. J. 428: 439-450.

Witzel, I., S. Loibl, G. von Minckwitz, C. Mundhenke, J. Huober, C. Hanusch, S. Henschen, M. Hauschild, T. Lantzsch, H. Tesch, K. Latos, M. Just, J. Hilfrich, J. Barinoff, C.Z. Eulenburg, M. Roller, M. Untch, and V. Muller. 2010. Monitoring serum HER2 levels during neoadjuvant trastuzumab treatment within the GeparQuattro trial. Breast Cancer Res. Treat. 123: 437-445.

Wolfsberg, T.G., P. Primakoff, D.G. Myles, and J.M. White. 1995. ADAM, a novel family of membrane proteins containing A Disintegrin And Metalloprotease domain: multipotential functions in cell-cell and cell-matrix interactions. J. Cell Biol. 131: 275-278.

Wong, S.T., L.F. Winchell, B.K. McCune, H.S. Earp, J. Teixido, J. Massague, B. Herman, and D.C. Lee. 1989. The TGF-alpha precursor expressed on the cell surface binds to the EGF receptor on adjacent cells, leading to signal transduction. Cell. 56: 495-506.

Wright, C.M., J.E. Larsen, N.K. Hayward, M.U. Martins, M.E. Tan, M.R. Davidson, S.M. Savarimuthu, R.E. McLachlan, L.H. Passmore, M.N. Windsor, B.E. Clarke, E.E. Duhig, I.A. Yang, R.V. Bowman, and K.M. Fong. 2010. ADAM28: a potential oncogene involved in asbestos-related lung adenocarcinomas. Genes Chromosomes Cancer. 49: 688-698.

Wu, E., P.I. Croucher, and N. McKie. 1997. Expression of members of the novel membrane linked metalloproteinase family ADAM in cells derived from a range of haematological malignancies. Biochem. Biophys. Res. Commun. 235: 437-442.

Wu, X., A. Blanck, G. Norstedt, L. Sahlin, and A. Flores-Morales. 2002. Identification of genes with higher expression in human uterine leiomyomas than in the corresponding myometrium. Mol. Hum. Reprod. 8: 246-254.

Xu, P. and R. Derynck. 2010. Direct activation of TACE-mediated ectodomain shedding by p38 MAP kinase regulates EGF receptor-dependent cell proliferation. Mol. Cell. 37: 551-566.

Xu, Q., X. Liu, Y. Cai, Y. Yu, and W. Chen. 2010. RNAi-mediated ADAM9 gene silencing inhibits metastasis of adenoid cystic carcinoma cells. Tumour Biol. 31: 217-224.

Xu, Q., X. Liu, W. Chen, and Z. Zhang. 2010. Inhibiting adenoid cystic carcinoma cells growth and metastasis by blocking the expression of ADAM 10 using RNA interference. J. Transl. Med. 8: 136.

Yamada, D., K. Ohuchida, K. Mizumoto, S. Ohhashi, J. Yu, T. Egami, H. Fujita, E. Nagai, and M. Tanaka. 2007. Increased expression of ADAM 9 and ADAM 15 mRNA in pancreatic cancer. Anticancer Res. 27: 793-799.

Yamauchi, H., A. O'Neill, R. Gelman, W. Carney, D.Y. Tenney, S. Hosch, and D.F. Hayes. 1997. Prediction of response to antiestrogen therapy in advanced breast cancer patients by pretreatment circulating levels of extracellular domain of the HER-2/c-neu protein. J. Clin. Oncol. 15: 2518-2525.

Yang, J.J., L.Y. Cho, S.H. Ma, K.P. Ko, A. Shin, B.Y. Choi, D.S. Han, K.S. Song, Y.S. Kim, S.H. Chang, H.R. Shin, D. Kang, K.Y. Yoo, and S.K. Park. 2011. Oncogenic CagA promotes gastric cancer risk via activating ERK signaling pathways: a nested case-control study. PLoS One. 6: e21155.

Yang, M., Y. Li, K. Chilukuri, O.A. Brady, M.I. Boulos, J.C. Kappes, and D.S. Galileo. 2011. L1 stimulation of human glioma cell motility correlates with FAK activation. J. Neurooncol. 105: 27-44.

Yang, M.H., P.Y. Chu, S.C. Chen, T.W. Chung, W.C. Chen, L.B. Tan, W.C. Kan, H.Y. Wang, S.B. Su, and Y.C. Tyan. 2011. Characterization of ADAM28 as a biomarker of bladder transitional cell carcinomas by urinary proteome analysis. Biochem. Biophys. Res. Commun. 411: 714-720.

Yasuda, H., S. Hirata, K. Inoue, H. Mashima, H. Ohnishi, and M. Yoshiba. 2007. Involvement of membrane-type bile acid receptor M-BAR/TGR5 in bile acid-induced activation of epidermal growth factor receptor and mitogen-activated protein kinases in gastric carcinoma cells. Biochem. Biophys. Res. Commun. 354: 154-159.

Young, J., K.G. Biden, L.A. Simms, P. Huggard, R. Karamatic, H.J. Eyre, G.R. Sutherland, N. Herath, M. Barker, G.J. Anderson, D.R. Fitzpatrick, G.A. Ramm, J.R. Jass, and B.A. Leggett. 2001. HPP1: a transmembrane protein-encoding gene commonly methylated in colorectal polyps and cancers. Proc. Natl. Acad. Sci. U. S. A. 98: 265-270.

Yu, Q., B.P. Toole, and I. Stamenkovic. 1997. Induction of apoptosis of metastatic mammary carcinoma cells in vivo by disruption of tumor cell surface CD44 function. J. Exp. Med. 186: 1985-1996.

Zawadzki, V., A. Perschl, M. Rosel, A. Hekele, and M. Zoller. 1998. Blockade of metastasis formation by CD44-receptor globulin. Int. J. Cancer. 75: 919-924.

Zhang, Q., S.M. Thomas, V.W. Lui, S. Xi, J.M. Siegfried, H. Fan, T.E. Smithgall, G.B. Mills, and J.R. Grandis. 2006. Phosphorylation of TNF-alpha converting enzyme by gastrin-releasing peptide induces amphiregulin release and EGF receptor activation. Proc. Natl. Acad. Sci. U.S.A. 103: 6901-6906.

Zhang, X.P., T. Kamata, K. Yokoyama, W. Puzon-McLaughlin, and Y. Takada. 1998. Specific interaction of the recombinant disintegrin-like domain of MDC-15 (metargidin, ADAM-15) with integrin alphavbeta3. J. Biol. Chem. 273: 7345-7350.

Zhang, Z., B.D. Humphreys, and J.V. Bonventre. 2007. Shedding of the urinary biomarker kidney injury molecule-1 (KIM-1) is regulated by MAP kinases and juxtamembrane region. J. Am. Soc. Nephrol. 18: 2704-2714.

Zhang, Z., P. Oliver, J.R. Lancaster Jr, P.O. Schwarzenberger, M.S. Joshi, J. Cork, and J.K. Kolls. 2001. Reactive oxygen species mediate tumor necrosis factor alpha-converting, enzyme-dependent ectodomain shedding induced by phorbol myristate acetate. FASEB J. 15: 303-305.

Zhao, B.J., S.N. Tan, Y. Cui, D.G. Sun, and X. Ma. 2008. Aberrant promoter methylation of the TPEF gene in esophageal squamous cell carcinoma. Dis. Esophagus. 21: 582-588.

Zhao, X.Y., D. Schneider, S.L. Biroc, R. Parry, B. Alicke, P. Toy, J.A. Xuan, C. Sakamoto, K. Wada, M. Schulze, B. Muller-Tiemann, G. Parry, and H. Dinter. 2005. Targeting tomoregulin for radioimmunotherapy of prostate cancer. Cancer Res. 65: 2846-2853.

Zheng, Y., J. Schlondorff, and C.P. Blobel. 2002. Evidence for regulation of the tumor necrosis factor alpha-convertase (TACE) by protein-tyrosine phosphatase PTPH1. J. Biol. Chem. 277: 42463-42470.

Zhou, B.B., M. Peyton, B. He, C. Liu, L. Girard, E. Caudler, Y. Lo, F. Baribaud, I. Mikami, N. Reguart, G. Yang, Y. Li, W. Yao, K. Vaddi, A.F. Gazdar, S.M. Friedman,

D.M. Jablons, R.C. Newton, J.S. Fridman, J.D. Minna, and P.A. Scherle. 2006. Targeting ADAM-mediated ligand cleavage to inhibit HER3 and EGFR pathways in non-small cell lung cancer. Cancer. Cell. 10: 39-50.

Zhou, M., R. Graham, G. Russell, and P.I. Croucher. 2001. MDC-9 (ADAM-9/Meltrin gamma) functions as an adhesion molecule by binding the alpha(v)beta(5) integrin. Biochem. Biophys. Res. Commun. 280: 574-580.

Zigrino, P., C. Mauch, J.W. Fox, and R. Nischt. 2005. Adam-9 expression and regulation in human skin melanoma and melanoma cell lines. Int. J. Cancer. 116: 853-859.

Zöller, M. 2011. CD44: can a cancer-initiating cell profit from an abundantly expressed molecule? Nat. Rev. Cancer. 11: 254-267.

CHAPTER 6

The BRCT Signaling 'Landscape': Emerging Concepts and Implications in Tumorigenesis and Cancer Diagnostics

Dimitris Mastellos[1] and George Nounesis[2*]

[1]Protein Chemistry Laboratory, IRRP, National Centre for Scientific Research Demokritos, 153 10 Aghia Paraskevi, Greece

[2]Biomolecular Physics Laboratory, IRRP, National Centre for Scientific Research Demokritos, 153 10 Aghia Paraskevi, Greece

ABSTRACT

This chapter provides a comprehensive overview of the complex signaling landscape guided by BRCT-mediated protein interactions and discusses their implications in multiple cellular processes that regulate genomic stability, the DNA damage response and tumorigenesis. It will focus primarily on discussing critical perspectives of the biology of BRCT signaling modules, with emphasis on their phosphopeptide binding capacity and involvement in the assembly of high-order multiprotein complexes that mediate ionizing radiation (IR)-induced DNA repair. The main objective of this chapter is to integrate the fascinating biology lying behind BRCT signaling modules and their versatile protein interactions with cancer predisposition and diagnostics. Particular emphasis will also be placed on emerging concepts that elucidate structure-function elements of the BRCT domains found in the tumor suppressors BRCA1 and BARD1. The impact of cancer-linked BRCT missense mutations on the overall stability and function of this signaling module will be integrated with 3-D structure-guided and stability-based biophysical approaches. Through the integrative, multi-platform evaluation of the pathogenicity of BRCT-missense variants isolated from cancer patients, this chapter will also provide a conceptual framework for the selection of novel cancer

**Corresponding author*: E-mail: nounesis@rrp.demokritos.gr

biomarkers that might facilitate the rational development and testing of novel, patient-oriented cancer diagnostic platforms based on BRCT interactions.

THE 'BUILDING BLOCKS' OF THE DNA DAMAGE RESPONSE SIGNALING APPARATUS

In an era that has witnessed the impressive growth of high-throughput genomic and proteomic approaches in cancer research it is becoming ever more appreciable that carcinogenesis reflects a dynamic, multi-layered process of cellular transformation that is highly stochastic in nature (Balazsi et al. 2011; Fletcher and Houlston 2010). The transition of a quiescent somatic cell to malignancy and uncontrollable cell growth does not appear to abide by the rules of a deterministic model that depends on the linear succession of molecular events and on a fixed set of cancer-triggering mutations in single genes (Michor et al. 2011; Fisher 1958; Fletcher and Houlston 2010). Indeed, neoplastic transformation is a multifaceted process influenced by a multitude of genetic and epigenetic-environmental factors, including elements of the tumor microenvironment (e.g. angiogenic signals, nutrients) (Chouaib et al. 2010; Hu and Polyak 2008), the metabolic/redox state of precancerous cells (Cairns et al. 2011), the extent of immune surveillance exerted on tumor cells (Gonda et al. 2009) and a broad spectrum of stroma-derived agents that either promote or suppress the outgrowth of malignant cells (Kryczek et al. 2007; Chouaib et al. 2010).

In this context, any perturbations that can destabilize the molecular circuitry that is responsible for maintaining cellular integrity and securing finely regulated cell growth may promote malignant transformation of otherwise quiescent, terminally differentiated cells. The subversion of genomic stability is a common denominator in tumorigenesis and a 'hallmark' in multiple pathways leading to malignant transformation and cancer growth (Hanahan and Weinberg 2011; Shiloh 2003). An essential 'sentinel' of genomic stability and tumor suppression, positioned at the interface of cell cycle regulation and transcriptional activation, is the **DNA damage checkpoint response pathway** (Bakkenist and Kastan 2004).

The *DNA damage checkpoint response* is a pivotal 'gatekeeper' of genomic integrity and cellular homeostasis that is rapidly activated upon exposure to several environmental 'stress' cues and genotoxic agents, such as *ionizing radiation* and *chemical carcinogens* (e.g. DNA-alkylating agents) (Squatrito and Holland 2011). It culminates through the orchestrated assembly of high-order multiproteinaceous complexes that mediate chromatin remodeling to increase accessibility to areas of DNA damage (e.g. DSBs) (Xu and Price 2011) and catalyze the downstream transcriptional activation or repression of diverse gene targets that regulate cell cycle progression and DNA repair and apoptosis (Costes et al. 2010; Reinhardt and Yaffe 2009). The DNA damage response depends on the concerted activity of protein serine/threonine kinases and modular phosphoserine/threonine-binding domains to relay the damage signal and recruit repair proteins from various intracellular compartments (Reinhardt and Yaffe 2009).

Modular protein domains that recognize specific phosphopeptide motifs in target proteins and also protein kinases/phosphatases that modify target substrates by reversible phosphorylation are the key players in the molecular network regulating the DNA damage response (Mohammad and Yaffe 2009). A crucial step in the activation of an effective DNA repair circuit is the assembly of multi-protein complexes and molecular scaffolds that will forge high-order chromatin remodeling complexes at ionizing radiation-induced DNA damage foci (IRIF) (Bakkenist and Kastan 2003; Rogakou et al. 1998). Among the modular protein domains that have been recognized to play an important role in the DNA repair process are three families of evolutionarily conserved protein domains: the **14-3-3 proteins**, the **FHA (*Forkhead associated*)** and **BRCT (*Breast Cancer 1 carboxyl- terminal*)** domains (Mohammad and Yaffe 2009). These protein modules act in a highly coordinated manner to regulate the DNA damage response and cell cycle checkpoint by recruiting protein complexes that assemble via reversible phosphopeptide recognition and thereby modulate transcriptional activation and chromatin remodeling at foci of IR-induced DNA damage (Mailand et al. 2007; Qi and Martinez 2003).

These phosphopeptide-binding protein modules act as efficient molecular 'transducers' in the cytoplasm-nucleus interface. They relay the activation signal produced by protein kinases in response to genotoxic damage (e.g. IR-induced DNA breaks) to downstream effectors and phosphopeptide signaling complexes that will shape the appropriate gene expression signature underlying cell cycle checkpoint control and DNA repair (Li et al. 2000). This is achieved through the specific recognition of short aminoacid sequence motifs that surround a central phosphorylated Serine or Threonine residue in several target proteins (Manke et al. 2003).

Phosphoserine/threonine-binding domain-containing proteins affect multiple levels of the DNA repair molecular circuitry. They play instrumental roles in establishing the DNA damage checkpoints in the G1, S and G2 phases of the cell cycle, in safeguarding checkpoint maintenance and most importantly in guiding the assembly of multi-protein signaling complexes at the foci of DNA damage (DNA double strand breaks) and also mobilizing the large protein complexes that will modulate the transcriptional activation of gatekeeping, 'genome guardian' genes.

THE SIGNALING 'LANDSCAPE' OF DNA DAMAGE AND REPAIR WITHIN THE CELL

There are two classes of DNA damage–activated protein kinases, depending on the sequence of the peptide recognition motif found on their target substrates (Abraham 2001). Positioned upstream in the DNA damage response signaling cascade are the phosphatidylinositol-3-OH-kinase-like (**PI(3)KK**) kinases ATM, ATR and DNA-PK (Shiloh 2003; Bakkenist and Kastan 2004). These kinases preferentially phosphorylate target motifs containing serines/threonines immediately followed by a glutamine residue (Abraham 2001). The second major class of DNA damage-induced kinases includes the cell cycle checkpoint kinases ChK1, Chk2 and ChK3(MK2) (Chen et al. 2006; Manke et al. 2005; Reinhardt et al. 2007; Li et al. 2008). These two classes of specialized kinases are responsible for the polarization of the DNA damage response and the localization of different complexes to discrete cell compartments during the DNA damage response and repair. They act upon protein substrates that participate in the assembly of distinct macromolecular complexes in the DNA damage response-signaling landscape. ATM-dependent phosphorylated-targets participate in the assembly of protein complexes at foci of DNA damage (Yu et al. 2003; Rodriguez et al. 2003) while cell cycle checkpoint control is mediated mainly by Chk1/2/MK2-responsive phosphopeptide binding domains and Chk1/2 phosphorylated targets (Chen et al. 2006; Manke et al. 2005). In line with this functional segregation profile of these kinases, the respective phosphopeptide-binding specificity of the **14-3-3** proteins, and **FHA** and **BRCT** domains perfectly accommodates the compartmentalization of the DNA damage response-induced kinases. Indeed, the modular domains shared by signaling proteins that localize to foci of DNA damage (double-strand breaks) (e.g. BRCT domains) exhibit phosphopeptide binding specificity for ATM kinase motifs (Yu et al. 2003), while protein domains involved in cell cycle checkpoint control and maintenance (e. g, 14-3-3 proteins) show preferential binding specificity for phospho-peptide motifs recognized by Chk1/2/MK2 kinases (Peng et al. 1997; Dalal et al. 2004; Hermeking 2006; Kumagai et al. 1998).

The initial step upon activation of the DNA damage response signaling cascade involves the coordinated interaction between specific ATR/ATM substrates (e.g. H2 histone variant H2AX) and BRCT- and/or FHA-containing proteins (e.g. the tumor suppressors BRCA1, 53BP1 and the DNA damage modulator MDC1) (Rogakou et al. 1998; Fernandez-Capetillo et al. 2002). A crucial event in the initial stages of the activation of the DNA damage response to double-stranded breaks is the phosphorylation of histone H2AX by the IR-induced kinase ATM (Burma et al. 2001). ATM-dependent H2AX phosphorylation generates a C-terminal phosphopeptide motif that is specifically recognized by the downstream scaffold protein MDC1 through contact with its tandem BRCT repeat (Stucki et al. 2005). This phosphorylation-dependent protein engagement induces the formation of large protein complexes that act as multi-docking scaffolds and localize to nuclear foci of radiation-induced DNA damage (Stucki et al. 2005). The combined recruitment of chromatin remodeling factors, DNA repair proteins and transcriptional coactivators to DNA double-strand breaks amplifies significantly the damage signal, leading to the downstream activation of cell cycle checkpoint kinases Chk1,Chk2 and Mk2, which are the main signaling 'effectors' of genome stability that regulate both the G1/S phase and G2/M cell cycle transition (Manke et al. 2005). These kinases will in turn phosphorylate Chk1/2 peptide motifs in target substrates and thus promote the recruitment of Chk1/2-dependent phosphopeptide-binding 14-3-3- proteins to these phosphorylated targets (Peng et al. 1997). The assembly of these phosphorylation-dependent protein complexes helps establish the G2/M cell cycle checkpoint, that is the last and most crucial genome-stabilizing (gatekeeping) 'sentinel' which enables damaged cells to repair chromatin aberrations (DNA double-strand breaks) or activate programmed cell death -as a last resort- before entering into mitosis with an irreversibly damaged genome.

It is of particular note that in a large array of tumors the molecular machinery regulating the maintenance of the G2/M checkpoint is severely deregulated at multiple 'entry points' (kinases and/or kinase-specific substrates), thus severely affecting genomic stability and cancer predisposition (Lobrich and Jeggo 2007). Furthermore, radioenhancing anti-tumor agents that have been shown to modulate more efficiently the radioresistance of tumor cells tend to interfere with the activity of crucial effectors of the signaling network that establishes the G2/M checkpoint in the cell cycle (Koniaras et al. 2001). Genetic profiling of several tumors has revealed a growing cluster of mutations that localize to genes regulating the G2/M checkpoint and also aberrant G2/M gene expression through copy number variations (Lobrich and Jeggo 2007; Talos et al. 2007). In the light of these observations, it is becoming ever more urgent to better understand the complex dynamics and fine specificities behind the protein interactions that are mediated by phosphopeptide-binding protein modules involved in the DNA damage response, such as the *FHA*, *14-3-3* and *BRCT* domains.

PHOSPHOPEPTIDE-BINDING MODULES INVOLVED IN DNA DAMAGE SENSING

BRCT domain-containing proteins: The BRCT (BRca1 C-Terminal) domain is a protein signaling module broadly encountered among DNA damage responsive and cell cycle checkpoint control proteins (Callebaut and Mornon 1997; Bork et al. 1997). It is mainly associated with phosphorylation-dependent protein-protein interactions that lead to the assembly of large macromolecular scaffolds and phosphopeptide signaling complexes that localize to nuclear foci of IR-induced DNA damage (Manke et al. 2003; Yu et al. 2003). BRCT domains share high evolutionary conservation across diverse species (ranging from yeast cells to mammals) and exhibit a high binding affinity for phosphorylated ligands that contain a conserved structural motif around a central pSer/Thr residue (Manke et al. 2003; Yu et al. 2003; Rappas et al. 2011).

A comprehensive search across available nucleotide sequence databases from different species, coupled to rigorous bioinformatics analysis, have revealed the presence of a multitude of proteins containing BRCT domains in tandem repeats or as single modules (Bork et al. 1997). The first indication of BRCT domain involvement in phospho-dependent protein-protein interactions came from studies in the budding yeast *Saccharomyces cerevisiae* involving the cell cycle checkpoint protein RAD9 (Soulier and Lowndes 1999). RAD9 was shown to homo-oligomerize preferentially with hyperphosphorylated RAD9 monomers via its C-terminal BRCT domains (Soulier and Lowndes 1999). There are approximately 30 BRCT-domain-containing proteins in mammalian cells, the majority of which appear to play essential roles in pathways regulating genomic stability, DNA damage sensing and DNA replication and repair (ref.). BRCA1, BARD1, 53BP1, DNA ligase IV and MDC1 (mediator of DNA damage checkpoint 1) contain 2 BRCT domains in tandem repeats, while microcephalin (MCPH1) contains 3 BRCT domains; a single domain being close to the N-terminus and spaced far apart from the C-terminus BRCT tandem repeat (Wood et al. 2008). Pax-2 transactivation domain-interacting protein (PTIP), a large protein involved in DNA double-strand break repair, consists of 6 BRCT domains (Wang et al. 2010). Interestingly, Topoisomerase IIβ-binding protein 1 (TopBP1), a protein involved in the regulation of DNA replication checkpoint consists of 8 BRCT modules, thus providing several docking surfaces for phosphoprotein interactions and the assembly of higher order multi-protein complexes (Rappas et al. 2011).

XRCC1, REV1 and poly-(ADP)ribose-polymerase-1 (PARP-1) -a set of proteins acting as DNA damage sensors and repair 'mechanics'- are among only a few proteins that contain a single BRCT domain (Mohammad and Yaffe 2009).

Despite the fact that phosphopeptide binding is a promiscuous property within the BRCT-domain containing protein family, not all BRCT domains bind to phosphopeptide targets (Leung and Glover 2011). To function as intact phosphopeptide-binding modules, at least two BRCT domains have to be organized in a structural tandem (Williams et al. 2001). A class of tandem BRCT domains bind efficiently to target phosphopeptides with a strong preference for aromatic residues (e.g. Phenylalanine) in the +3 position downstream of the pSer residue (Manke et al. 2003; Yu et al. 2003; Rodriguez et al. 2003). An example of such as phosphopeptide binding specificity is provided by the BRCT signaling module of the BRCA1 tumor suppressor protein (Clapperton et al. 2004; Shiozaki et al. 2004). The BRCT tandem of BRCA1 recognizes the phosphorylated peptide motif **-pSer-P-T-F** in a wide array of protein targets, including the BRCA1-associated C-terminal helicase/Fanconi Anemia group J protein (BACH1/FANCJ/BRIP1) (Manke et al. 2003; Yu et al. 2003); CtIP (Yu and Chen 2004) and Abraxas/Coiled-coil domain containing protein 98 (CCDC98) (Wang et al. 2007; Kim et al. 2007b; Liu et al. 2007). Insights into the overall conformation of the BRCT tandem of BRCA1 and the way by which the two individual domains fold into a functional signaling module were first provided by the crystal structure of the BRCA1-BRCT tandem in complex with a BACH1 phosphopeptide (Clapperton et al. 2004; Shiozaki et al. 2004). X-ray diffraction analysis revealed that the N-terminal BRCT domain mainly stabilizes the phosphorylated Serine and the overall folding of the BRCT tandem repeat contributes to the formation of a hydrophobic pocket that stabilizes the aromatic residue in the 3+ position of the phosphopeptide (Clapperton et al. 2004; Shiozaki et al. 2004). The ability of BRCA1 to form such BRCT-mediated interactions with diverse phosphoproteins (Rodriguez et al. 2003) that participate in the DNA damage response and repair pathways is instrumental for the assembly of higher order macromolecular complexes and the formation of protein scaffolds that recruit signaling proteins to nuclear foci of DNA damage. Apart from their well-appreciated role in phosphopeptide recognition, recent studies have indicated that BRCT domains exert novel functions that are independent of their phosphopeptide binding capacity (Leung and Glover 2011; Mohammad and Yaffe 2009). These phosphorylation-independent protein interactions appear to modulate p53-dependent transcriptional co-activation, DNA binding, poly(ADP-ribose) (PAR) binding, regulation of translational efficiency in the ribosome and mRNA maturation and stability (Derbyshire et al. 2002; Yamane and Tsuruo 1999; Chai et al. 1999; Dizin et al. 2006). A schematic outline of the complex signaling network that regulates DNA damage sensing and repair and comprises both phospho-dependent and phospho-independent interactions of the BRCT repeat of human BRCA1 tumor suppressor is shown in Figure 1.

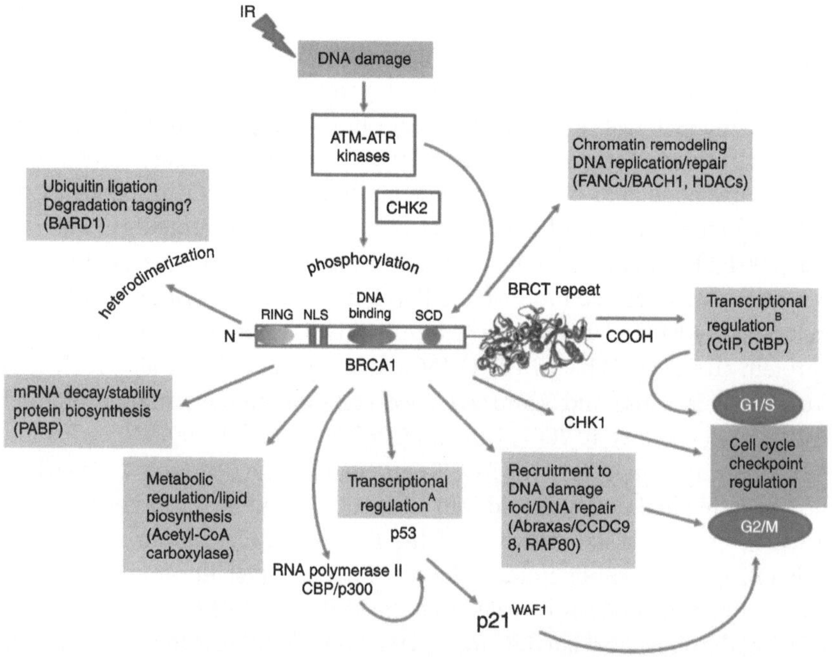

Fig. 1 Protein interactions mediated by the tandem BRCT repeat of BRCA1 tumor suppressor: A schematic outline of the multiple BRCT-mediated interactions that modulate DNA damage sensing and repair and ionizing radiation (IR)-dependent cell cycle checkpoint regulation.

Color image of this figure appears in the color plate section at the end of the book.

Recent studies have highlighted a novel role of BRCT-containing proteins, such as BRCA1, in DNA damage and repair responses by pointing to another research direction with important clinical implications; the role of BRCT-containing proteins in mitigating oxidative stress-induced DNA damage (Georgakilas 2011). It appears that BRCA1 plays an essential role in safeguarding the genome from accumulating oxidatively-generated clustered DNA lesions -other than DSBs- that are usually associated with the individual's sensitivity to environmental and toxicological stress cues (e.g. exposure to environmental γ-radiation doses or DNA-oxidizing chemical agents) (Georgakilas et al. 2009). Dysfunctional or mutated BRCA1 may lead to accumulation of such oxidative DNA lesions and thereby promote mutagenesis and malignant transformation through misrepair of DNA lesions and defective apoptotic pathways within the damaged cells (Hair et al. 2010). The role of BRCA1 as a potential biomarker for non-DSB oxidative DNA damage merits further investigation.

NOVEL INSIGHTS INTO BRCT DOMAIN STRUCTURE

Overall, the BRCT domains identified in several proteins are characterized by low sequence conservation. Nevertheless they possess a generally conserved structural organization. As shown in Table 1, BRCTs are found in single, tandem and even triple architectural motifs, determining the basic protein functions they are part of. The crystal structure of a BRCT domain was first obtained for the N-terminal BRCT of the X-ray repair cross-complementing protein 1 (XRCC1). The fold has as central core a four-stranded β-sheet that is surrounded by two α-helices from one side and one α-helix from the other, (Zhang et al. 1998). Such BRCT domains existing in single copies exhibit conserved residues localized mainly at the hydrophobic core. Their interaction with phosphorylated targets does not appear to be straightforward; the domains most likely participate in dynamic signaling processes involving contacts with other non-BRCT domains required to stabilize the interactions (Leung and Glover 2011).

The crystal structures of the tandem-BRCT-repeat domains (each repeat containing ~ 90 aminoacids) have been solved for several human proteins including BRCA1 (Williams et al. 2001), 53BP1 (Derbyshire et al. 2002) and BARD1 (Birrane et al. 2007). A more complete list of tandem-BRCT containing proteins appears in Table 1. The tandem-BRCT structures reveal common organization and architecture. Exactly as in the case of single BRCTs, the BRCT repeats participating in tandem formations, consist of the parallel four-stranded β-sheet located at the central part of the fold surrounded by

Table 1 Human BRCT-domain containing proteins.

BRCT-domain-containing Human proteins	*Number of BRCT repeats*	*PDB ID*	*Reference*
BRCA1	Tandem	1JNX	(Williams et al. 2001)
MDC1	Tandem	2ADO	(Lee et al. 2005)
53BP1	Tandem	1KZY	(Joo et al. 2002)
BARD1	Tandem	2NTE	(Birrane et al. 2007)
LigIV	Tandem	3II6	(Wu et al. 2009)
PARP1	Single	2COK[1]	(Ahel et al. 2009)
PARP4	Single	–	(Kickhoefer et al. 1999)
XRCC1	Two distanced repeats packed as single	1CDZ	(Zhang et al. 1998)
Lig3	Single	1IN1	(Krishnan et al. 2001)
DNA POLL	Single	2JW5[2]	(Garcia-Diaz et al. 2004)
DNA POLM	Single	2HTF	(Derose et al. 2007)
ECT2	Two distanced repeats packed as single	3L46[3]	(Kim et al. 2005)
DNTT (DNA)	Single	2COE[4]	(Mahajan and Mitchell 2003)
MCPH1	Multiple (three)	2WT8	(Richards et al. 2010)
PAXIP1	Multiple (six)	–	(Jowsey et al. 2004)
PES1	Single	2EP8[5]	(Holzel et al. 2007)
REV1	Single	2EBW[6]	(Hara et al. 2010)
RFC1	Single	2EBU[7]	(Mossi et al. 1997)
TopBP1	Multiple (eight) containing single, tandem and triple BRCT architecture	2XNK	(Rappas et al. 2011)

1. Nagashima T., Hayashi F. and Yokoyama S. Solution structure of BRCT domain of poly(ADP-ribose) polymerase-1. (to be published)
2. Mueller G. A., Moon A. F., DeRose E. F., Davis B. J., Ramsden D. A., Pedersen L. C. and London R. E. Polymerase Lambda BRCT domain. (to be published)
3. Nedyalkova L., Tempel W., Tong Y., Crombet L., Wang H., Guan X., Arrowsmith C. H., Edwards A. M., Bountra C., Weigelt J., Bochkarev A. and Park H. Crystal structure of the second BRCT domain of epithelial cell transforming 2 (ECT2). (to be published)
4. Nagashima T., Hayashi F. and Yokoyama S. Solution structure of BRCT domain of terminal deoxynucleotidyltransferase. (to be published)
5. Nagashima T., Hayashi F. and Yokoyama S. Solution structure of the BRCT domain from human Pescadillo homolog 1. (to be published)
6. Nagashima T., Hayashi F. and Yokoyama S. Solution structure of the BRCT domain from human DNA repair protein REV1. (to be published)
7. Nagashima T., Hayashi F. and Yokoyama S. Solution structure of the BRCT domain from human replication factor C large subunit 1. (to be published)

three α-helices. The two repeats pack together in a very specific head-to-tail manner, giving rise to the formation of a conserved, almost all-hydrophobic inter-repeat interface which consists of the α2 helix of the N-terminal BRCT repeat and the α1' and α3' helices of the C-terminal BRCT repeat. Mutation-induced destabilization of the inert-BRCT interface leads to loss of function and has been associated with hereditary breast/ovarian cancer. Flexible, BRCT-repeat-interconnecting linkers are also involved in the interface formation, usually packing right against the inter-BRCT interface. The inter-BRCT linkers are variable in length, sequence and structure. Linkers may separate and control the relative position of the BRCT repeats as well as protect and stabilize the inter-BRCT hydrophobic interface. They can also play a direct role in protein-protein interactions such as in the case of the BRCT domain of 53BP1 (53BP1-BRCT) binding to p53. Their importance for the domain's function is further highlighted by their implication in human disease. Indeed, inherited missense mutations that are located at the inter-BRCT linker of BRCT-BRCA1 such as G1738R, have been associated with breast-ovarian cancer (Chenevix-Trench et al. 2006). Alterations of the tandem-BRCT packing occur when linkers are significantly long as in the case of the BRCT repeats of DNA ligase IV (LigIV). In such cases the two repeats pack as singles (Dore et al. 2006; Sibanda et al. 2001; Wu et al. 2009). Figure 2 displays the superposition of three tandem-BRCT domains as found in BRCA1, 53BP1 and BARD1, to illustrate the highly conserved structural organization of this integral signaling module.

The structural integrity of the tandem repeats is essential for the molecule's interaction with phosphorylated protein targets including the DNA helicase BACH1 and the CtBP-interacting protein CtIP. Crystallographic studies of BRCA1-BRCT domains bound to BRIP1 (Williams et al. 2004; Clapperton et al. 2004; Shiozaki et al. 2004), CtIP (Varma et al. 2005) and ACC1 ligands (Shen and Tong 2008), have provided the molecular basis for this functional selectivity by identifying pSer-X-X-Phe as the binding recognition motif of these interactions, whereas the structure of the MDC1-BRCT complex with γ-H2AX revealed the binding mechanism with a tyrosine residue substituting phenylalanine in the phosphopeptide motif (Stucki et al. 2005). Specifically, a conserved shallow pocket in the N-terminal BRCT receives the phosphoserine at position 0 of the ligand, whereas a deeper hydrophobic pocket in the groove between the BRCT repeats selects Phe/Tyr residues at position +3. Recent comparative crystallographic studies involving the BRCT domains of BRCA1 and BARD1 revealed that these phosphoprotein recognition sites

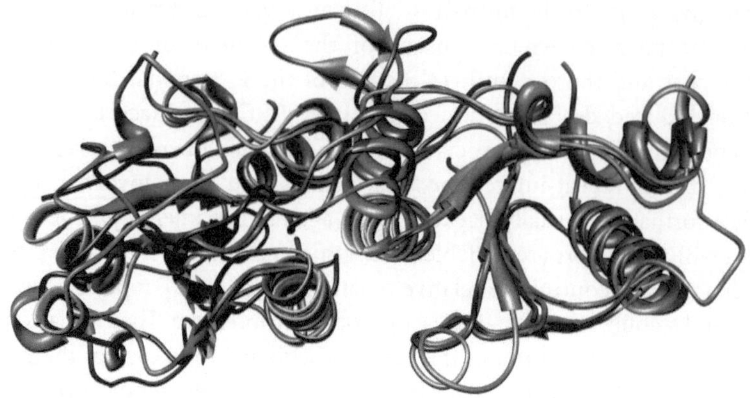

Fig. 2 Superposition of the crystal structures of 53BP1-BRCT, (green color, PDB ID: 1GZH, (Derbyshire et al. 2002)), BRCA1-BRCT (blue color, PDB ID: 1JNX, (Williams et al. 2001)), and BARD1-BRCT (purple color, PDB ID: 2NTE, (Birrane et al. 2007)). The different structural organization of the inter-BRCT linker of 53BP1 is shown in the middle-top part of the figure with the characteristic β-ribbon structure. The helical insertion -also an exclusive characteristic of the 53BP1-BRCT structure- can be observed at the bottom-right part of the figure (by permission from Thanassoulas et al. 2010).

Color image of this figure appears in the color plate section at the end of the book.

are also present in BARD1 (Birrane et al. 2007). However, oriented peptide library studies (Rodriguez et al. 2003) and isothermal titration calorimetry experiments (Birrane et al. 2007) failed to record an interaction between BARD1 BRCT and pSer-X-X-Phe phosphopeptides, suggesting that either aminoacids outside the binding pockets, not present in the case of BARD1, are essential for the high affinity binding or that the BARD1-BRCT domain recognizes macromolecular ligands with entirely different sequence motifs and structural characteristics. Missense mutations within the BRCT domain have been proposed to induce the molecule's loss of function by directly altering the binding sites and/or by destabilization. Characteristically, NMR structural results of the tandem-BRCT-domain suggest that for biologically active conformations the proper alignment of the N and C-terminal BRCT repeats is essential (Botuyan et al. 2004).

A triple BRCT domain has been identified in TopBP1 (Rappas et al. 2011; Huo et al. 2010). The packing of the each inter-repeat interface in this case is significantly altered from the canonical packing. This may indeed be the result of shorter than usual inter-BRCT linkers, paving the way for an interaction in twisted BRCT orientation.

THE PROTEIN INTERACTION NETWORK MODULATED BY BRCT-CONTAINING PROTEINS

BRCT domains are signaling modules that exhibit remarkable versatility in terms of selecting diverse protein partners and contact surfaces for interaction. In some proteins, ***single BRCT domains*** can specifically interact with BRCT domains found in other proteins through the formation of distinct interfaces that do not share the hydrophobic nature of the inter BRCT interface in tandem BRCT repeats (Leung and Glover 2011; Glover et al. 2004). Single BRCT domains are occasionally present as single copies in proteins, with Poly(ADP-ribose) polymerase 1 and DNA ligase III being typical examples of this domain architecture (Yu et al. 2003; Masson et al. 1998). They can also be found in multiple -yet separately spaced- copies in other proteins, having a stretch of more than 130 aminoacid residues in between each other (e.g. in the scaffolding protein XRCC1) (Zhang et al. 1998). The two single BRCT modules of the scaffolding protein XRCC1 appear to interact with distinct protein targets that participate in the signaling network that mediates DNA damage and base excision repair: the N-terminal BRCT domain of XRCC1 interacts with the BRCT repeat of poly(ADP-ribose) polymerase 1 (PARP1) (Masson et al. 1998), whereas the C-terminal BRCT repeat associates with the single BRCT repeat of DNA ligase III (Taylor et al. 1998). Furthermore, the BRCT domain of the *S. cerevisiae* checkpoint protein Rad9 mediates a homologous Rad9–Rad9 interaction after DNA damage through its homotypic contact with other BRCT domains in monomeric Rad9 molecules (Soulier and Lowndes 1999).

The presence of **multiple BRCT modules** within the same protein constitutes a striking paradigm of how these protein signaling modules may form extended contact surfaces with greater plasticity, accommodating higher-order macromolecular scaffold assemblies and dynamic phosphoprotein interactions. The presence of multiple BRCT copies furnishes the protein with a higher degree of freedom, adding complexity and further regulatory elements to its functional 'crosstalk' with other pathways (Leung and Glover 2011). TopBP1, a protein involved in the regulation of DNA replication checkpoint consists of 8 BRCT modules, thus presenting several docking surfaces for phosphoprotein interactions and the assembly of higher order multi-protein complexes.

The recently solved crystal structure of the N-terminus of human TopBP1 revealed an unconventional packing mode involving the three N-terminal BRCT domains (Huo et al. 2010). The overall assembly and packing of the TopBP1-BRCT domains provides evidence for the formation of a triple BRCT repeat module that can interact with Rad9, thereby inducing ATR activation and subsequent DNA replication checkpoint activation (Huo et al. 2010). Furthermore, it has been suggested that a broader phosphopeptide selectivity profile may be accommodated by TopBP1, partly due to the striking flexibility and variable hydrophobicity of the various BRCT-BRCT interfaces across the protein (Leung and Glover 2011) .

The case of NBS1 (*Nijmegen break-age syndrome protein 1*), a protein involved in DNA double-strand break repair, illustrates a paradigm of how a tandem BRCT repeat may form a tripartite signaling module in close interaction with an FHA domain, another phosphopeptide-binding module. This hybrid 'FHA-BRCT-BRCT' module in NBS1 may serve as a versatile structural scaffold for the cross-linking of two basic phospho-peptide signaling pathways involved in DNA damage and repair (Lloyd et al. 2009).

However, to this date the most common form of interaction is between **BRCT tandem repeats** and protein domains of different structure. In the following sections we provide an overview of such heterologous BRCT-mediated interactions with other target proteins.

The BRCA1-BRCT paradigm:

1. Phospho-dependent BRCA1 BRCT-interactions

1.1 FANCJ/BACH1/BRIP1 (Role in DNA replication and repair): The Fanconi Anemia group J protein FANCJ, also known as BACH1 or BRIP1, is a DNA helicase involved in DNA replication and repair (Hiom 2010; Cantor and Guillemette 2011). FANCJ has also been shown to serve a crucial role in the maintenance of genomic stability and tumor suppression (Cantor and Guillemette 2011). Its association with pathways regulating genome stability and malignant transformation was highlighted by the identification of clinically relevant mutations not only in breast cancer, but also in the childhood cancer syndrome, Fanconi anemia (Kennedy and D'Andrea 2006). The crucial role of FANCJ in tumorigenesis was further underscored by studies documenting a direct physical interaction between FANCJ/BACH1 and the tumor suppressor BRCA 1 (Clapperton et al. 2004; Cantor and Guillemette 2011). Genetic ablation/mutation of FANCJ DNA helicase, or abrogation of its physical interaction with BRCA1, have been associated with severe defects in key effector pathways of the DNA damage response (Shiozaki et al. 2004). Indeed, it is now well established that the BRCA1-FANCJ interaction plays an instrumental role in promoting DNA replication and proofreading capacity, checkpoint control and DNA crosslink repair and recombination (Leung and Glover 2011).

FANCJ/BACH1 was originally identified as a BRCA1-BRCT interacting partner in GST-pull down studies probing 293T and U20S cell lysates for novel BRCT-binding molecules (Cantor et al. 2001). The dependence of this novel BRCT-mediated interaction on protein phosphorylation was demonstrated by studies showing that phosphorylation of BACH1 at Ser990 is a prerequisite for its association with the BRCT tandem repeat of BRCA1 (Rodriguez et al. 2003; Yu et al. 2003). It was also shown that this phosphopeptide-mediated interaction is cell cycle-regulated and absolutely essential for the establishment of the G2/M checkpoint control in response to DNA damage (Cantor et al. 2001). Mutant BRCA1 that ether lacks one BRCT domain or contains missense mutations (P1749R and M1775R) fails to interact with BACH1. These data suggest that an intact BRCA1-BRCT domain structure is required for its interaction with FANCJ and their coordinated function in IR-induced DNA damage response. As mentioned above, Phe at position P+3 is also critical for phosphorylation-dependent interactions between BACH1 and BRCA1 (Clapperton et al. 2004). Taken together, these studies and the characterization of other BRCT-containing proteins indicate that phospho-dependent binding constitutes a crucial aspect of BRCT function, and demonstrate the versatility of this signaling module across various BRCT-containing proteins. The various modes of interaction (such as BRCT-homodimerization and non-phospho dependent binding) should complement each other functionally and allow the BRCT-containing proteins to act in diverse pathways that coordinate the DNA damage response and DSB repair process (Leung and Glover 2011).

Recently it was shown that another BRCT-containing protein, ToBP1, interacts with FANCJ in the signaling network regulating the DNA damage response (Leung et al. 2011). The recognition of the BACH1/FANCJ helicase by TopBP1 is critical for the activation of the DNA replication checkpoint at stalled replication forks and is facilitated by the C-terminal tandem BRCT7/8 domains of TopBP1 and a phosphorylated Thr(1133) binding motif in FANCJ/BRIP1/BACH1 (Leung et al. 2011).

1.2 CtIP (Role in transcriptional regulation of gene expression): Recent studies have indicated that the tumor suppressor protein BRCA1 regulates gene expression at the level of mRNA transcription (Yu and Chen 2004; Yu et al. 1998). This novel role of BRCA1 is supported by evidence that the C-terminal BRCT domains of BRCA1 can activate transcription of reporter genes in yeast cells when expressed as fusion proteins with a heterologous GAL4 DNA-binding domain (Monteiro et al. 1996). In addition, wild type BRCA1 has been shown to co-purify from whole cell extracts with the holoenzyme of RNA polymerase II, suggesting a physical interaction with the RNA polymerase II-dependent transcriptional complex (Scully et al. 1997). CtBP-interacting protein (CtIP) is a polypeptide that was recently shown to specifically bind CtBP (Schaeper et al. 1998), a transcriptional co-repressor used by the adenoviral E1A oncoprotein (Meloni et al. 1999) and several diverse DNA-binding, transcription factors, including the *Drosophila* Knirps, Snail, and Hairy proteins (Meloni et al. 1999).

The direct interaction of the transcriptional suppressor CtBP binding partner CtIP with the BRCT tandem of BRCA1 was revealed through a two-hybrid screening in *Saccharomyces cerevisiae* cells (Yu et al. 1998). Within the BRCA1-binding region of CtIP, the sequence surrounding Ser327 resembles the phosphorylation motif on BACH1/BRIP1 and undergoes transient phosphorylation during the G2 phase of the cell cycle (Yu et al. 1998). Furthermore, substitution of Ser327 with Ala abolished the in vitro and in vivo interaction of CtIP with BRCT BRCA1 (Yu and Chen 2004). The BRCA1/CtIP complex, which only assembles during the G2 phase, is required for the G2/M transition checkpoint and DNA damage-induced Chk1 activation, suggesting that CtIP cooperates with BRCA1 in cell cycle checkpoint control (Yu and Chen 2004). However, the BRCA1/CtIP complex is not required for prolonged G2 accumulation after DNA damage, a process controlled by a separate BRCA1/BACH1 complex. Notably, the in vivo interaction between BRCA1 and CtIP is completely abrogated by the tumor-associated mutations of BRCA1 (A1708E and P1749R) as well as the nonsense mutation that eliminates the C-terminal 11 amino acids of BRCA1 (Y1853delta) (Yu et al. 1998). In fact, cancer- associated mutations on any BRCA1 BRCT repeats or its linker region disrupt the BRCA1/CtIP interaction (Li et al. 1999). Clearly, the BRCT domain assists BRCA1 function by its ability to interact with diverse partners at different cell cycle stages.

1.3 Acetyl-CoA carboxylase (Role in metabolic regulation-lipid biosynthesis): One of the pivotal hallmarks of malignant transformation and a feature that defines to a great extent the aggressive behavior of tumor cells is the high rate of energy-consuming anabolic processes driving the synthesis of DNA, proteins and lipids (all of which are regulated by the energy status of the cell) (Hanahan and Weinberg 2011). In this context, the BRCA1 tumor suppressor has been recently linked to the metabolic regulation of tumor cells by studies that have identified a novel interacting partner of the BRCA1-BRCT domain that is an essential component of the metabolic machinery within the cell (Magnard et al. 2002; Moreau et al. 2006).

ACCA (Acetyl Coenzyme A (CoA) Carboxylase-alpha) was identified as a novel BRCA1-BRCT interacting protein through a glutathione-S-transferase (GST) pull-down screening in murine cells (Magnard et al. 2002). ACCA is the rate-limiting enzyme catalyzing de novo fatty acid biogenesis (Tong and Harwood, Jr. 2006). More specifically, ACCA plays a central role in fatty acid metabolism by catalyzing the carboxylation of Acetyl CoA to malonyl-CoA for the synthesis of long-chain fatty acids (Towle et al. 1997; Moreau et al. 2006). A short-term regulatory mechanism for ACCA function involves the reversible phosphorylation and dephosphorylation of its Ser79 that result in its inactivation and activation respectively (Tong and Harwood, Jr. 2006). BRCA1 interacts only with the phosphorylated (i.e. inactive) form of ACCA. Sequences surrounding residue Ser1263 on ACCA strongly resemble those surrounding residue Ser990 of BACH1, and likely mediate the recognition of ACCA by BRCT repeats of BRCA1. BRCA1 modulates lipid synthesis through its phospho-dependent binding to ACCA, thereby preventing ACCA from dephosphorylation and its subsequent activation (Moreau et al. 2006). Importantly, this interaction appears to have clinical implications for tumorigenesis, since it is severely disrupted by five germ-line BRCA1 deleterious mutations that destabilize the tandem BRCT repeat (e.g., A1708E, M1775R, P1749R, R1835X and Y1853X) (Magnard et al. 2002). A number of studies have linked lipid biogenesis with cancer predisposition. For example, ACCA and another major enzyme involved in lipogenesis FAS (Fatty Acid Synthase), are highly expressed in several human malignancies, including breast cancer (Milgraum et al. 1997). Collectively, these findings have shed light into novel pathways and mechanisms by which BRCA1 may exert its tumor suppressor function, and provide a conceptual framework for understanding at the molecular level how BRCT domains may mediate critical signaling events in metabolic pathways within the cell.

1.4 Abraxas/CCDC98 and RAP80 (Role in DNA repair and G2/M checkpoint control): A critical function of BRCA1 is its DNA damage 'sensor' capability that is exerted though multiple phosphopeptide-dependent interactions and subsequent recruitment to nuclear foci of DNA damage (double strand breaks) in response to ionizing radiation (Yu et al. 2003). The DNA damage response relies on the assembly of high order multi-protein complexes in the vicinity of damaged DNA foci. The identification of novel proteins that participate in these large molecular scaffolds is crucial for understanding the biology of DNA repair and designing more effective cancer therapeutics.

Phosphopeptide affinity proteomic analysis of the BRCA1 complex led to the identification of Abraxas/CCDC98 as a novel phosphoprotein interacting with the BRCA1 BRCT domains through a pSer-X-X-Phe motif (Wang et al. 2007). Abraxas/CCDC98 was shown to promote the recruitment of BRCA1 to foci of DNA damage and also mediate BRCA1-dependent G2/M checkpoint activation. The binding of Abraxas/CCDC98 to BRCA1 requires phosphorylation, and is mutually exclusive with BACH1 or CtIP interaction with BRCA1 (Wang et al. 2007). This is consistent with the observation that each of these proteins interacts with BRCA1 through the BRCT domains. Another ubiquitin-binding protein, RAP80, was found to associate with the Abraxas/CCDC98-BRCA1 complex (Wang et al. 2007; Liu et al. 2007; Kim et al. 2007a). RAP80 and BRCA1 interact in a BRCA1-BRCT dependent manner, as two clinically relevant missense mutations of BRCA1 BRCT domains (V1696L and P1749R) led to disrupted RAP80-BRCA1 binding (Wang et al. 2007; Sobhian et al. 2007). However, RAP80 does not contain a pSXXF motif, suggesting an indirect interaction between RAP80 and BRCA1. Indeed, several groups have shown that Abraxas/CCDC98 bridges the interaction between RAP80 and BRCA1 (Wang et al. 2007; Kim et al. 2007b; Liu et al. 2007). After DNA damage, the Abraxas/CCDC98-RAP80 complex translocates to sites of damage, followed by the recruitment of BRCA1 via its binding to the phosphorylated C-terminus of Abraxas/CCDC98. Both Abraxas/CCDC98 and RAP80 are required for DNA damage resistance, DNA repair and G2/M checkpoint control. The biochemical understanding of the interactions of these proteins with BRCA1 therefore not only helps to illustrate signaling pathways that recruit BRCA1 to sites of DNA damage, but also sheds light on the function and activity of other BRCT-containing proteins.

2. Phospho-independent BRCA1-BRCT interactions

2.1 Poly(A)-Binding Protein (Role in mRNA decay/stability and protein biosynthesis): RNA translation, stability and protein biosynthesis are highly regulated processes occurring in the cytoplasm-ribosome interface within the cell. Eukaryotic pre-messenger RNA molecules undergo a series of important chemical modifications after their biosynthesis by the RNA polymerase II holoenzyme. These modifications include the addition of a 5' 7-methyl-G cap and a 3' poly(A) tail (Deo et al. 1999). Pre-mRNA polyadenylation involves recognition of a 5'-AAUAAA-3' consensus sequence followed by an endonucleolytic cleavage and the addition of a great number of adenine nucleotides (60-80 nts in yeast cells and up to 250 in mammals) to the 3' end of the pre-mRNA by the action of a poly(A) polymerase (Wahle and Keller 1996). The poly(A) tail of messenger RNAs plays important roles in both translation initiation and mRNA stabilization/degradation (Sachs and Wahle 1993). Crucial to its function in translation initiation is its physical interaction with a 71 kDa poly(A)-binding protein, PABP, in a highly ordered nucleoprotein complex (Blobel 1973). It has been shown that multiple copies of PABP bind to the polyA tail of eukaryotic mRNA transcripts. Translation initiation is stimulated by the poly(A) tail/PABP complex through interactions between PABP and eIF4G, a translation initiation factor and an effector of mRNA recruitment to the ribosome (Deo et al. 1999). This multi-protein interaction results in the formation of a "closed loop" structure that promotes translation initiation at the ribosome. (Jacobson and Peltz 1996). A recent study has provided evidence for a novel role of BRCA1 in tumorigenesis through its interaction with elements of the protein biosynthesis machinery. Eva Dizin and colleagues identified poly(A)-binding protein 1 (PABP) as a novel BRCA1-BRCT interacting protein in a yeast two-hybrid screen (Dizin et al. 2006). They went on to validate this interaction by GST-pull down assays and coimmunoprecipitation studies in mammalian cells. The biological significance of this novel interaction and its potential role in tumor suppression pathways were further implied by the finding that germ-line BRCA1 cancer-associated mutations (both truncations and missense variants), destabilizing the BRCT tandem repeat, could abrogate the physical interaction with PABP (Dizin et al. 2006). Deletion mapping demonstrated that the RNA recognition motifs 1–4 region of PABP is required to mediate the interaction with BRCA1. Inhibition of endogenous BRCA1 expression using a small interfering (si)-RNA-based approach decreased overall rates of protein synthesis, suggesting that the PABP-BRCA1 interaction may modulate protein translation efficiency. In contrast, overexpression of BRCA1 in mammalian cells activated translation suggesting that PABP may be a specific BRCT-interacting target responsible for this effect. Using a RNA transfection approach, it was shown that the BRCA1-BRCT domain modulates translation, at a post-transcriptional level. These findings -albeit far from being recapitulated in different cell-based systems and in vivo tumor settings- provide a new 'twist' to the complex BRCT signaling landscape by suggesting that BRCA1 might modulate mRNA stability and protein synthesis via its interaction with PABP. This novel BRCT-mediated protein interaction might further explain how BRCA1 exerts its tumor suppressor function through multiple non-redundant pathways.

2.2 CREB-binding Protein (CBP) (Role in transcriptional regulation): The transcriptional coactivators cAMP response element (CREB)–binding protein (CBP) and p300 bind to several sequence-specific transcription factors and thereby regulate RNA polymerase II-dependent transcription (Kalkhoven 2004; Johannessen et al. 2004). Recruitment of these coactivators after signal-induced phosphorylation of the CREB kinase inducible domain at serine-133 strongly enhances CREB-dependent transcription (Johannessen et al. 2004). CBP and p300 have been implicated in diverse cellular pathways that modulate DNA repair, growth, differentiation and apoptosis. They exert their functions through acetylation of histones and other regulatory proteins (e.g. p53) (Cui et al. 1998). Germ-line mutations in these genes have been linked to the development of human diseases including various malignancies and several studies have implicated these coactivators in tumor suppressor pathways (Kalkhoven 2004). BRCA1 has previously been shown to interact with the RNA polymerase II complex and to activate transcription when fused to the GAL4 DNA-binding domain (Monteiro et al. 1996). A recent study employing a GST-pull down approach and in vitro biochemical and immunoprecipitation assays revealed that the second BRCT domain of BRCA1 associates with the CBP/p300 coactivator complex and also binds to p53, acting as a coactivator that increases p53-dependent transcription from the $p21^{WAF1/CIP1}$ promoter (Chai et al. 1999). These findings elaborate on the physical association of BRCA1 with the RNA polymerase II holoenzyme and provide new insights into the mechanism by which BRCA1 might stabilize the structure of this multiprotein complex and thereby modulate the transcriptional activity of RNA polymerase II. They also suggest that BRCA1 might modulate the transcriptional activation of a wide array of gene targets through its physical interaction with the CBP/p300 coactivator complex and the subsequent acetylation of p53. Whether cancer-linked BRCA1 mutations that lead to truncation of the BRCT domains or overall destabilization of their structure (e.g. BRCT missense variants) have any effect on the coactivator function of BRCA1 in gene transcription and affect the recruitment of CBP/p300 and p53 to the RNA polymerase II-transcriptional complex remains to be elucidated.

CANCER-LINKED MISSENSE MUTATIONS IN THE BRCT DOMAIN OF BRCA1 AND UNCLASSIFIED VARIANTS

Genetic screening for predisposition to hereditary breast ovarian/cancer has proven to be essential for early diagnosis and successful treatment. The widespread testing of the BRCA1 gene for individuals with a family history of breast/ovarian cancer has over the last fifteen years, uncovered a large number of missense **V**ariants of **U**ncertain clinical **S**ignificance (VUSs) concentrated at the protein's BRCT domain. It is essential for cancer patients and their families that these variants are successfully classified in order to improve the interpretation of cancer susceptibility test results. This classification task has proven to be very challenging since most of the missense variants are rare worldwide. A considerable research effort is thus being placed towards using genetic epidemiology and bioinformatics methodologies as well as biochemical and biophysical approaches, aiming at accurately determining whether a missense mutation is pathogenic or neutral.

To provide an updated status of the classifications of all the variants entered in the BIC (breast cancer mutation) database, concerning the BRCT domain, a Table listing is presented (Table 2). Color coding is used to highlight the currently proposed pathogenicity Class in order to facilitate the reading. The Table incorporates all of the published assigned grades from studies that used integrative genetic methods to assess the missense-variant pathogenicity as well as grades assigned from comprehensive analyses based on experimental structural and functional data at the protein level. Results from smaller-scale studies focused on specific variants are also included.

In 2007, a systematic risk assessment for 1,433 VUSs was based on data that included co-occurrence with known deleterious mutations, personal and family history of the patients and cosegregation of the variant with the disease within pedigrees (Easton et al. 2007). The risks reported in the study can then be assigned to Classes according to the classification scheme proposed by the International Agency for Research in Cancer (IARC) (Plon et al. 2008). Class 1 corresponds to the classification "not pathogenic" or "no clinical significance" with likelihood of less than 0.1% for pathogenicity; Class 2 is "likely not pathogenic" with 0.1 – 0.5% likelihood of pathogenicity; Class 3 is "uncertain" (5 to 95%); Class 4 is "likely pathogenic" (95 to 99%) and Class 5 is "pathogenic" (>99%). Carriers of Class 4 and 5 variants are advised to undergo surveillance and cancer prevention treatments. Class 4 and Class 2 variants may change to Class 5 and Class 3 respectively as more genetic data

	[1]	[2]		
N1647K				
S1651F				
M1652T				
M1652I			[3]	[13]
V1653M				
S1655F				
G1656D				
F1662S				
M1663K				[13]
M1663L				[13]
L1664P				[13]
V1665M				[13]
A1669S			[4]	[13]
E1682K				
E1682V				
T1685I				
T1685A				
M1689R				
M1689T				
T1691K				
T1691I				
D1692Y				[13]
D1692N				[13]
D1692H				
F1695L				
V1696L				[14]
C1697R				
R1699W				[15]
R1699Q			[5]	[13]
R1699L				[13]
T1700A				
G1706E				
G1706A			[6]	[13]
A1708E				[13
A1708V				
V1713A				
V1714G				
S1715R				[13]
S1715N				[13]
S1715C				[13]
W1718C				
W1718S				
T1720A				
S1722F				
R1726G				

N1730S				
D1733G			[3]	
F1734S				
V1736A				[13]
V1736G				
G1738R				
G1738E				
D1739Y				
D1739V				
D1739G				
D1739E				
V1741G				
H1746N				
P1749R				
R1751P				
R1751Q				[13]
A1752P				
A1752V				
R1753T				
F1761S				
G1763V				
L1764P				[13]
I1766S				[13]
P1771R				
P1771L				
T1773S				[13]
T1773I				
M1775R			[7]	[15]
M1775K			[8]	[14]
D1778G				
D1778Y				
D1778N				[13]
L1780P				[13]
M1783T				[14]
M1783I				
M1783L				
Q1785H				
C1787S				[13]
G1788V				[13]
G1788D				[13]
A1789S				
E1794D				
G1803A				
V1804D				
H1805P				
P1806A				[13]

V1808A				[13]
V1809A				
V1809F				[14]
V1810G				
Q1811R				
P1812A			[9]	[14]
D1818G			[10]	
N1819S				
A1823T			[11]	
Q1826H				
A1830T				
V1833M				[13]
R1835P				
E1836K				
W1837G				[13]
W1837R				[13]
W1837C				
V1838E			[12]	
S1841N				[13]
S1841R				
A1843P				[13]
L1844R				
D1851E				
Y1853C				[13]
L1854P				
P1856S				
P1859R				

Table 2 Missense variants of the BRCT domain of BRCA1.

List of missense mutations identified in the BRCT domain of human BRCA1 and recorded in the BIC database. Different highlight coloring is used to indicate the pathogenicity Class of the mutation, as assessed by either integrative genetic methods or by other experimental approaches including biochemical and biophysical assays. Mutations that appear colored in column 1 are those for which consent already exists between epidemiological and experimental methodologies classifying them as most likely pathogenic (red) or most likely neutral (blue). Non colored variants remain unclassified. In column 2, the results from the study of Easton et al., (Easton et al. 2007) are presented. The coloring code follows the five grade IARC classification scheme: Red (5): Pathogenic; Orange (4): Likely Pathogenic; Yellow (3): Unclassified; Green (2): Likely Neutral and Blue (1): Neutral. In column 3 the results from

Table 2 Contd. ...

are accumulated with time, although it is highly unlikely that Class 2 will turn to Class 4 or 5. Class 3 variants carry sufficient uncertainty that no predictive testing should be advised for relatives and the clinical advice should depend solely on the pattern of cancer in the family and tumor histology and should not concern the presence of the variant.

Of very high value for risk assessment and the interpretation of genetic testing results is the comprehensive analysis of BRCT missense variants by Le et al. (Lee et al. 2010). For this study *in vitro* protein folding, phosphopeptide binding and cell-based transcriptional assays were used to assess the functional consequences from missense variants of the BRCT domain of BRCA1. Based on the results the variants are grouped in five categories: Strong Functional Effect, Moderate Functional Effect, Uncertain Result, Low and finally No Functional Effect. The results are then correlated with the Classes defined by the genetic assessment of Easton et al. (Easton et al. 2007) and other genetic assessment studies to produce a validation set of likely pathogenic and likely neutral variants. These variants have been highlighted by red or blue respectively in the list presented in column one of the Missense Variant Table (Table 2).

Proposed classification schemes of the BRCA1-BRCT VUS, based on biophysical assays involving the changes in the thermodynamic stability of the wild-type and the calorimetric assessment of the binding ability of the variants to phosphopeptides are also included in the Table. Based on the results by Rowling et al., (Rowling et al. 2010) the following Classification scheme was proposed: Very Destabilizing, Moderately Destabilizing, Mildly Destabilizing and No Destabilizing Effect.

Table 2 Contd. ...

the comprehensive experimental analysis by Lee et al., (Lee et al. 2010) are presented. The coloring code follows the scheme: Strong Functional Effect (red), Moderate Functional Effect (orange), Uncertain Result (yellow), Low Functional Effect (green) and No Functional Effect (blue). Column 4 hosts the results from genetic assessment studies of selected variants. As in column 2, the IARC classification scheme is also used here. Finally, in column 5 results from biophysical studies are presented that are based on stability measurements and protein/phosphopeptide interactions. For the results from Rowling et al., (Rowling et al. 2010), the color code follows the proposed classification scheme: Very Destabilizing (red), Moderately Destabilizing (orange), Mildly Destabilizing (green) and No Destabilizing Effect (blue). The numbers in brackets indicate the following references: [1] (Easton et al. 2007); [2] (Lee et al. 2010); [3] (Tavtigian et al. 2006); [4] (Judkins et al. 2005); [5] (Chenevix-Trench et al. 2006); [6] (Lovelock et al. 2006); [7] (Miki et al. 1994); [8] (Tischkowitz et al. 2008); [9] (Gaildrat et al. 2010); [10] (Rouleau et al. 2010); [11] (Ladopoulou et al. 2002); [12] (Spurdle et al. 2008); [13] (Rowling et al. 2010); [14] (Drikos et al. 2009); [15] (Nikolopoulos et al. 2007).

Color image of this table appears in the color plate section at the end of the book.

It is important to concentrate on the variants (color-highlighted in column 1) for which there exists a consent between genetic assessment and structural/functional effects. There are fourteen variants high-lighted in red, indicating pathogenicity: T1685I, T1685A, M1689R, R1699W, G1706E, A1708E, S1715R, G1738R, L1746P, I1766S, M1775R, M1775K, G1788V and V1838E. All the mutations appear in highly conserved aminoacids in the BRCT sequence. With the exception of R1699W, M1775K and M177R, they all show exhibit demonstrate destabilizing effects of BRCT fold as well as functional problems related to no-specific binding of phosphopeptides. R1699W as well as M1775R and M1775K even though they do not produce significantly destabilized molecules they severely alter the ability to bind phosphopeptides due to structural changes in the +3 binding pocket. Ten mutations are characterized as neutral (blue). They are: M1652T, M1652I, F1662S, A1669S, E1682K, T1720K, D1733G, R1751Q, V1804D and P1859R. Of these only R1751 and A1669 residues are highly conserved. Yet R1751Q and A1669S mutations cause no destabilizing effects and have no connection with the phosphopeptide binding sites. Special attention must be paid to the variants P1812A, D1818G and A1823T. Independent studies have demonstrated that these three variants can cause skipping of Exon 23 in the generated BRCA1 transcript and thus carry high probability of being deleterious.

THE BARD1-BRCT PARADIGM: STATE-OF-THE-ART AND CHALLENGES

BARD1 (BRCA1-associated RING Domain Protein 1) was originally identified as a BRCA1-interacting partner by means of a yeast-two-hybrid screening approach (Wu et al. 1996). BARD1 shares remarkable structural similarity with the BRCA1 protein, comprising both an N-terminal RING domain and a C-terminal BRCT tandem repeat (Glover et al. 2004). The interaction of BARD1 with the tumor suppressor BRCA1 is effected through contact of the respective RING-finger domains and involvement of adjacent helical regions of both proteins (Edwards et al. 2008). The resulting heterodimer has E3 ubiquitin ligase activity and plays crucial roles in cell cycle checkpoint control and DNA repair. The RING domain-mediated BARD1/BRCA1 heterodimer has been implicated in tumorigenesis and BRCA1-mediated tumor suppression (Irminger-Finger and Jefford 2006). Indeed, missense mutations in BRCA1 that disrupt binding to BARD1 clearly segregate with human breast cancer cases and are strongly linked to cancer predisposition (Wu et al. 1996). Tumor suppression functions have also been attributed to BARD1 independently of its interaction with BRCA1 (Irminger-Finger and Jefford 2006).

BARD1 contains a tandem BRCT repeat at its C-terminus, as well as a set of ankyrin repeats immediately N-terminal to the BRCT region (Edwards et al. 2008). In vitro peptide binding studies have indicated that the BARD1 BRCT repeats may bind serine-phosphorylated peptides (Rodriguez et al. 2003). However, attempts to isolate bonafide phosphorylation-dependent BARD1-BRCT binding partners from human cells have thus far been unsuccessful (Kim et al. 2007a). A recent study performing a thermodynamic survey of the overall stability of the BARD1-BRCT interaction with a BRIP1/BACH1 phosphopeptide suggested that BRIP1 may be a putative BARD1-binding partner. A saturable binding of BARD1-BRCT to the BRIP1 phosphopeptide was observed at lower temperatures via an enthalpy-driven interaction, which appeared specific to the pSer-X-X-Phe peptide-binding motif (Thanassoulas et al. 2010).

The recent solving of the crystal structure of the BARD1 BRCT tandem repeat (Birrane et al. 2007) has provided important insights into the structural elements of the BRCT domain that may contribute to phosphopeptide binding. The structure of the BARD1 phosphoserine-binding pocket P1 appears strikingly similar to those of the BRCA1 and MDC1 BRCT domains, implying a similar mode of interaction with the phosphate group of the putative peptide substrate (Birrane et al. 2007). However, specificity pocket P2 exhibits distinct structural features, including two histidine residues, His685 and His686, which may contribute to ligand binding, depending on their pH-driven protonation state and the overall electrostatic potential of the adjacent region. The available BARD1 BRCT structure lays the groundwork for better understanding the impact of the cancer-associated missense mutations C645R, V695L, and S761N on the overall folding, stability and function of BARD1. A recent BARD1 mutation analysis in hereditary breast/ovarian cancer families (De et al. 2010) revealed the presence of a deleterious frameshift mutation in BARD1 that results in a truncated protein lacking the second BRCT domain. In line with previous studies, this recent report provides further evidence for a crucial role of the BARD1-BRCT domain in tumorigenesis and cancer predisposition.

CONCLUDING REMARKS AND PERSPECTIVES

Since the initial discovery of the BRCA1 tumor suppressor and the determination of the crystal structures of various BRCT domains in complex with their model phosphopeptide targets, numerous studies have provided evidence for the crucial role of this integral signaling module in diverse cellular pathways that modulate the DNA damage response, the cell cycle checkpoint control and ultimately the cell's 'ill-fated' transition to malignancy.

It is now well-appreciated that the remarkable versatility of BRCT domains in choosing binding partners stems from the structural plasticity of the domain and the distinct structural elements that regulate its phosphopeptide-binding selectivity. Besides their phosphopeptide recognition capacity, BRCT domains have also been implicated in phosphorylation-independent protein interactions, including binding to initiation factors of the ribosomal translation machinery and DNA binding. This plethora of interactions is attributed not only to the sequence and structure variations within the BRCT domains but also to the ability of this signaling module to engage in higher-order macromolecular assemblies involving other phosphoprotein recognition modules and scaffolding proteins. The growing number of BRCT mutations identified in cancer patients and their detrimental impact on the domain's stability and phosphopeptide-binding function underscore the importance of BRCT-mediated interactions in maintaining genomic stability and tumor suppression. The integrative evaluation of the pathogenicity of novel unclassified BRCT mutations using combinatorial and multidisciplinary approaches is anticipated to provide further insights into the mechanisms by which BRCT modules affect genomic stability and homeostatic repair pathways that uphold a tumor suppressive environment within the cell. The discovery of novel BRCT targets by means of high-throughput screening and validation of their tumor suppressive properties in preclinical cancer models is anticipated to thrust the development of patient-oriented cancer diagnostic protocols.

REFERENCES

Abraham, R.T. 2001. Cell cycle checkpoint signaling through the ATM and ATR kinases. Genes Dev. 15: 2177-2196.

Ahel, D., Z. Horejsi, N. Wiechens, S.E. Polo, E. Garcia-Wilson, I. Ahel, H. Flynn, M. Skehel, S.C. West, S.P. Jackson, T. Owen-Hughes and S.J. Boulton. 2009. Poly(ADP-ribose)-dependent regulation of DNA repair by the chromatin remodeling enzyme ALC1. Science 325: 1240-1243.

Bakkenist, C.J. and M.B. Kastan. 2003. DNA damage activates ATM through intermolecular autophosphorylation and dimer dissociation. Nature 421: 499-506.

Bakkenist, C.J. and M.B. Kastan. 2004. Initiating cellular stress responses. Cell 118: 9-17.

Balazsi, G., A. van Oudenaarden and J.J. Collins. 2011. Cellular decision making and biological noise: from microbes to mammals. Cell 144: 910-925.

Birrane, G., A.K. Varma, A. Soni and J.A. Ladias. 2007. Crystal structure of the BARD1 BRCT domains. Biochemistry 46: 7706-7712.

Blobel, G. 1973. A protein of molecular weight 78,000 bound to the polyadenylate region of eukaryotic messenger RNAs. Proc. Natl. Acad. Sci. U.S.A. 70: 924-928.

Bork, P., K. Hofmann, P. Bucher, A.F. Neuwald, S.F. Altschul and E.V. Koonin. 1997. A superfamily of conserved domains in DNA damage-responsive cell cycle checkpoint proteins. FASEB J. 11: 68-76.

Botuyan, M.V., Y. Nomine, X. Yu, N. Juranic, S. Macura, J. Chen and G. Mer. 2004. Structural basis of BACH1 phosphopeptide recognition by BRCA1 tandem BRCT domains. Structure. 12: 1137-1146.

Burma, S., B.P. Chen, M. Murphy, A. Kurimasa and D.J. Chen. 2001. ATM phosphorylates histone H2AX in response to DNA double-strand breaks. J. Biol. Chem. 276: 42462-42467.

Cairns, R.A., I.S. Harris and T.W. Mak. 2011. Regulation of cancer cell metabolism. Nat. Rev. Cancer 11: 85-95.

Callebaut, I. and J.P. Mornon. 1997. From BRCA1 to RAP1: a widespread BRCT module closely associated with DNA repair. FEBS Lett. 400: 25-30.

Cantor, S.B., D.W. Bell, S. Ganesan, E.M. Kass, R. Drapkin, S. Grossman, D.C. Wahrer, D.C. Sgroi, W.S. Lane, D.A. Haber and D.M. Livingston. 2001. BACH1, a novel helicase-like protein, interacts directly with BRCA1 and contributes to its DNA repair function. Cell 105: 149-160.

Cantor, S.B. and S. Guillemette. 2011. Hereditary breast cancer and the BRCA1-associated FANCJ/BACH1/BRIP1. Future. Oncol. 7: 253-261.

Chai, Y.L., J. Cui, N. Shao, E. Shyam, P. Reddy and V.N. Rao. 1999. The second BRCT domain of BRCA1 proteins interacts with p53 and stimulates transcription from the p21WAF1/CIP1 promoter. Oncogene 18: 263-268.

Chen, J.S., S.Y. Lin, W.L. Tso, G.C. Yeh, W.S. Lee, H. Tseng, L.C. Chen and Y.S. Ho. 2006. Checkpoint kinase 1-mediated phosphorylation of Cdc25C and bad proteins are involved in antitumor effects of loratadine-induced G2/M phase cell-cycle arrest and apoptosis. Mol. Carcinog. 45: 461-478.

Chenevix-Trench, G., S. Healey, S. Lakhani, P. Waring, M. Cummings, R. Brinkworth, A.M. Deffenbaugh, L.A. Burbidge, D. Pruss, T. Judkins, T. Scholl, A. Bekessy, A. Marsh, P. Lovelock, M. Wong, A. Tesoriero, H. Renard, M. Southey, J.L. Hopper, K. Yannoukakos, M. Brown, D. Easton, S.V. Tavtigian, D. Goldgar and A.B. Spurdle. 2006. Genetic and histopathologic evaluation of BRCA1 and BRCA2 DNA sequence variants of unknown clinical significance. Cancer Res. 66: 2019-2027.

Chouaib, S., C. Kieda, H. Benlalam, M.Z. Noman, F. Mami-Chouaib and C. Ruegg. 2010. Endothelial cells as key determinants of the tumor microenvironment: interaction with tumor cells, extracellular matrix and immune killer cells. Crit Rev. Immunol. 30: 529-545.

Clapperton, J.A., I.A. Manke, D.M. Lowery, T. Ho, L.F. Haire, M.B. Yaffe and S.J. Smerdon. 2004. Structure and mechanism of BRCA1 BRCT domain recognition of phosphorylated BACH1 with implications for cancer. Nat. Struct. Mol. Biol. 11: 512-518.

Costes, S.V., I. Chiolo, J.M. Pluth, M.H. Barcellos-Hoff and B. Jakob. 2010. Spatiotemporal characterization of ionizing radiation induced DNA damage foci and their relation to chromatin organization. Mutat. Res. 704: 78-87.

Cui, J.Q., N. Shao, Y. Chai, H. Wang, E.S. Reddy and V.N. Rao. 1998. BRCA1 splice variants BRCA1a and BRCA1b associate with CBP co-activator. Oncol. Rep. 5: 591-595.

Dalal, S.N., M.B. Yaffe and J. A. DeCaprio. 2004. 14-3-3 family members act coordinately to regulate mitotic progression. Cell Cycle 3: 672-677.

De, B.S., G.J. De, R. Loris, N. Janin, W. Lissens, E. Sermijn and E. Teugels. 2010. Cancer predisposing missense and protein truncating BARD1 mutations in non-BRCA1 or BRCA2 breast cancer families. Hum. Mutat. 31: E1175-E1185.

Deo, R.C., J.B. Bonanno, N. Sonenberg and S.K. Burley. 1999. Recognition of polyadenylate RNA by the poly(A)-binding protein. Cell 98: 835-845.

Derbyshire, D.J., B.P. Basu, L.C. Serpell, W.S. Joo, T. Date, K. Iwabuchi and A.J. Doherty. 2002. Crystal structure of human 53BP1 BRCT domains bound to p53 tumour suppressor. EMBO J. 21: 3863-3872.

Derose, E.F., M.W. Clarkson, S.A. Gilmore, C.J. Galban, A. Tripathy, J.M. Havener, G.A. Mueller, D.A. Ramsden, R.E. London and A.L. Lee. 2007. Solution structure of polymerase mu's BRCT Domain reveals an element essential for its role in nonhomologous end joining. Biochemistry 46: 12100-12110.

Dizin, E., C. Gressier, C. Magnard, H. Ray, D. Decimo, T. Ohlmann and V.N. Dalla. 2006. BRCA1 interacts with poly(A)-binding protein: implication of BRCA1 in translation regulation. J. Biol. Chem. 281: 24236-24246.

Dore, A.S., N. Furnham, O.R. Davies, B.L. Sibanda, D.Y. Chirgadze, S.P. Jackson, L. Pellegrini and T.L. Blundell. 2006. Structure of an Xrcc4-DNA ligase IV yeast ortholog complex reveals a novel BRCT interaction mode. DNA Repair (Amst) 5: 362-368.

Drikos, I., G. Nounesis and C.E. Vorgias. 2009. Characterization of cancer-linked BRCA1-BRCT missense variants and their interaction with phosphoprotein targets. Proteins 77: 464-476.

Easton, D.F., A.M. Deffenbaugh, D. Pruss, C. Frye, R.J. Wenstrup, K. Allen-Brady, S.V. Tavtigian, A.N. Monteiro, E.S. Iversen, F.J. Couch and D.E. Goldgar. 2007. A systematic genetic assessment of 1,433 sequence variants of unknown clinical significance in the BRCA1 and BRCA2 breast cancer-predisposition genes. Am. J. Hum. Genet. 81: 873-883.

Edwards, R.A., M.S. Lee, S.E. Tsutakawa, R.S. Williams, I. Nazeer, F.E. Kleiman, J.A. Tainer and J.N. Glover. 2008. The BARD1 C-terminal domain structure and interactions with polyadenylation factor CstF-50. Biochemistry 47: 11446-11456.

Fernandez-Capetillo, O., H.T. Chen, A. Celeste, I. Ward, P.J. Romanienko, J.C. Morales, K. Naka, Z. Xia, R.D. Camerini-Otero, N. Motoyama, P.B. Carpenter, W. M. Bonner, J. Chen and A. Nussenzweig. 2002. DNA damage-induced G2-M checkpoint activation by histone H2AX and 53BP1. Nat. Cell Biol. 4: 993-997.

Fisher, J.C. 1958. Multiple-mutation theory of carcinogenesis. Nature 181: 651-652.

Fletcher, O. and R.S. Houlston. 2010. Architecture of inherited susceptibility to common cancer. Nat. Rev. Cancer 10: 353-361.

Gaildrat, P., S. Krieger, J.C. Thery, A. Killian, A. Rousselin, P. Berthet, T. Frebourg, A. Hardouin, A. Martins and M. Tosi. 2010. The BRCA1 c.5434C->G (p.Pro1812Ala) variant induces a deleterious exon 23 skipping by affecting exonic splicing regulatory elements. J. Med. Genet. 47: 398-403.

Garcia-Diaz, M., K. Bebenek, J.M. Krahn, L. Blanco, T.A. Kunkel and L.C. Pedersen. 2004. A structural solution for the DNA polymerase lambda-dependent repair of DNA gaps with minimal homology. Mol. Cell 13: 561-572.

Georgakilas, A.G. 2011. From chemistry of DNA damage to repair and biological significance. Comprehending the future. Mutat. Res. 711: 1-2.

Georgakilas, A.G., K. Aziz, D. Ziech, S. Georgakila and M.I. Panayiotidis. 2009. BRCA1 involvement in toxicological responses and human cancer etiology. Toxicol. Lett. 188: 77-83.

Glover, J.N., R.S. Williams and M.S. Lee. 2004. Interactions between BRCT repeats and phosphoproteins: tangled up in two. Trends Biochem. Sci. 29: 579-585.

Gonda, T.A., S. Tu and T.C. Wang. 2009. Chronic inflammation, the tumor microenvironment and carcinogenesis. Cell Cycle 8: 2005-2013.

Hair, J.M., G.I. Terzoudi, V.I. Hatzi, K.A. Lehockey, D. Srivastava, W. Wang, G.E. Pantelias and A.G. Georgakilas. 2010. BRCA1 role in the mitigation of radiotoxicity and chromosomal instability through repair of clustered DNA lesions. Chem. Biol. Interact. 188: 350-358.

Hanahan, D. and R.A. Weinberg. 2011. Hallmarks of cancer: the next generation. Cell 144: 646-674.

Hara, K., H. Hashimoto, Y. Murakumo, S. Kobayashi, T. Kogame, S. Unzai, S. Akashi, S. Takeda, T. Shimizu and M. Sato. 2010. Crystal structure of human REV7 in complex with a human REV3 fragment and structural implication of the interaction between DNA polymerase zeta and REV1. J. Biol. Chem. 285: 12299-12307.

Hermeking, H. 2006. 14-3-3 proteins and cancer biology. Semin. Cancer Biol. 16: 161.

Hiom, K. 2010. FANCJ: solving problems in DNA replication. DNA Repair (Amst) 9: 250-256.

Holzel, M., T. Grimm, M. Rohrmoser, A. Malamoussi, T. Harasim, A. Gruber-Eber, E. Kremmer and D. Eick. 2007. The BRCT domain of mammalian Pes1 is crucial for nucleolar localization and rRNA processing. Nucleic Acids Res. 35: 789-800.

Hu, M. and K. Polyak. 2008. Microenvironmental regulation of cancer development. Curr. Opin. Genet. Dev. 18: 27-34.

Huo, Y.G., L. Bai, M. Xu and T. Jiang. 2010. Crystal structure of the N-terminal region of human Topoisomerase IIbeta binding protein 1. Biochem. Biophys. Res. Commun. 401: 401-405.

Irminger-Finger, I. and C.E. Jefford. 2006. Is there more to BARD1 than BRCA1? Nat. Rev. Cancer 6: 382-391.

Jacobson, A. and S.W. Peltz. 1996. Interrelationships of the pathways of mRNA decay and translation in eukaryotic cells. Annu. Rev. Biochem. 65: 693-739.

Johannessen M., M.P. Delghandi and U. Moens. 2004. What turns CREB on? Cell Signal. 16: 1211-1227.

Joo, W.S., P.D. Jeffrey, S.B. Cantor, M.S. Finnin, D.M. Livingston and N.P. Pavletich. 2002. Structure of the 53BP1 BRCT region bound to p53 and its comparison to the BRCA1 BRCT structure. Genes Dev. 16: 583-593.

Jowsey, P.A., A.J. Doherty and J. Rouse. 2004. Human PTIP facilitates ATM-mediated activation of p53 and promotes cellular resistance to ionizing radiation. J. Biol. Chem. 279: 55562-55569.

Judkins, T., B.C. Hendrickson, A.M. Deffenbaugh, K. Eliason, B. Leclair, M.J. Norton, B.E. Ward, D. Pruss and T. Scholl. 2005. Application of embryonic lethal or other obvious phenotypes to characterize the clinical significance of genetic variants found in trans with known deleterious mutations. Cancer Res. 65: 10096-10103.

Kalkhoven, E. 2004. CBP and p300: HATs for different occasions. Biochem. Pharmacol. 68: 1145-1155.

Kennedy, R.D. and A.D. D'Andrea. 2006. DNA repair pathways in clinical practice: lessons from pediatric cancer susceptibility syndromes. J. Clin. Oncol. 24: 3799-3808.

Kickhoefer, V.A., A.C. Siva, N.L. Kedersha, E.M. Inman, C. Ruland, M. Streuli and L.H. Rome. 1999. The 193-kD vault protein, VPARP, is a novel poly(ADP-ribose) polymerase. J. Cell Biol. 146: 917-928.

Kim, H., J. Chen and X. Yu. 2007a. Ubiquitin-binding protein RAP80 mediates BRCA1-dependent DNA damage response. Science 316: 1202-1205.

Kim, H., J. Huang and J. Chen. 2007b. CCDC98 is a BRCA1-BRCT domain-binding protein involved in the DNA damage response. Nat. Struct. Mol. Biol. 14: 710-715.

Kim, J.E., D.D. Billadeau and J. Chen. 2005. The tandem BRCT domains of Ect2 are required for both negative and positive regulation of Ect2 in cytokinesis. J. Biol. Chem. 280: 5733-5739.

Koniaras, K., A.R. Cuddihy, H. Christopoulos, A. Hogg and M.J. O'Connell. 2001. Inhibition of Chk1-dependent G2 DNA damage checkpoint radiosensitizes p53 mutant human cells. Oncogene 20: 7453-7463.

Krishnan, V.V., K.H. Thornton, M.P. Thelen and M. Cosman. 2001. Solution structure and backbone dynamics of the human DNA ligase IIIalpha BRCT domain. Biochemistry 40: 13158-13166.

Kryczek, I., S. Wei, E. Keller, R. Liu and W. Zou. 2007. Stroma-derived factor (SDF-1/CXCL12) and human tumor pathogenesis. Am. J. Physiol Cell Physiol 292: C987-C995.

Kumagai, A., P.S. Yakowec and W.G. Dunphy. 1998. 14-3-3 proteins act as negative regulators of the mitotic inducer Cdc25 in Xenopus egg extracts. Mol. Biol. Cell 9: 345-354.

Ladopoulou, A., I. Konstantopoulou, S. Armaou, E. Efstathiou, M. Mihalatos, G. Nasioulas, G. Bardi, N. Pandis, D. Yannoukakos and Athens; Hellenic Cooperative Oncology Group. 2002. A change in the last base of BRCA1 exon 23, 5586G-->A, results in abnormal RNA splicing. Cancer Genet. Cytogenet. 134: 175-177.

Lee, M.S., R.A. Edwards, G.L. Thede and J.N. Glover. 2005. Structure of the BRCT repeat domain of MDC1 and its specificity for the free COOH-terminal end of the gamma-H2AX histone tail. J. Biol. Chem. 280: 32053-32056.

Lee, M.S., R. Green, S.M. Marsillac, N. Coquelle, R.S. Williams, T. Yeung, D. Foo, D.D. Hau, B. Hui, A.N. Monteiro and J.N. Glover. 2010. Comprehensive analysis of missense variations in the BRCT domain of BRCA1 by structural and functional assays. Cancer Res. 70: 4880-4890.

Leung, C.C. and J.N. Glover. 2011. BRCT domains: Easy as one, two, three. Cell Cycle 10: 2461-2470.

Leung, C.C., Z. Gong, J. Chen and J.N. Glover. 2011. Molecular basis of BACH1/FANCJ recognition by TopBP1 in DNA replication checkpoint control. J. Biol. Chem. 286: 4292-4301.

Li, J., G.I. Lee, S.R. Van Doren and J.C. Walker. 2000. The FHA domain mediates phosphoprotein interactions. J. Cell Sci. 113 Pt 23: 4143-4149.

Li, J., I.A. Taylor, J. Lloyd, J.A. Clapperton, S. Howell, D. MacMillan and S.J. Smerdon. 2008. Chk2 oligomerization studied by phosphopeptide ligation: implications for regulation and phosphodependent interactions. J. Biol. Chem. 283: 36019-36030.

Li, S., P.L. Chen, T. Subramanian, G. Chinnadurai, G. Tomlinson, C.K. Osborne, Z.D. Sharp and W.H. Lee. 1999. Binding of CtIP to the BRCT repeats of BRCA1 involved in the transcription regulation of p21 is disrupted upon DNA damage. J. Biol. Chem. 274: 11334-11338.

Liu, Z., J. Wu and X. Yu. 2007. CCDC98 targets BRCA1 to DNA damage sites. Nat. Struct. Mol. Biol. 14: 716-720.

Lloyd, J., J.R. Chapman, J.A. Clapperton, L.F. Haire, E. Hartsuiker, J. Li, A.M. Carr, S.P. Jackson and S.J. Smerdon. 2009. A supramodular FHA/BRCT-repeat architecture mediates Nbs1 adaptor function in response to DNA damage. Cell 139: 100-111.

Lobrich, M. and P.A. Jeggo. 2007. The impact of a negligent G2/M checkpoint on genomic instability and cancer induction. Nat. Rev. Cancer 7: 861-869.

Lovelock, P.K., S. Healey, W. Au, E.Y. Sum, A. Tesoriero, E.M. Wong, S. Hinson, R. Brinkworth, A. Bekessy, O. Diez, L. Izatt, E. Solomon, M. Jenkins, H. Renard, J. Hopper, P. Waring, S.V. Tavtigian, D. Goldgar, G.J. Lindeman, J.E. Visvader, F.J. Couch, B.R. Henderson, M. Southey, G. Chenevix-Trench, A.B. Spurdle and M.A. Brown. 2006. Genetic, functional, and histopathological evaluation of two C-terminal BRCA1 missense variants. J. Med. Genet. 43: 74-83.

Magnard, C., R. Bachelier, A. Vincent, M. Jaquinod, S. Kieffer, G.M. Lenoir and N.D. Venezia. 2002. BRCA1 interacts with acetyl-CoA carboxylase through its tandem of BRCT domains. Oncogene 21: 6729-6739.

Mahajan, K.N. and B.S. Mitchell. 2003. Role of human Pso4 in mammalian DNA repair and association with terminal deoxynucleotidyl transferase. Proc. Natl. Acad. Sci. U.S.A. 100: 10746-10751.

Mailand, N., S. Bekker-Jensen, H. Faustrup, F. Melander, J. Bartek, C. Lukas and J. Lukas. 2007. RNF8 ubiquitylates histones at DNA double-strand breaks and promotes assembly of repair proteins. Cell 131: 887-900.

Manke, I.A., D.M. Lowery, A. Nguyen and M.B. Yaffe. 2003. BRCT repeats as phosphopeptide-binding modules involved in protein targeting. Science 302: 636-639.

Manke, I.A., A. Nguyen, D. Lim, M.Q. Stewart, A.E. Elia and M.B. Yaffe. 2005. MAPKAP kinase-2 is a cell cycle checkpoint kinase that regulates the G2/M transition and S phase progression in response to UV irradiation. Mol. Cell 17: 37-48.

Masson, M., C. Niedergang, V. Schreiber, S. Muller, M.J. Menissier-de and Murcia G. de. 1998. XRCC1 is specifically associated with poly(ADP-ribose) polymerase and negatively regulates its activity following DNA damage. Mol. Cell Biol. 18: 3563-3571.

Meloni, A.R., E.J. Smith and J.R. Nevins. 1999. A mechanism for Rb/p130-mediated transcription repression involving recruitment of the CtBP corepressor. Proc. Natl. Acad. Sci. U. S. A. 96: 9574-9579.

Michor, F., J. Liphardt, M. Ferrari and J. Widom. 2011. What does physics have to do with cancer? Nat. Rev. Cancer 11: 657-670.

Miki, Y., J. Swensen, D. Shattuck-Eidens, P.A. Futreal, K. Harshman, S. Tavtigian, Q. Liu, C. Cochran, L.M. Bennett, W. Ding, et al. 1994. A strong candidate for the breast and ovarian cancer susceptibility gene BRCA1. Science 266: 66-71.

Milgraum, L.Z., L.A. Witters, G.R. Pasternack and F.P. Kuhajda. 1997. Enzymes of the fatty acid synthesis pathway are highly expressed in in situ breast carcinoma. Clin. Cancer Res. 3: 2115-2120.

Mohammad, D.H. and M.B. Yaffe. 2009. 14-3-3 proteins, FHA domains and BRCT domains in the DNA damage response. DNA Repair (Amst) 8: 1009-1017.

Monteiro, A.N., A. August and H. Hanafusa. 1996. Evidence for a transcriptional activation function of BRCA1 C-terminal region. Proc. Natl. Acad. Sci. U.S.A. 93: 13595-13599.

Moreau, K., E. Dizin, H. Ray, C. Luquain, E. Lefai, F. Foufelle, M. Billaud, G.M. Lenoir and N.D. Venezia. 2006. BRCA1 affects lipid synthesis through its interaction with acetyl-CoA carboxylase. J. Biol. Chem. 281: 3172-3181.

Mossi, R., Z.O. Jonsson, B.L. Allen, S.H. Hardin and U. Hubscher. 1997. Replication factor C interacts with the C-terminal side of proliferating cell nuclear antigen. J. Biol. Chem. 272: 1769-1776.

Nikolopoulos, G., S. Pyrpassopoulos, A. Thanassoulas, P. Klimentzou, C. Zikos, M. Vlassi, C.E. Vorgias, D. Yannoukakos and G. Nounesis. 2007. Thermal unfolding of human BRCA1 BRCT-domain variants. Biochim. Biophys. Acta 1774: 772-780.

Peng, C.Y., P.R. Graves, R.S. Thoma, Z. Wu, A.S. Shaw and H. Piwnica-Worms. 1997. Mitotic and G2 checkpoint control: regulation of 14-3-3 protein binding by phosphorylation of Cdc25C on serine-216. Science 277: 1501-1505.

Plon, S.E., D.M. Eccles, D. Easton, W.D. Foulkes, M. Genuardi, M.S. Greenblatt, F.B. Hogervorst, N. Hoogerbrugge, A.B. Spurdle and S.V. Tavtigian. 2008. Sequence variant classification and reporting: recommendations for improving the interpretation of cancer susceptibility genetic test results. Hum. Mutat. 29: 1282-1291.

Qi, W. and J.D. Martinez. 2003. Reduction of 14-3-3 proteins correlates with increased sensitivity to killing of human lung cancer cells by ionizing radiation. Radiat. Res. 160: 217-223.

Rappas, M., A.W. Oliver and L.H. Pearl. 2011. Structure and function of the Rad9-binding region of the DNA-damage checkpoint adaptor TopBP1. Nucleic Acids Res. 39: 313-324.

Reinhardt, H.C., A.S. Aslanian, J.A. Lees and M.B. Yaffe. 2007. p53-deficient cells rely on ATM- and ATR-mediated checkpoint signaling through the p38MAPK/MK2 pathway for survival after DNA damage. Cancer Cell 11: 175-189.

Reinhardt, H.C. and M.B. Yaffe. 2009. Kinases that control the cell cycle in response to DNA damage: Chk1, Chk2, and MK2. Curr. Opin. Cell Biol. 21: 245-255.

Richards, M.W., J.W. Leung, S.M. Roe, K. Li, J. Chen and R. Bayliss. 2010. A pocket on the surface of the N-terminal BRCT domain of Mcph1 is required to prevent abnormal chromosome condensation. J. Mol. Biol. 395: 908-915.

Rodriguez, M., X. Yu, J. Chen and Z. Songyang. 2003. Phosphopeptide binding specificities of BRCA1 COOH-terminal (BRCT) domains. J. Biol. Chem. 278: 52914-52918.

Rogakou, E.P., D.R. Pilch, A.H. Orr, V.S. Ivanova and W.M. Bonner. 1998. DNA double-stranded breaks induce histone H2AX phosphorylation on serine 139. J. Biol. Chem. 273: 5858-5868.

Rouleau, E., C. Lefol, V. Moncoutier, L. Castera, C. Houdayer, S. Caputo, I. Bieche, M. Buisson, S. Mazoyer, D. Stoppa-Lyonnet, C. Nogues and R. Lidereau. 2010. A missense variant within BRCA1 exon 23 causing exon skipping. Cancer Genet. Cytogenet. 202: 144-146.

Rowling, P.J., R. Cook and L.S. Itzhaki. 2010. Toward classification of BRCA1 missense variants using a biophysical approach. J. Biol. Chem. 285: 20080-20087.

Sachs, A. and E. Wahle. 1993. Poly(A) tail metabolism and function in eucaryotes. J. Biol. Chem. 268: 22955-22958.

Schaeper, U., T. Subramanian, L. Lim, J.M. Boyd and G. Chinnadurai. 1998. Interaction between a cellular protein that binds to the C-terminal region of adenovirus E1A

(CtBP) and a novel cellular protein is disrupted by E1A through a conserved PLDLS motif. J. Biol. Chem. 273: 8549-8552.

Scully, R., S.F. Anderson, D.M. Chao, W. Wei, L. Ye, R.A. Young, D.M. Livingston and J.D. Parvin. 1997. BRCA1 is a component of the RNA polymerase II holoenzyme. Proc. Natl. Acad. Sci. U. S. A. 94: 5605-5610.

Shen, Y. and L. Tong. 2008. Structural evidence for direct interactions between the BRCT domains of human BRCA1 and a phospho-peptide from human ACC1. Biochemistry 47: 5767-5773.

Shiloh, Y. 2003. ATM and related protein kinases: safeguarding genome integrity. Nat. Rev. Cancer 3: 155-168.

Shiozaki, E.N., L. Gu, N. Yan and Y. Shi. 2004. Structure of the BRCT repeats of BRCA1 bound to a BACH1 phosphopeptide: implications for signaling. Mol. Cell 14: 405-412.

Sibanda, B.L., S.E. Critchlow, J. Begun, X.Y. Pei, S.P. Jackson, T.L. Blundell and L. Pellegrini. 2001. Crystal structure of an Xrcc4-DNA ligase IV complex. Nat. Struct. Biol. 8: 1015-1019.

Sobhian, B., G. Shao, D.R. Lilli, A.C. Culhane, L.A. Moreau, B. Xia, D.M. Livingston and R.A. Greenberg. 2007. RAP80 targets BRCA1 to specific ubiquitin structures at DNA damage sites. Science 316: 1198-1202.

Soulier, J. and N.F. Lowndes. 1999. The BRCT domain of the S. cerevisiae checkpoint protein Rad9 mediates a Rad9-Rad9 interaction after DNA damage. Curr. Biol. 9: 551-554.

Spurdle, A.B., S.R. Lakhani, S. Healey, S. Parry, L.M. Da Silva, R. Brinkworth, J.L. Hopper, M.A. Brown, D. Babikyan, G. Chenevix-Trench, S.V. Tavtigian and D.E. Goldgar. 2008. Clinical classification of BRCA1 and BRCA2 DNA sequence variants: the value of cytokeratin profiles and evolutionary analysis--a report from the kConFab Investigators. J. Clin. Oncol. 26: 1657-1663.

Squatrito, M. and E.C. Holland. 2011. DNA damage response and growth factor signaling pathways in gliomagenesis and therapeutic resistance. Cancer Res. 71: 5945-5949.

Stucki, M., J.A. Clapperton, D. Mohammad, M.B. Yaffe, S.J. Smerdon and S.P. Jackson. 2005. MDC1 directly binds phosphorylated histone H2AX to regulate cellular responses to DNA double-strand breaks. Cell 123: 1213-1226.

Talos, F., A. Nemajerova, E.R. Flores, O. Petrenko and U.M. Moll. 2007. p73 suppresses polyploidy and aneuploidy in the absence of functional p53. Mol. Cell 27: 647-659.

Tavtigian, S.V., A.M. Deffenbaugh, L. Yin, T. Judkins, T. Scholl, P.B. Samollow, S.D. De, A. Zharkikh and A. Thomas. 2006. Comprehensive statistical study of 452 BRCA1 missense substitutions with classification of eight recurrent substitutions as neutral. J. Med. Genet. 43: 295-305.

Taylor, R.M., B. Wickstead, S. Cronin and K.W. Caldecott. 1998. Role of a BRCT domain in the interaction of DNA ligase III-alpha with the DNA repair protein XRCC1. Curr. Biol. 8: 877-880.

Thanassoulas, A., M. Nomikos, M. Theodoridou, D. Yannoukakos, D. Mastellos and G. Nounesis. 2010. Thermodynamic study of the BRCT domain of BARD1 and its interaction with the -pSER-X-X-Phe- motif-containing BRIP1 peptide. Biochim. Biophys. Acta 1804: 1908-1916.

Tischkowitz, M., N. Hamel, M.A. Carvalho, G. Birrane, A. Soni, E.H. van Beers, S.A. Joosse, N. Wong, D. Novak, L.A. Quenneville, S.A. Grist, P.M. Nederlof, D.E. Goldgar, S.V. Tavtigian, A.N. Monteiro, J.A. Ladias and W.D. Foulkes. 2008. Pathogenicity of the BRCA1 missense variant M1775K is determined by the disruption of the BRCT phosphopeptide-binding pocket: a multi-modal approach. Eur. J. Hum. Genet. 16: 820-832.

Tong, L. and H.J. Harwood, Jr. 2006. Acetyl-coenzyme A carboxylases: versatile targets for drug discovery. J. Cell Biochem. 99: 1476-1488.

Towle, H.C., E.N. Kaytor and H.M. Shih. 1997. Regulation of the expression of lipogenic enzyme genes by carbohydrate. Annu. Rev. Nutr. 17: 405-433.

Varma, A.K., R.S. Brown, G. Birrane and J.A. Ladias. 2005. Structural basis for cell cycle checkpoint control by the BRCA1-CtIP complex. Biochemistry 44: 10941-10946.

Wahle, E. and W. Keller. 1996. The biochemistry of polyadenylation. Trends Biochem. Sci. 21: 247-250.

Wang, B., S. Matsuoka, B.A. Ballif, D. Zhang, A. Smogorzewska, S.P. Gygi and S.J. Elledge. 2007. Abraxas and RAP80 form a BRCA1 protein complex required for the DNA damage response. Science 316: 1194-1198.

Wang, X., K. Takenaka and S. Takeda. 2010. PTIP promotes DNA double-strand break repair through homologous recombination. Genes Cells.

William, R.S., R. Green and J.N. Glover. 2001. Crystal structure of the BRCT repeat region from the breast cancer-associated protein BRCA1. Nat. Struct. Biol. 8: 838-842.

Williams, R.S., M.S. Lee, D.D. Hau and J.N. Glover. 2004. Structural basis of phosphopeptide recognition by the BRCT domain of BRCA1. Nat. Struct. Mol. Biol. 11: 519-525.

Wood, J.L., Y. Liang, K. Li and J. Chen. 2008. Microcephalin/MCPH1 associates with the Condensin II complex to function in homologous recombination repair. J. Biol. Chem. 283: 29586-29592.

Wu, L.C., Z.W. Wang, J.T. Tsan, M.A. Spillman, A. Phung, X.L. Xu, M.C. Yang, L.Y. Hwang, A.M. Bowcock and R. Baer. 1996. Identification of a RING protein that can interact in vivo with the BRCA1 gene product. Nat. Genet. 14: 430-440.

Wu, P.Y., P. Frit, S. Meesala, S. Dauvillier, M. Modesti, S.N. Andres, Y. Huang, J. Sekiguchi, P. Calsou, B. Salles and M.S. Junop. 2009. Structural and functional interaction between the human DNA repair proteins DNA ligase IV and XRCC4. Mol. Cell Biol. 29: 3163-3172.

Xu, Y. and B.D. Price. 2011. Chromatin dynamics and the repair of DNA double strand breaks. Cell Cycle 10: 261-267.

Yamane, K. and T. Tsuruo. 1999. Conserved BRCT regions of TopBP1 and of the tumor suppressor BRCA1 bind strand breaks and termini of DNA. Oncogene 18: 5194-5203.

Yu, X. and J. Chen. 2004. DNA damage-induced cell cycle checkpoint control requires CtIP, a phosphorylation-dependent binding partner of BRCA1 C-terminal domains. Mol. Cell Biol. 24: 9478-9486.

Yu, X., C.C. Chini, M. He, G. Mer and J. Chen. 2003. The BRCT domain is a phospho-protein binding domain. Science 302: 639-642.

Yu X., L.C. Wu, A.M. Bowcock, A. Aronheim and R. Baer. 1998. The C-terminal (BRCT) domains of BRCA1 interact in vivo with CtIP, a protein implicated in the CtBP pathway of transcriptional repression. J. Biol. Chem. 273: 25388-25392.

Zhang, X., S. Morera, P.A. Bates, P.C. Whitehead, A.I. Coffer, K. Hainbucher, R.A. Nash, M.J. Sternberg, T. Lindahl and P.S. Freemont. 1998. Structure of an XRCC1 BRCT domain: a new protein-protein interaction module. EMBO J. 17: 6404-6411.

CHAPTER 7

Cellular Senescence and Cancer Development: Antagonistic and Synergistic Relations

Dimitris Kletsas, PhD
Laboratory of Cell Proliferation & Ageing, Institute of Biology
National Centre for Scientific Research "Demokritos" Athens, Greece
E-mail: dkletsas@bio.demokritos.gr

ABSTRACT

Cellular senescence is a major feature of normal cells. It is provoked by continuous cell proliferation or as a cells' response to genotoxic stresses and it is considered to be a potent anticancer mechanism. On the other hand, several anticancer treatments, such as ionizing radiation or DNA-damaging anticancer drugs, can drive cells into senescence. Senescent cells express a pro-inflammatory phenotype, characterized by the secretion of several cytokines, growth factors and proteases and several lines of evidence indicate that the presence of senescent cells in stroma, via this secretome, facilitate tumor growth. In this book chapter we intend to describe the characteristics of senescent cells, their presence and role in vivo and especially the antagonistic and synergistic relations between cellular senescence and tumor development.

INTRODUCTION

Cancer development is a multistep process, during which cells acquire various biological capabilities, i.e. sustained proliferative signaling, evading growth suppressors, resisting cell death, inducing angiogenesis, activation of invasion and metastasis, reprogramming of energy metabolism, evading immune destruction, as well as replicative immortality (Hanahan and Weinberg 2011). This latter trait represents the requirement of cancer cells for an unlimited proliferative potential, and it is in contrast to the majority of normal cells which are able for only a limited number of successive cycles of cell growth and division (Hanahan and Weinberg 2011). Accordingly, this feature of normal cells, called cellular senescence, has been proposed to be an antitumor mechanism, and several lines of evidence support this hypothesis. However, it has been reported that senescent cells can also enhance tumor development. This article will focus on the mechanisms of senescence and its ambiguous role in tumorigenesis.

REPLICATIVE AND STRESS-INDUCED SENESCENCE

Until the early 1960's it was generally accepted that normal cells are immortal and they could proliferate indefinitely if cultured under appropriate conditions. However, L. Hayflick and his colleagues have challenged this dogma and have shown that normal human embryonic lung fibroblasts posses a limited in vitro replicative lifespan, i.e. they can perform a limited number of cell divisions, after which they enter a state called (replicative) senescence, characterized by their inability to proliferate even in the presence of adequate mitogenic stimuli (Hayflick 1965). These initial studies have been reproduce for fibroblasts from several tissues and developmental stages, as well as for different cell types, such as lymphocytes (Effros et al. 2003), keratinocytes (Kang et al. 2000), melanocytes (Bandyopadhyay et al. 2001), epithelial, endothelial and smooth muscle cells (Romanov et al. 2001; Gorenne et al. 2006; Matthews et al. 2006; Erusalimsky 2009; Erusalimsky and Skene 2009), as well as mesenchymal stem cells (Shibata et al. 2007; Sikora et al. 2011).

Senescent cells, and especially fibroblasts, are characterized by specific morphological and functional characteristics: they loose their elongated, spindle-like appearance with a near-rounded, elliptical nucleus and become enlarged, flattened with a cytoplasm filled with lysosomal autofluorescent granule, while their nuclei are highly lobed (Lipetz and Cristofalo 1972; Kletsas 2003). They also remain viable and metabolically active for many months in culture. Even more, they express an increased resistance to several apoptotic stimuli (Green and Evan 2002; Hampel et al. 2004). However, the major characteristic of senescent cells is their inability to proliferate. They are usually found arrested in the late G1 phase of the cell cycle. This is due to the over-expression of cyclin-dependent kinase (CDK) inhibitors, such as $p21^{WAF1}$ and $p16^{INK4a}$, which lead to the hypophosporylation of the pRb protein and the subsequent growth arrest. Both CDK inhibitors can alone or in synergy establish and maintain the growth arrest of senescent cells (Herbig et al. 2004; Campisi and d'Adda di Fagagna 2007).

Each normal cell line can undergo a constant number of population doublings before entering into senescence. This gave rise to the idea of a "mitotic clock", an inherent mechanism of cellular lifespan. In this line, Olovnikov has predicted that with each round of cell division chromosome telomeres are becoming shorter due to the inability of standard DNA polymerases to replicate DNA ends (Olovnikov 1973), and this has been proved experimentally several years later (Harley et al. 1990). This telomere shortening is crucial for replicative senescence, as the majority of cancer cells and of immortal germ cells express telomerase, an enzyme responsible for the maintenance of telomere length (Shay and Wright 2007). Furthermore, the ectopic expression of catalytic subunit of telomerase (TERT) extends the lifespan of normal cells and can even lead to immortalization (Bodnar et al. 1998). It is now believed that shortening of few telomeres beyond a certain threshold is perceived by the cells as a DNA damage, thus activating DNA damage response (DDR) pathways, such as the ATM-Chk2-p53-p21^{WAF1}-pRb pathway, leading to a permanent growth arrest. From the above, it seems that two of the major tumor suppressor genes, i.e. p53 and pRb, are involved in replicative senescence. This is further substantiated as several viral proteins, such as SV40 Large T antigen (that can bind and inactivate both p53 and pRb), or the combination of adenovirus E1A and E1B or of papilloma virus E6 and E7 (which separately inactivate these proteins), can also drive normal cells to immortalization (Shay et al. 1991). Interestingly, a combination of TERT delivery (which maintains telomere length), SV40 Large T antigen (which inactivates p53 and pRb) and an activated oncogene (H-rasV12), found to mutated in many tumors, can convert a normal cell into a tumor cell (Hahn et al. 1999), thus indicating that these genetic events are important in cellular senescence and carcinogenesis, as well.

Beyond serial subculturing (leading to replicative senescence), cells can senesce prematurely after a serial exposure to subcytotoxic doses of noxes, usually in telomere shortening-independent mode, a process called stress-induced premature senescence (SIPS) (Toussaint et al. 2000). These stresses include among others the exposure to oxidative stress, ionizing radiation, signaling molecules, various chemicals and chemotherapeutic drugs, all provoking (directly or indirectly) damage to DNA and other cellular macromolecules. Another stress provoking SIPS is the overexpression of several oncogenes. The prototypic oncogenes linked to senescence were members of the Ras family (Serrano et al. 1997; Collado et al. 2005; Michaloglou et al. 2005; Collado et al.

2007), while there is a long list of oncogenes whose overexpression in normal cells leads to oncogene-induced senescence (OIS) (Gorgoulis and Halazonetis 2010). This further supports the role of senescence as a major antitumor mechanism (Halazonetis et al. 2008). Interestingly, OIS was linked to an activation of a DNA damage response (Bartkova et al. 2006), indicating that DDR is a common motif for replicative and stress-induced senescence.

As mentioned above, one of the stresses that can lead to premature senescence is oxidative insult, provoked either exogenously (by ionizing radiation, oxidizing chemicals or UV light) or endogenously (as a side-effect of metabolic processes or during inflammation) (Toussaint et al. 2000; Kryston et al. 2011). This stress can induce DNA damage, such as single stand breaks (SSBs), double strand breaks (DSBs), as well as oxidatively induced clustered DNA lesions (OCDLs) (Francisco et al. 2008; Kryston et al. 2011). Interestingly, stress-induced, as well as replicative senescence, are characterized by a decline in DNA repair capability. In particular, it has been shown that senescent cells exhibit a reduced efficiency for the removal of 8-hydroxy-2' deoxyguanosine (8-OHdG) (the latter being a marker of oxidative DNA damage), compared to young cells (Wolf et al. 2002). In addition, cellular senescence is characterized by a less efficient DNA end joining (Seluanov et al. 2004), while a delayed DNA DSBs processing was observed during normal and pathologic ageing (Sedelnikova et al. 2008). The persistence of these DNA lesions obviously activates continuously DDR, thus provoking and securing premature senescence. It has been shown that oxidative stress can activate ATM, an upstream kinase in the DDR pathway. Interestingly, it has been shown that ATM phosphorylation can occur directly by H_2O_2-treatment, independently of DSB formation, indicating that oxidative stress and DSBs can synergize towards an increase in the level of ATM activation and a robust DNA damage response (Guo et al. 2010).

PRESENCE OF SENESCENT CELLS IN VIVO: IN SEARCH OF A RELIABLE MARKER

Although cellular senescence was intensely studied in vitro by many laboratories for several decades, there was always a skepticism concerning if it represents a real biological phenomenon or a laboratory artifact due to inappropriate culture conditions. Accordingly, a reliable senescence marker was more than necessary. Several markers have been proposed, such as the thymidine labeling index (Cristofalo and Sharf 1973), lipofuscin deposition and lysosomal enzyme activity (Cristofalo et al. 1967; Cristofalo 2005). However, by using these markers it was not possible to discriminate senescent cells from simply growth arrested cells or cells exposed to various exogenous stresses. In the mid 90's a specific histochemical procedure of β-galactosidase expression detectable at pH 6, termed "senescence associated-β galactosidase staining" (SA-β gal staining) was developed (Dimri et al. 1995). This marker was found to be expressed in the perinuclear region of senescent, but not-presenescent, human skin fibroblasts and keratinocytes, and was absent from quiescent fibroblasts and terminally differentiated keratinocytes. By using this technique the authors have shown for the first time the presence of senescent cells in vivo and an age-related accumulation of senescent cells (Dimri et al. 1995). SA β-gal staining for the identification of senescent cell in vivo has been, and still is, used by several laboratories. However, early on a criticism has been evolved. In particular, it has been shown that SA β-gal staining increased in replicating senescent, as well as in quiescent, serum starved or over confluent cells. This staining was irreversible as an outcome of replicative senescence or the exposure to H_2O_2, but was reversible after serum starvation or confluence (Severino et al. 2000; Cristofalo 2005; Yang and Hu 2005). The above indicate that SA β-gal staining must be used in cautious and in parallel to other markers.

Another feature of senescent cells that has been used to develop specific markers for their presence in vivo is the senescent associated heterochromatin foci (SAHF). The formation of these foci of heterochromatin was considered to be crucial for the acceleration of the senescence process via the silencing of genes that are responsible for a normal cell cycle progression (Narita et al. 2003). Heterochomatinization was evident with DAPI staining, the presence of histone H3 methylated on Lys9 or increased expression of the heterochromatic protein HP1γ (Collado et al. 2005; Bartkova et al. 2006). However, it has been recently shown that SAHF formation is not a universal feature of senescence. While premature senescence induced by oncogenic Ras is characterized by SAHF presence, this is not the case in cells that became senescent in response to telomere erosion or DNA damage (Kocylowski and Halazonetis 2011; Di Micco et al. 2011; Kosar et al. 2011). In particular, when several senescence-provoking stresses have been tested in a series of normal fibroblasts and keratinocytes, it has been found that although all cell types formed SAHF in response to activated H-Ras, only one fibroblast cell line formed these foci in response to other stresses, such as etoposide, doxorubicin, hydroxyurea, bacterial toxins or telomere erosion (Kosar et al. 2011). Even more, inactivation of ATM or p53 allows the proliferation of oncogene-expressing cells that retain senescent associated heterochromatin foci (Di Micco et al. 2011).

As mentioned above, DNA damage response is a central motif in replicative and stress-induced senescence. An early response to DNA damage and especially to DSBs is the phopshorylation of H2AX – a minor histone H2A variant – to produce γ-H2AX (Rogakou et al. 1998; Redon et al. 2002; Sedelnikova et al. 2002). H2AX is phosphorylated in Ser-139 by members of the phosphatidyl-inositol-3-kinase (PIKK) family, such as ataxia telangiectasia mutated (ATM) and ATM and RAD3-related (ATR). ATM and/or ATR can phosphorylate H2AX after exposure to ionizing radiation, UV or reactive oxygen species (Mah et al. 2010; Ward et al. 2004; Falck et al. 2005; Stiff et al. 2006; Guo et al. 2010). Interestingly, telomere-initiated senescence is characterized by the presence of γ-H2AX foci (d'Adda di Fagagna et al. 2003). Similarly, ionizing radiation-induced premature senescence is characterized by the presence of a limited number (2-3 per nucleus) of large γ-H2AX foci, representing probably sites of unrepairable DNA damage (Suzuki et al. 2006; Papadopoulou and Kletsas 2011). Finally, γ-H2AX foci appear in the tissue of aged mice, as well as of mice exposed to ionizing radiation (Hudson et al. 2011; Wang et al. 2009). However, the formation of γ-H2AX foci is implicated in several other pathologies too, such as cancer, chronic inflammatory diseases or ischemia-reperfusion injury, indicating that it does not represent a unique marker for senescence and can be used as such only in combination with other established senescence markers.

According to the above, a number of morphological and molecular markers must be used to identify the senescent cells in vivo, as to investigate their role in ageing and age-related pathologies. Indeed, senescent cells were discovered in sites of various diseases, e.g. atherosclerosis (Vasile et al. 2001; Gorgoulis et al. 2005), chronic hepatitis C (Paradis et al. 2001), venous ulcers (Stanley and Osler 2001), intervertebral disc herniation (Roberts et al. 2006), prostatic hyperplasia (Choi et al. 2000), to name a few. Concerning normal tissues, it was found that there is an accumulation of dermal fibroblasts expressing senescence markers in primates with advanced age. In contrast, in a post-mitotic tissue (skeletal muscle) only a small percentage of cells expressed these markers in the elderly (Jeyapalan et al. 2007).

As mentioned already, several markers used to characterize senescent cells are not exclusive for these cells as they are also expressed under other circumstances, e.g. in growth arrested cells. However, the cyclin-dependent kinase inhibitor $p16^{INK4a}$, that reinforces cell cycle arrest by activating the pRb tumor suppressor protein, was found to be a robust in vivo biomarker of cellular ageing. This protein was found to be overexpressed in almost all rodent tissues with advanced age, while its expression was attenuated in kidney, ovary and heart after calorie restriction, an established anti-ageing treatment (Krishnamurthy et al. 2004). In addition, in the dermis and epidermis of healthy human skin an increase of the $p16^{INK4a}$-positive cells was observed with age, while in immunofluorescence co-expression studies Ki-67-positive (proliferating) cells were found to be negative for $p16^{INK4a}$ (Ressler et al. 2006). Interestingly, very recently, the use of this senescence biomarker has been used to provide important information of the role of senescent cells in vivo. Baker and collaborators (Baker et al. 2011) have managed to eliminate $p16^{INK4a}$-positive cells in mice, by activating a drug-inducible caspase. In young mice, with no signs of senescence, the removal of $p16^{INK4a}$-positive cells delayed age-related deficits in several tissues (adipose tissue, skeletal muscle and eye), i.e. the appearance of kyfosis and cataract was delayed, the muscle fibers became thicker and the exercise fitness and fat deposit were increased. In addition, in older animals senescent cells elimination led to a decrease of senescent markers and to an increase of muscle fiber diameter, enlarged fat deposits and improved fitness performance. The above data, strongly indicate that $p16^{INK4a}$ is a reliable marker for senescent cells. Even more, they robustly support that the presence of senescent cells in vivo affects considerably tissue homeostasis and contributes to age-related pathologies and that their clearance can delay these changes and can be a future anti-ageing therapeutic approach.

AGED STROMA AND CANCER DEVELOPMENT

The majority of human tumors – especially in advanced age – are classified as carcinomas and are of epithelial origin (Elenbaas and Weinberg 2001; Radisky and Bissell 2004). Accordingly, the traditional focus in the studies on tumorigenesis was for many years epitheliocentric, based on the concept that tumor generation and development is a cell-autonomous process, driven by the genetic alterations of epithelial cells (Elenbaas and Weinberg 2001; Tlsty and Hein 2001; Baglole et al. 2006). However, in many carcinomas the surrounding tumor microenvironment or "stroma" can represent more than the 90% or the tumor mass (Schurch et al. 1981; Elenbaas and Weinberg 2001; Baglole et al. 2006; Orimo and Weinberg 2006). Nevertheless, for many years stroma was seen as a passive structural support for tumor growth. This concept has been challenged by a large body of evidence indicating that the symbiosis of cancer cells with their microenvironment regulates the process of tumorigenesis (Beacham and Cukierman 2005). Numerous studies imply that a normal stroma is able to constrain cancer development, indicating that it can be dominant over the malignant phenotype, while an "activated" tumor microenvironment can promote tumor growth (Park et al. 2000; Bissell et al. 2002; Kenny and Bissell 2003; Beacham and Cukierman 2005). Moreover, a hormonally altered stromal environment can stimulate carcinogenesis of non-tumorigenic epithelial cells (Cunha et al. 2003). Several factors are responsible for these cell-cell or cell-matrix interactions. As an example, alterations in matrix metalloproteases (MMPs) - the enzymes that can catabolise extracellular matrix components - can affect tissue homeostasis towards cancer development. Lack of MMPs can suppress tumorigenesis, whereas an environment ectopically overexpressing these enzymes can produce carcinomas derived from adjacent epithelial cells (D'Armiento et al. 1995; Sympson et al. 1995; Wilson et al. 1997; Masson et al. 1998; Rudolph-Owen et al. 1998; Thomasset et al. 1998; Sternlicht et al. 1999; Sternlicht et al. 2000; Tlsty and Hein 2001; Coussens and Werb 2002; Egeblad and Werb 2002). Fibroblasts represent the predominat stromal cell type that play a key role in tumor development (Elenbaas and Weinberg 2001; Tlsty and Hein 2001; Bhowmick et al. 2004; Beacham and Cukierman 2005; Baglole et al. 2006). They are responsible for the deposition of extracellular matrix (ECM) components, ECM remodeling via the secretion of MMPs and other enzymes, as well as for the release of cytokines and growth factors acting in a paracrine manner on cancer cells (Elenbaas and Weinberg 2001; Bhowmick et al. 2004; Kiaris et al. 2008). It has been shown that normal fibroblasts can inhibit tumor growth, while "activated" or cancer-associated fibroblasts (CAFs) can promote it (Picard et al. 1986; Horgan et al. 1987; Camps et al. 1990; Noel et al. 1993; Olumi et al. 1999; Kuperwasser et al. 2004; Maffini et al. 2004; Beacham and Cukierman 2005; Baglole et al. 2006).

Advanced age is considered to be the largest risk factor for a variety of diseases, including cancer. This is, at least partly, due to the accumulation of somatic mutations over a lifetime that leads to the activation of oncogenes and the silencing of tumor-suppressor genes, which drive cancer development and progression (DePinho 2000). However, several lines of evidence indicate that the aged stroma must play a crucial role in this process. This stems from the observed increased number of senescent cells over time (Dimri 1995; DePinho 2000). Furthermore, it has been hypothesized that the accumulation of senescent cells in the tissues of the elderly can create a permissive environment for the growth of neoplastic cells and this has been strongly supported by experimental evidence (Krtolica and Campisi 2002). In particular, it has been reported that human fibroblasts senescent by replicative exhaustion, expression of oncogenic RAS or after exposure to hydrogen peroxide can promote epithelial cell growth and tumorigenesis (Krtolica et al. 2001). Similarly, human fibroblasts prematurely senescent by DNA damaging agents, such as bleomycin or ionizing radiation, can also increase the growth of xenograft tumors in immunocompromised mice, (Liu and Hornsby 2007; Papadopoulou and Kletsas 2011). These data provide a direct link between cellular senescence and cancer. They also support the theory of "antagonistic pleiotropy", proposing that some tumor-suppressor mechanisms, such as replicative or premature senescence, can be beneficial early in life, while in older organisms (probably after the reproductive period) they can be deleterious for homeostasis (Kirkwood and Austad 2000). Even more, they reinforce the need for reliable senescence markers to be used in stromal cells, as they can imply the existence of a microenvironment able to support and promote tumor development.

THE SENESCENCE-ASSOCIATED SECRETORY PHENOTYPE

In order to understand the significance of senescent cell in vivo and their role in age-related disorders, including cancer, many laboratories have tried in the past to characterize the senescent phenotype by using classical biochemical approaches. On the other hand, more recently the use of cDNA microarrays showed that senescent cells express a specific pro-inflammatory phenotype, as these cells overexpress several genes linked to inflammatory states, such as matrix metalloproteases, growth factors, inflammatory cutokines and other molecules linked to inflammation (Shelton et al. 1999). Later on, protein arrays have been used as to describe the molecules secreted by senescent cells. A series of studies have shown that senescent cells secrete increased amounts of interleukins (e.g. IL-6, IL-1 and IL-7), chemokines (e.g. IL-8 and GRO-a), other inflammatory factors (e.g. GM-CSF and IFN-γ), growth factors (e.g. VEGF, HGF and SDF-1), proteases (e.g. MMPS and cathepsin B), several receptors (e.g. ICAM-1, Fas and EGF-R), non-protein soluble factors (e.g. PGE2) and ECM components (e.g. fibronectin) (Coppe et al. 2008; Rodier et al. 2009; Davalos et al. 2010). The sum of these factors is called senescence-associated secretory phenotype (SASP). The SASP factors, independently or collectively, seem to work in a paracrine manner and to affect tissue homeostasis. Interestingly, they can also reinforce cellular senescence in an autocrine manner (Coppe et al. 2008). Among these secreted factors that have been found to be overexpressed in replicative- or oncogene induced-senescence are several inflammatory mediators, such as IL-6 and IL-8, GROα, insulin-like growth factor binding protein 7 (IGFBP7) or plasminogen activator inhibitor-1 (PAI-1). Interestingly, these agents can strengthen senescence via distinct mechanisms (Acosta et al. 2008). For example, IGFBP7 exerts its effect by regulating the MEK/ERK pathway (Wajapeyee et al.), while PAI-1 via the PI3K activity (Kortlever 2006). On the other hand, IL-6 upregulates the cell cycle inhibitor $p15^{INK4b}$, while the CXCR2 ligands IL-8 and GROα activate reactive oxygen species and eventually trigger senescence (Acosta et al. 2008). It must be mentioned that SASP can be provoked by several senescence-inducers and requires several days to develop after DNA damage, indicating that it does not represent a DNA damage response. However, a persistent DDR seems to be necessary for a robust SASP (Davalos et al. 2010).

Among the proteins mentioned above that characterize SASP are factors known to be involved in cancer development, as well as in invasion and metastasis, such as VEGF, MMPs and SDF-1 (Coppe et al. 2008). In vitro and in vivo experiments have clearly shown that senescent cells can promote carcinogenesis through the secretion of these factors (Coppe et al. 2006; Liu and Hornsby 2007; Papadopoulou and Kletsas 2011). However, these effects seem to be specific for each cell type and the senescence inducer. For example, the ionizing radiation-mediated senescence, increase of MMP expression and enhancement of cancer cells' growth in vitro and in vivo has been reported for human lung but not skin fibroblasts (Liu and Hornsby 2007; Papadopoulou and Kletsas 2011). On the other hand, senescent fibroblasts can provoke an epithelial-to-mesenchymal transition (EMT) – a crucial step in cancer development – in breast but not lung cancer cells (Coppe et al. 2006; Papadopoulou and Kletsas 2011).

The expression of p53 and $p16^{INK4a}$ is crucial in senescence, while the majority of cancers harbor mutations in p53 and the $p16^{INK4a}$/pRb pathway. Accordingly, their role in SASP is of extreme interest. Recently, it has been shown that the ectopic expression of $p16^{INK4a}$ can provoke premature senescence but without the appearance of an inflammatory phenotype. Even more, $p16^{INK4a}$ expression by cells undergoing replicative senescence limited the accumulation of DNA damage and cytokine secretion. This indicates an indirect role for this inhibitor in suppressing the inflammatory secretory phenotype and suggests that the SASP is not a consequence of the activation or senescence *per se*, but rather is a damage response that is separable from the growth arrest (Coppe et al. 2011). In addition, p53 expression is not needed for the establishment and maintenance of SASP and furthermore it can also restrain it (Coppe et al. 2008). These data indicate that these senescence mediators can play an additional role in tumor suppression, i.e. by inducing senescence but in the same time by inhibiting the cancer supporting effects of SASP.

CLEARANCE OF SENESCENT CELLS IN VIVO

As senescent normal cells express a pro-inflammatory phenotype, it is expected that these cells could be eliminated by the action of the immune system. This has been tested recently in irradiated mouse tissues (Le et al. 2010). The animals were exposed to sublethal doses of ionizing radiation and senescent cells were identified by DNA damage foci and the increase in $p16^{INK4a}$ expression. Their presence in liver, brain and lung tissues was followed for a period of 45 weeks and has been found that there is an active clearance of ionizing-radiation DNA damaged cells in a tissue-specific manner, as senescent liver cells are preferentially eliminated, while lung cells were more persistent. Interestingly, the elimination of cells with senescent-like characteristics is independent of T, B and NK cells. Even more, a similar phenomenon was observed in p53-/-mice, suggesting that p53 does not play a crucial role in the persistence or clearance of cells after exposure to ionizing radiation (Le et al. 2010). Similarly, we found that skin fibroblasts from patients exposed to radiotherapy overexpress features of senescent cells (our unpublished observations). On the other hand, it has been shown that in mouse liver injured hepatocytes activate stellate cells causing fibrosis, but becoming senescent, as well. These senescent cells attract natural killer cells, thus activating senescent cells' clearance and consequently limiting tissue damage (Krizhanovsky et al. 2008). Even more, it has been found that in liver carcinomas of p53-deficient mouse, that the reactivation of p53 leads to an activation of a premature senescence program and an upregulation of inflammatory cytokines, leading to the triggering of an innate immune response and a tumor regression (Xue et al. 2007). Recently, it has been shown that senescent hepatocytes secrete soluble factors that trigger their immune-mediated clearance (characterized as "senescence surveillance"). Moreover, impaired clearance lead to the development of carcinomas, indicating that this type of surveillance can function as a part of the general anti-tumor immune barrier (Kang et al. 2011). These data indicate that the combination of senescence with the immune system can limit tumor development, suggesting new therapeutic modalities.

ANTI-CANCER TREATMENT AND CELLULAR SENESCENCE

Treatment-induced Premature Senescence in Normal Cells

Major anti-cancer treatment regimes are based on the use of agents inducing DNA damage, such as ionizing radiation or various chemotherapeutic drugs. However, these treatments inevitably affect the stroma surrounding the tumor. As DNA damage can provoke accelerated senescence, several studies have focus on the biological effect of chemotherapeutic drugs on stromal fibroblasts. Doxorubicin and etoposide are topoisomerase II poisons that provoke DNA double stand breaks (DSBs) and single stand breaks (SSBs) (Vamvakas et al. 1997). Both drugs can induce a senescent-like permanent growth arrest in normal human fibroblasts (Robles et al. 1999). On the other hand, camptothecin, a topoisomerase I poison that causes SSBs which can be converted to DSBs when the cells traverse S phase (Ryan et al. 1991; Chen and Liu 1994), is unable to provoke senescence in these cells (Robles et al. 1999). However cisplatin, another agent causing SSBs that can also lead to DSBs, can induce premature senescence in human fibroblasts with a concomitant reduction of gap junctions (Zhao et al. 2004). Busulfan is an alkylating agent that can cause DNA-DNA and DNA-protein cross-links that may trigger a DNA damage response (Iwamoto et al. 2004; Mertins et al. 2004), and in contrast to the majority of chemotherapeutic drugs it affects mostly slow proliferating or non-proliferating cells (Probin et al. 2006). Busulfan was also found to provoke premature senescence in human fibroblasts, but in a mode independent of p53 activation (Probin et al. 2006). In contrast, busulfan depletes intracellular glutathione and increases reactive oxygen species, leading to the activation of ERK and p38 MAPK. Incubation with the potent antioxidant N-acetyl-cysteine can effectively attenuate busulfan-mediated premature senescence (Probin et al. 2007). Finally taxol, a chemotherapeutic agent that promotes the assembly of tubulin heterodimers and inhibits their depolymerization, was found to promote premature senescence of human stromal fibroblasts (our unpublished observations). This is probably due to the formation of reactive oxygen species and to alterations of mitochondrial membrane permeability (Varbiro et al. 2001).

Ionizing radiation by producing DSBs and SSBs in the DNA is used in radiotherapy as to produce cancer cells' arrest and/or death. It also affects stromal cells and several studies have shown that it can provoke premature senescence (Naka et al. 2004; Suzuki et al. 2006; Papadopoulou and Kletsas 2011). This is achieved by the immediate activation of the

ATM-Chk2-p53-p21^{WAF1}-pRb axis and the delayed overexpression and phosphorylation of p16^{INK4a} and p38 MAPK, respectively.

Anticancer Treatment-mediated Premature Senescence in Cancer Cells

Conventionally, aim of chemotherapy and radiotherapy was to inhibit tumor growth by provoking cell cycle arrest and/or death of cancer cells. However, another effect of these treatments is the induction of premature senescence. Doxorubicin was found to drive HCT116 colon carcinoma and HT1080 fibrosarcoma cells to a senescence-like phenotype. By manipulating the expression of p53 and p21^{WAF1} it has been shown that both are positive regulators of SIPS, but their function is not absolutely necessary for this response (Chang et al. 1999). Adriamycin can provoke also senescence in breast cancer MCF-7 cells, in a telomere length-independent manner; however, inactivation of p53 can turn these cells to apoptosis (Elmore et al. 2002). Low concentrations of topoisomerase I inhibitor camptothecin can drive HCT116 cells to premature senescence, in a p21^{WAF1}-dependent manner (Han et al. 2002). Cisplatin provokes a senescent-like growth arrest in CNE1 nasopharyngeal carcinoma cells in a dose-dependent and telomerase-independent manner (Wang et al. 1998). In addition, in vitro and in vivo experiments (by using xenografts) have also shown that cisplatin can induce a senescence-like phenotype in non-small cell lung cancer cells overexpressing p16^{INK4a}, indicating that the ectopic expression of this protein by gene transfer can improve tumor suppression by this drug (Fang et al. 2007). Finally discodermolide, that stabilizes microtubules more potently than Taxol, can provoke accelerated senescence in A549 non-small cell lung cancer cells, equally to doxorubicin (Klein et al. 2005).

Ionizing radiation can also lead to premature senescence of cancer cells. By using p53 wild-type MCF-7 cells, MDA-MB-231 cells with mutant p53 and MCF-7/E6 cells with attenuated p53 function, it has been shown that radiation can induce accelerated senescence in breast cancer cells in a p53-dependent manner (Jones et al. 2005). Similarly, the induction of accelerated senescence in several cancer cell lines with wild type p53 was found to be dependent on p21^{WAF1} function (Mirzayans et al. 2005).

Besides the direct effect of anticancer treatment on the induction of senescence in cancer cells it has been shown that they can also exert a bystander effect. When naive MCF-7 or HCT116 cells are exposed to conditioned medium from MCF-7 cells that became senescence by adriamycin-treatment they also undergo senescence (Di et al. 2008). In addition, cells treated with ionizing radiation, as well as senescent or malignant cells can release several mediators, such as TGF-β or nitric oxide, and produce DNA damage in nearby cells, as shown by the formation of γH2AX foci (Dickey et al. 2009). Even more interesting, these interactions can be not only local but also systemic, as tumors can induce DNA damage in distant tissues in vivo by the activation of circulating cytokines (Redon et al. 2010).

The induction of cancer cells' senescence by anticancer drugs was observed also in vivo (Roninson 2003). Studies on breast cancer patients who had or not received chemotherapy indicated that these drugs can induce senescence in tumor tissue. SA-β-gal staining was associated with low p53 and high $p16^{INK4a}$ expression (te Poele et al. 2002). Similarly, the presence of p53 and $p16^{INK4a}$ is crucial for the chemotherapy-mediated induction of senescence in primary murine lymphomas and, furthermore, tumors with the ability for senescence have a better prognosis following chemotherapy (Schmitt et al. 2002).

Work from many laboratories has clearly shown that cellular senescence (replicative or stress-induced) represents a potent anticancer mechanism. On the other hand, the pro-inflammatory phenotype of senescent cells and their specific secretome seems to modulate locally and possibly systemically the microenvironment and to facilitate tumor growth. The delineation of the complex molecular networks linking these events may lead to novel therapeutic approaches aiming at the regulation of the abundance of senescent cells and their deleterious effects on tissue homeostasis.

REFERENCES

Acosta, J.C., A. O'Loghlen, A. Banito, S. Raguz and J. Gil. 2008. Control of senescence by CXCR2 and its ligands. Cell Cycle 7(19): 2956-2959.

Baglole, C.J., D.M. Ray, S.H. Bernstein, S.E. Feldon, T.J. Smith, P.J. Sime and R.P. Phipps. 2006. More than structural cells, fibroblasts create and orchestrate the tumor microenvironment. Immunol Invest 35(3-4): 297-325.

Baker, D.J., T. Wijshake, T. Tchkonia, N.K. LeBrasseur, B.G. Childs, B. van de Sluis, J.L. Kirkland and J.M. van Deursen. 2011. Clearance of p16Ink4a-positive senescent cells delays ageing-associated disorders. Nature 479(7372): 232-236.

Bandyopadhyay, D., N. Timchenko, T. Suwa, P.J. Hornsby, J. Campisi and E.E. Medrano. 2001. The human melanocyte: a model system to study the complexity of cellular aging and transformation in non-fibroblastic cells. Exp Gerontol 36(8): 1265-1275.

Bartkova, J., N. Rezaei, M. Liontos, P. Karakaidos, D. Kletsas, N. Issaeva, L.V. Vassiliou, E. Kolettas, K. Niforou, V.C. Zoumpourlis, M. Takaoka, H. Nakagawa,

F. Tort, K. Fugger, F. Johansson, M. Sehested, C.L. Andersen, L. Dyrskjot, T. Orntoft, J. Lukas, C. Kittas, T. Helleday, T.D. Halazonetis, J. Bartek and V.G. Gorgoulis. 2006. Oncogene-induced senescence is part of the tumorigenesis barrier imposed by DNA damage checkpoints. Nature 444(7119): 633-637.

Beacham, D.A. and E. Cukierman. 2005. Stromagenesis: the changing face of fibroblastic microenvironments during tumor progression. Semin Cancer Biol 15(5): 329-341.

Bhowmick, N.A., E.G. Neilson and H.L. Moses. 2004. Stromal fibroblasts in cancer initiation and progression. Nature 432(7015): 332-337.

Bissell, M.J., D.C. Radisky, A. Rizki, V.M. Weaver and O.W. Petersen. 2002. The organizing principle: microenvironmental influences in the normal and malignant breast. Differentiation 70(9-10): 537-546.

Bodnar, A.G., M. Ouellette, M. Frolkis, S.E. Holt, C.P. Chiu, G.B. Morin, C.B. Harley, J.W. Shay, S. Lichtsteiner and W.E. Wright. 1998. Extension of life-span by introduction of telomerase into normal human cells. Science 279(5349): 349-352.

Campisi, J. and F. d'Adda di Fagagna. 2007. Cellular senescence: when bad things happen to good cells. Nat Rev Mol Cell Biol 8(9): 729-740.

Camps, J.L., S.M. Chang, T.C. Hsu, M.R. Freeman, S.J. Hong, H.E. Zhau, A.C. von Eschenbach and L.W. Chung. 1990. Fibroblast-mediated acceleration of human epithelial tumor growth in vivo. Proc Natl Acad Sci U S A 87(1): 75-79.

Chang, B.D., Y. Xuan, E.V. Broude, H. Zhu, B. Schott, J. Fang and I.B. Roninson. 1999. Role of p53 and p21waf1/cip1 in senescence-like terminal proliferation arrest induced in human tumor cells by chemotherapeutic drugs. Oncogene 18(34): 4808-4818.

Chen, A.Y. and L.F. Liu. 1994. DNA topoisomerases: essential enzymes and lethal targets. Annu Rev Pharmacol Toxicol 34: 191-218.

Choi, J., I. Shendrik, M. Peacocke, D. Peehl, R. Buttyan, E.F. Ikeguchi, A.E. Katz and M.C. Benson. 2000. Expression of senescence-associated beta-galactosidase in enlarged prostates from men with benign prostatic hyperplasia. Urology 56(1): 160-166.

Collado, M., M.A. Blasco and M. Serrano. 2007. Cellular senescence in cancer and aging. Cell 130(2): 223-233.

Collado, M., J. Gil, A. Efeyan, C. Guerra, A.J. Schuhmacher, M. Barradas, A. Benguria, A. Zaballos, J.M. Flores, M. Barbacid, D. Beach and M. Serrano. 2005. Tumour biology: senescence in premalignant tumours. Nature 436(7051): 642.

Coppe, J.P., K. Kauser, J. Campisi and C.M. Beausejour. 2006. Secretion of vascular endothelial growth factor by primary human fibroblasts at senescence. J Biol Chem 281(40): 29568-29574.

Coppe, J.P., C.K. Patil, F. Rodier, Y. Sun, D.P. Munoz, J. Goldstein, P.S. Nelson, P.Y. Desprez and J. Campisi. 2008. Senescence-associated secretory phenotypes reveal cell-nonautonomous functions of oncogenic RAS and the p53 tumor suppressor. PLoS Biol 6(12): 2853-2868.

Coppe, J.P., F. Rodier, C.K. Patil, A. Freund, P.Y. Desprez and J. Campisi. 2011. Tumor suppressor and aging biomarker p16(INK4a) induces cellular senescence without the associated inflammatory secretory phenotype. J Biol Chem 286(42): 36396-36403.

Coussens, L.M. and Z. Werb. 2002. Inflammation and cancer. Nature 420(6917): 860-870.

Cristofalo, V.J. 2005. SA beta Gal staining: biomarker or delusion. Exp Gerontol 40(10): 836-838.

Cristofalo, V.J., N. Parris and D. Kritchevsky. 1967. Enzyme activity during the growth and aging of human cells in vitro. J Cell Physiol 69(3): 263-271.

Cristofalo, V.J. and B.B. Sharf. 1973. Cellular senescence and DNA synthesis. Thymidineincorporationasameasureofpopulationageinhumandiploidcells.ExpCell Res 76(2): 419-427.

Cunha, G.R., S.W. Hayward, Y.Z. Wang and W.A. Ricke. 2003. Role of the stromal microenvironment in carcinogenesis of the prostate. Int J Cancer 107(1): 1-10.

d'Adda di Fagagna, F., P.M. Reaper, L. Clay-Farrace, H. Fiegler, P. Carr, T. Von Zglinicki, G. Saretzki, N.P. Carter and S.P. Jackson. 2003. A DNA damage checkpoint response in telomere-initiated senescence. Nature 426(6963): 194-198.

D'Armiento, J., T. DiColandrea, S.S. Dalal, Y. Okada, M.T. Huang, A.H. Conney and K. Chada. 1995. Collagenase expression in transgenic mouse skin causes hyperkeratosis and acanthosis and increases susceptibility to tumorigenesis. Mol Cell Biol 15(10): 5732-5739.

Davalos, A.R., J.P. Coppe, J. Campisi and P.Y. Desprez. 2010. Senescent cells as a sourceof inflammatory factors for tumor progression. Cancer Metastasis Rev 29(2): 273-283.

Di Micco, R., G. Sulli, M. Dobreva, M. Liontos, O.A. Botrugno, G. Gargiulo, R. dal Zuffo, V. Matti, G. d'Ario, E. Montani, C. Mercurio, W.C. Hahn, V. Gorgoulis, S. Minucci and F. d'Adda di Fagagna. 2011. Interplay between oncogene-induced DNA damage response and heterochromatin in senescence and cancer. Nat Cell Biol 13(3): 292-302.

Di, X., A.T. Bright, R. Bellott, E. Gaskins, J. Robert, S. Holt, D. Gewirtz and L. Elmore. 2008. A chemotherapy-associated senescence bystander effect in breast cancer cells. Cancer Biol Ther 7(6): 864-872.

Dickey, J.S., B.J. Baird, C.E. Redon, M.V. Sokolov, O.A. Sedelnikova and W.M. Bonner. 2009. Intercellular communication of cellular stress monitored by gamma-H2AX induction. Carcinogenesis 30(10): 1686-1695.

Dimri, G.P., X. Lee, G. Basile, M. Acosta, G. Scott, C. Roskelley, E.E. Medrano, M. Linskens, I. Rubelj, O. Pereira-Smith, M. Peacocke and J. Campisi. 1995. A biomarker that identifies senescent human cells in culture and in aging skin in vivo. Proc Natl Acad Sci U S A 92(20): 9363-9367.

Effros, R.B., M. Dagarag and H.F. Valenzuela. 2003. In vitro senescence of immune cells. Exp Gerontol 38(11-12): 1243-1249.

Egeblad, M. and Z. Werb. 2002. New functions for the matrix metalloproteinases in cancer progression. Nat Rev Cancer 2(3): 161-174.

Elenbaas, B. and R.A. Weinberg. 2001. Heterotypic signaling between epithelial tumor cells and fibroblasts in carcinoma formation. Exp Cell Res 264(1): 169-184.

Elmore, L.W., C.W. Rehder, X. Di, P.A. McChesney, C.K. Jackson-Cook, D.A. Gewirtz and S.E. Holt. 2002. Adriamycin-induced senescence in breast tumor cells involves functional p53 and telomere dysfunction. J Biol Chem 277(38): 35509-35515.

Erusalimsky, J.D. 2009. Vascular endothelial senescence: from mechanisms to pathophysiology. J Appl Physiol 106(1): 326-332.

Erusalimsky, J.D. and C. Skene. 2009. Mechanisms of endothelial senescence. Exp Physiol 94(3): 299-304.

Falck, J., J. Coates and S.P. Jackson. 2005. Conserved modes of recruitment of ATM, ATR and DNA-PKcs to sites of DNA damage. Nature 434(7033): 605-611.

Fang, K., C.C. Chiu, C.H. Li, Y.T. Chang and H.T. Hwang. 2007. Cisplatin-induced senescence and growth inhibition in human non-small cell lung cancer cells with ectopic transfer of p16INK4a. Oncol Res 16(10): 479-488.

Francisco, D. C., P. Peddi, J.M. Hair, B.A. Flood, A.M. Cecil, P.T. Kalogerinis, G. Sigounas and A.G. Georgakilas. 2008. Induction and processing of complex DNA damage in human breast cancer cells MCF-7 and nonmalignant MCF-10A cells. Free Radic Biol Med 44(4): 558-569.

Gorenne, I., M. Kavurma, S. Scott and M. Bennett. 2006. Vascular smooth muscle cell senescence in atherosclerosis. Cardiovasc Res 72(1): 9-17.

Gorgoulis, V.G. and T.D. Halazonetis. 2010. Oncogene-induced senescence: the bright and dark side of the response. Curr Opin Cell Biol 22(6): 816-827.

Gorgoulis, V.G., H. Pratsinis, P. Zacharatos, C. Demoliou, F. Sigala, P.J. Asimacopoulos, A.G. Papavassiliou and D. Kletsas. 2005. p53-dependent ICAM-1 overexpression in senescent human cells identified in atherosclerotic lesions. Lab Invest 85(4): 502-511.

Green, D.R. and G.I. Evan. 2002. A matter of life and death. Cancer Cell 1(1): 19-30.

Guo, Z., S. Kozlov, M.F. Lavin, M.D. Person and T.T. Paull. 2010. ATM activation by oxidative stress. Science 330(6003): 517-521.

Hahn, W.C., C.M. Counter, A.S. Lundberg, R.L. Beijersbergen, M.W. Brooks and R.A. Weinberg. 1999. Creation of human tumour cells with defined genetic elements. Nature 400(6743): 464-468.

Halazonetis, T.D., V.G. Gorgoulis and J. Bartek. 2008. An oncogene-induced DNA damage model for cancer development. Science 319(5868): 1352-1355.

Hampel, B., F. Malisan, H. Niederegger, R. Testi and P. Jansen-Durr. 2004. Differential regulation of apoptotic cell death in senescent human cells. Exp Gerontol 39(11-12): 1713-1721.

Han, Z., W. Wei, S. Dunaway, J.W. Darnowski, P. Calabresi, J. Sedivy, E.A. Hendrickson, K.V. Balan, P. Pantazis and J.H. Wyche. 2002. Role of p21 in apoptosis and senescence of human colon cancer cells treated with camptothecin. J Biol Chem 277(19): 17154-17160.

Hanahan, D. and R.A. Weinberg. 2011. Hallmarks of cancer: the next generation. Cell 144(5): 646-674.

Harley, C. B., A.B. Futcher and C.W. Greider. 1990. Telomeres shorten during ageing of human fibroblasts. Nature 345(6274): 458-460.

Hayflick, L. 1965. The limited in vitro lifetime of human diploid cell strains. Exp Cell Res 37: 614-636.

Herbig, U., W.A. Jobling, B.P. Chen, D.J. Chen and J.M. Sedivy. 2004. Telomere shortening triggers senescence of human cells through a pathway involving ATM, p53, and p21(CIP1), but not p16(INK4a). Mol Cell 14(4): 501-513.

Horgan, K., D.L. Jones and R.E. Mansel. 1987. Mitogenicity of human fibroblasts in vivo for human breast cancer cells. Br J Surg 74(3): 227-229.

Hudson, D., I. Kovalchuk, I. Koturbash, B. Kolb, O.A. Martin and O. Kovalchuk. 2011. Induction and persistence of radiation-induced DNA damage is more pronounced in young animals than in old animals. Aging (Albany NY) 3(6): 609-620.

Iwamoto, T., Y. Hiraku, S. Oikawa, H. Mizutani, M. Kojima and S. Kawanishi. 2004. DNA intrastrand cross-link at the 5'-GA-3' sequence formed by busulfan and its role in the cytotoxic effect. Cancer Sci 95(5): 454-458.

Jeyapalan, J.C., M. Ferreira, J.M. Sedivy and U. Herbig. 2007. Accumulation of senescent cells in mitotic tissue of aging primates. Mech Ageing Dev 128(1): 36-44.

Jones, K.R., L.W. Elmore, C. Jackson-Cook, G. Demasters, L.F. Povirk, S.E. Holt and D.A. Gewirtz. 2005. p53-Dependent accelerated senescence induced by ionizing radiation in breast tumour cells. Int J Radiat Biol 81(6): 445-458.

Kang, M.K., C. Bibb, M.A. Baluda, O. Rey and N.H. Park. 2000. In vitro replication and differentiation of normal human oral keratinocytes. Exp Cell Res 258(2): 288-297.

Kang, T.W., T. Yevsa, N. Woller, L. Hoenicke, T. Wuestefeld, D. Dauch, A. Hohmeyer, M. Gereke, R. Rudalska, A. Potapova, M. Iken, M. Vucur, S. Weiss, M. Heikenwalder, S. Khan, J. Gil, D. Bruder, M. Manns, P. Schirmacher, F. Tacke, M. Ott, T. Luedde, T. Longerich, S. Kubicka and L. Zender. 2011. Senescence surveillance of pre-malignant hepatocytes limits liver cancer development. Nature 479(7374): 547-551.

Kenny, P.A. and M.J. Bissell. 2003. Tumor reversion: correction of malignant behavior by microenvironmental cues. Int J Cancer 107(5): 688-695.

Kiaris, H., G. Trimis and A.G. Papavassiliou. 2008. Regulation of tumor-stromal fibroblast interactions: implications in anticancer therapy. Curr Med Chem 15(29): 3062-3067.

Kirkwood, T.B. and S.N. Austad. 2000. Why do we age? Nature 408(6809): 233-238.

Klein, L.E., B.S. Freeze, A.B. Smith, 3rd and S.B. Horwitz. 2005. The microtubule stabilizing agent discodermolide is a potent inducer of accelerated cell senescence. Cell Cycle 4(3): 501-507.

Kletsas, D. 2003. Aging of fibroblast, Kluwer Academic Publishers. Biology of Aging and its Modulations. Vol. 2 Aging of cells In and Outside the Body. pp. 27-46.

Kocylowski, M.K. and T.D. Halazonetis. 2011. SAHF, to senesce or not to senesce? Cell Cycle 10(5): 738-739.

Kosar, M., J. Bartkova, S. Hubackova, Z. Hodny, J. Lukas and J. Bartek. 2011. Senescence-associated heterochromatin foci are dispensable for cellular senescence, occur in a cell type- and insult-dependent manner and follow expression of p16(ink4a). Cell Cycle 10(3): 457-468.

Krishnamurthy, J., C. Torrice, M.R. Ramsey, G.I. Kovalev, K. Al-Regaiey, L. Su and N.E. Sharpless. 2004. Ink4a/Arf expression is a biomarker of aging. J Clin Invest 114(9): 1299-1307.

Krizhanovsky, V., M. Yon, R.A. Dickins, S. Hearn, J. Simon, C. Miething, H. Yee, L. Zender and S.W. Lowe. 2008. Senescence of activated stellate cells limits liver fibrosis. Cell 134(4): 657-667.

Krtolica, A., S. Parrinello, S. Lockett, P.Y. Desprez and J. Campisi. 2001. Senescent fibroblasts promote epithelial cell growth and tumorigenesis: a link between cancer and aging. Proc Natl Acad Sci U S A 98(21): 12072-12077.

Kryston, T.B., A.B. Georgiev, P. Pissis and A.G. Georgakilas. 2011. Role of oxidative stress and DNA damage in human carcinogenesis. Mutat Res 711(1-2): 193-201.

Kuperwasser, C., T. Chavarria, M. Wu, G. Magrane, J.W. Gray, L. Carey, A. Richardson and R.A. Weinberg. 2004. Reconstruction of functionally normal and malignant human breast tissues in mice. Proc Natl Acad Sci U S A 101(14): 4966-4971.

Le, O.N., F. Rodier, F. Fontaine, J.P. Coppe, J. Campisi, J. DeGregori, C. Laverdiere, V. Kokta, E. Haddad and C.M. Beausejour. 2010. Ionizing radiation-induced long-term expression of senescence markers in mice is independent of p53 and immune status. Aging Cell 9(3): 398-409.

Lipetz, J. and V.J. Cristofalo. 1972. Ultrastructural changes accompanying the aging of human diploid cells in culture. J Ultrastruct Res 39(1): 43-56.

Liu, D. and P.J. Hornsby. 2007. Senescent human fibroblasts increase the early growth of xenograft tumors via matrix metalloproteinase secretion. Cancer Res 67(7): 3117-3126.

Maffini, M.V., A.M. Soto, J.M. Calabro, A.A. Ucci and C. Sonnenschein. 2004. The stroma as a crucial target in rat mammary gland carcinogenesis. J Cell Sci 117(Pt 8): 1495-1502.

Mah, L.J., A. El-Osta and T.C. Karagiannis. 2010. GammaH2AX as a molecular marker of aging and disease. Epigenetics 5(2): 129-136.

Masson, R., O. Lefebvre, A. Noel, M.E. Fahime, M.P. Chenard, C. Wendling, F. Kebers, M. LeMeur, A. Dierich, J.M. Foidart, P. Basset and M.C. Rio. 1998. In vivo evidence that the stromelysin-3 metalloproteinase contributes in a paracrine manner to epithelial cell malignancy. J Cell Biol 140(6): 1535-1541.

Matthews, C., I. Gorenne, S. Scott, N. Figg, P. Kirkpatrick, A. Ritchie, M. Goddard and M. Bennett. 2006. Vascular smooth muscle cells undergo telomere-based senescence in human atherosclerosis: effects of telomerase and oxidative stress. Circ Res 99(2): 156-164.

Mertins, S.D., T.G. Myers, S.L. Holbeck, W. Medina-Perez, E. Wang, G. Kohlhagen, Y. Pommier and S.E. Bates. 2004. In vitro evaluation of dimethane sulfonate analogues with potential alkylating activity and selective renal cell carcinoma cytotoxicity. Mol Cancer Ther 3(7): 849-860.

Michaloglou, C., L.C. Vredeveld, M.S. Soengas, C. Denoyelle, T. Kuilman, C.M. van der Horst, D.M. Majoor, J.W. Shay, W.J. Mooi and D.S. Peeper. 2005. BRAFE600-associated senescence-like cell cycle arrest of human naevi. Nature 436(7051): 720-724.

Mirzayans, R., A. Scott, M. Cameron and D. Murray. 2005. Induction of accelerated senescence by gamma radiation in human solid tumor-derived cell lines expressing wild-type TP53. Radiat Res 163(1): 53-62.

Naka, K., A. Tachibana, K. Ikeda and N. Motoyama. 2004. Stress-induced premature senescence in hTERT-expressing ataxia telangiectasia fibroblasts. J Biol Chem 279(3): 2030-2037.

Narita, M., S. Nunez, E. Heard, A.W. Lin, S.A. Hearn, D.L. Spector, G.J. Hannon and S.W. Lowe. 2003. Rb-mediated heterochromatin formation and silencing of E2F target genes during cellular senescence. Cell 113(6): 703-716.

Noel, A., M.C. De Pauw-Gillet, G. Purnell, B. Nusgens, C.M. Lapiere and J.M. Foidart. 1993. Enhancement of tumorigenicity of human breast adenocarcinoma cells in nude mice by matrigel and fibroblasts. Br J Cancer 68(5): 909-915.

Olovnikov, A.M. 1973. A theory of marginotomy. The incomplete copying of template margin in enzymic synthesis of polynucleotides and biological significance of the phenomenon. J Theor Biol 41(1): 181-190.

Olumi, A.F., G.D. Grossfeld, S.W. Hayward, P.R. Carroll, T.D. Tlsty and G.R. Cunha. 1999. Carcinoma-associated fibroblasts direct tumor progression of initiated human prostatic epithelium. Cancer Res 59(19): 5002-5011.

Orimo, A. and R.A. Weinberg. 2006. Stromal fibroblasts in cancer: a novel tumor-promoting cell type. Cell Cycle 5(15): 1597-1601.

Papadopoulou, A. and D. Kletsas. 2011. Human lung fibroblasts prematurely senescent after exposure to ionizing radiation enhance the growth of malignant lung epithelial cells in vitro and in vivo. Int J Oncol 39(4): 989-999.

Paradis, V., N. Youssef, D. Dargere, N. Ba, F. Bonvoust, J. Deschatrette and P. Bedossa. 2001. Replicative senescence in normal liver, chronic hepatitis C, and hepatocellular carcinomas. Hum Pathol 32(3): 327-332.

Park, C.C., M.J. Bissell and M.H. Barcellos-Hoff. 2000. The influence of the microenvironment on the malignant phenotype. Mol Med Today 6(8): 324-329.

Picard, O., Y. Rolland and M.F. Poupon. 1986. Fibroblast-dependent tumorigenicity of cells in nude mice: implication for implantation of metastases. Cancer Res 46(7): 3290-3294.

Probin, V., Y. Wang, A. Bai and D. Zhou. 2006. Busulfan selectively induces cellular senescence but not apoptosis in WI38 fibroblasts via a p53-independent but extracellular signal-regulated kinase-p38 mitogen-activated protein kinase-dependent mechanism. J Pharmacol Exp Ther 319(2): 551-560.

Probin, V., Y. Wang and D. Zhou. 2007. Busulfan-induced senescence is dependent on ROS production upstream of the MAPK pathway. Free Radic Biol Med 42(12): 1858-1865.

Radisky, D.C. and M.J. Bissell. 2004. Cancer. Respect thy neighbor! Science 303(5659): 775-777.

Redon, C., D. Pilch, E. Rogakou, O. Sedelnikova, K. Newrock and W. Bonner. 2002. Histone H2A variants H2AX and H2AZ. Curr Opin Genet Dev 12(2): 162-169.

Redon, C.E., J.S. Dickey, A.J. Nakamura, I.G. Kareva, D. Naf, S. Nowsheen, T.B. Kryston, W.M. Bonner, A.G. Georgakilas and O.A. Sedelnikova. 2010. Tumors induce complex DNA damage in distant proliferative tissues in vivo. Proc Natl Acad Sci U S A 107(42): 17992-17997.

Ressler, S., J. Bartkova, H. Niederegger, J. Bartek, K. Scharffetter-Kochanek, P. Jansen-Durr and M. Wlaschek. 2006. p16INK4A is a robust in vivo biomarker of cellular aging in human skin. Aging Cell 5(5): 379-389.

Roberts, S., E.H. Evans, D. Kletsas, D.C. Jaffray and S.M. Eisenstein. 2006. Senescence in human intervertebral discs. Eur Spine J 15 Suppl 3: S312-316.

Robles, S.J., P.W. Buehler, A. Negrusz and G.R. Adami. 1999. Permanent cell cycle arrest in asynchronously proliferating normal human fibroblasts treated with doxorubicin or etoposide but not camptothecin. Biochem Pharmacol 58(4): 675-685.

Rodier, F., J.P. Coppe, C.K. Patil, W.A. Hoeijmakers, D.P. Munoz, S.R. Raza, A. Freund, E. Campeau, A.R. Davalos and J. Campisi. 2009. Persistent DNA damage signalling triggers senescence-associated inflammatory cytokine secretion. Nat Cell Biol 11(8): 973-979.

Rogakou, E.P., D.R. Pilch, A.H. Orr, V.S. Ivanova and W.M. Bonner. 1998. DNA double-stranded breaks induce histone H2AX phosphorylation on serine 139. J Biol Chem 273(10): 5858-5868.

Romanov, S.R., B.K. Kozakiewicz, C.R. Holst, M.R. Stampfer, L.M. Haupt and T.D. Tlsty. 2001. Normal human mammary epithelial cells spontaneously escape senescence and acquire genomic changes. Nature 409(6820): 633-637.

Roninson, I.B. 2003. Tumor cell senescence in cancer treatment. Cancer Res 63(11): 2705-2715.

Rudolph-Owen, L.A., R. Chan, W.J. Muller and L.M. Matrisian. 1998. The matrix metalloproteinase matrilysin influences early-stage mammary tumorigenesis. Cancer Res 58(23): 5500-5506.

Ryan, A.J., S. Squires, H.L. Strutt and R.T. Johnson. 1991. Camptothecin cytotoxicity in mammalian cells is associated with the induction of persistent double strand breaks in replicating DNA. Nucleic Acids Res 19(12): 3295-3300.

Schmitt, C.A., J.S. Fridman, M. Yang, S. Lee, E. Baranov, R.M. Hoffman and S.W. Lowe. 2002. A senescence program controlled by p53 and p16INK4a contributes to the outcome of cancer therapy. Cell 109(3): 335-346.

Schurch, W., T.A. Seemayer and R. Lagace. 1981. Stromal myofibroblasts in primary invasive and metastatic carcinomas. A combined immunological, light and electron microscopic study. Virchows Arch A Pathol Anat Histol 391(2): 125-139.

Sedelnikova, O.A., I. Horikawa, C. Redon, A. Nakamura, D.B. Zimonjic, N.C. Popescu and W.M. Bonner. 2008. Delayed kinetics of DNA double-strand break processing in normal and pathological aging. Aging Cell 7(1): 89-100.

Sedelnikova, O.A., E.P. Rogakou, I.G. Panyutin and W.M. Bonner. 2002. Quantitative detection of (125)IdU-induced DNA double-strand breaks with gamma-H2AX antibody. Radiat Res 158(4): 486-492.

Seluanov, A., D. Mittelman, O.M. Pereira-Smith, J.H. Wilson and V. Gorbunova. 2004. DNA end joining becomes less efficient and more error-prone during cellular senescence. Proc Natl Acad Sci U S A 101(20): 7624-7629.

Serrano, M., A.W. Lin, M.E. McCurrach, D. Beach and S.W. Lowe. 1997. Oncogenic ras provokes premature cell senescence associated with accumulation of p53 and p16INK4a. Cell 88(5): 593-602.

Severino, J., R.G. Allen, S. Balin, A. Balin and V.J. Cristofalo. 2000. Is beta-galactosidase staining a marker of senescence in vitro and in vivo? Exp Cell Res 257(1): 162-171.

Shay, J.W., O.M. Pereira-Smith and W.E. Wright. 1991. A role for both RB and p53 in the regulation of human cellular senescence. Exp Cell Res 196(1): 33-9.

Shay, J.W. and W.E. Wright. 2007. Hallmarks of telomeres in ageing research. J Pathol 211(2): 114-123.

Shelton, D.N., E. Chang, P.S. Whittier, D. Choi and W.D. Funk. 1999. Microarray analysis of replicative senescence. Curr Biol 9(17): 939-945.

Shibata, K.R., T. Aoyama, Y. Shima, K. Fukiage, S. Otsuka, M. Furu, Y. Kohno, K. Ito, S. Fujibayashi, M. Neo, T. Nakayama, T. Nakamura and J. Toguchida. 2007. Expression of the p16INK4A gene is associated closely with senescence of human mesenchymal stem cells and is potentially silenced by DNA methylation during in vitro expansion. Stem Cells 25(9): 2371-2382.

Sikora, E., T. Arendt, M. Bennett and M. Narita. 2011. Impact of cellular senescence signature on ageing research. Ageing Res Rev 10(1): 146-152.

Stanley, A. and T. Osler. 2001. Senescence and the healing rates of venous ulcers. J Vasc Surg 33(6): 1206-1211.

Sternlicht, M. D., M.J. Bissell and Z. Werb. 2000. The matrix metalloproteinase stromelysin-1 acts as a natural mammary tumor promoter. Oncogene 19(8): 1102-1113.

Sternlicht, M.D., A. Lochter, C.J. Sympson, B. Huey, J.P. Rougier, J.W. Gray, D. Pinkel, M.J. Bissell and Z. Werb. 1999. The stromal proteinase MMP3/stromelysin-1 promotes mammary carcinogenesis. Cell 98(2): 137-146.

Stiff, T., S.A. Walker, K. Cerosaletti, A.A. Goodarzi, E. Petermann, P. Concannon, M. O'Driscoll and P.A. Jeggo. 2006. ATR-dependent phosphorylation and activation of ATM in response to UV treatment or replication fork stalling. EMBO J 25(24): 5775-5782.

Suzuki, M., K. Suzuki, S. Kodama and M. Watanabe. 2006. Phosphorylated histone H2AX foci persist on rejoined mitotic chromosomes in normal human diploid cells exposed to ionizing radiation. Radiat Res 165(3): 269-276.

Sympson, C.J., M.J. Bissell and Z. Werb. 1995. Mammary gland tumor formation in transgenic mice overexpressing stromelysin-1. Semin Cancer Biol 6(3): 159-163.

te Poele, R.H., A.L. Okorokov, L. Jardine, J. Cummings and S.P. Joel. 2002. DNA damage is able to induce senescence in tumor cells in vitro and in vivo. Cancer Res 62(6): 1876-1883.

Thomasset, N., A. Lochter, C.J. Sympson, L.R. Lund, D.R. Williams, O. Behrendtsen, Z. Werb and M.J. Bissell. 1998. Expression of autoactivated stromelysin-1 in mammary glands of transgenic mice leads to a reactive stroma during early development. Am J Pathol 153(2): 457-467.

Tlsty, T.D. and P.W. Hein. 2001. Know thy neighbor: stromal cells can contribute oncogenic signals. Curr Opin Genet Dev 11(1): 54-59.

Toussaint, O., E.E. Medrano and T. von Zglinicki. 2000. Cellular and molecular mechanisms of stress-induced premature senescence (SIPS) of human diploid fibroblasts and melanocytes. Exp Gerontol 35(8): 927-945.

Vamvakas, S., E.H. Vock and W.K. Lutz. 1997. On the role of DNA double-strand breaks in toxicity and carcinogenesis. Crit Rev Toxicol 27(2): 155-174.

Varbiro, G., B. Veres, F. Gallyas, Jr. and B. Sumegi. 2001. Direct effect of Taxol on free radical formation and mitochondrial permeability transition. Free Radic Biol Med 31(4): 548-558.

Vasile, E., Y. Tomita, L.F. Brown, O. Kocher and H.F. Dvorak. 2001. Differential expression of thymosin beta-10 by early passage and senescent vascular endothelium is modulated by VPF/VEGF: evidence for senescent endothelial cells in vivo at sites of atherosclerosis. FASEB J 15(2): 458-466.

Wajapeyee, N., R.W. Serra, X. Zhu, M. Mahalingam and M.R. Green. 2010. Role for IGFBP7 in senescence induction by BRAF. Cell 141(5): 746-747.

Wang, C., D. Jurk, M. Maddick, G. Nelson, C. Martin-Ruiz and T. von Zglinicki. 2009. DNA damage response and cellular senescence in tissues of aging mice. Aging Cell 8(3): 311-323.

Wang, X., S.C. Wong, J. Pan, S.W. Tsao, K.H. Fung, D.L. Kwong, J.S. Sham and J.M. Nicholls. 1998. Evidence of cisplatin-induced senescent-like growth arrest in nasopharyngeal carcinoma cells. Cancer Res 58(22): 5019-5022.

Ward, I.M., K. Minn and J. Chen. 2004. UV-induced ataxia-telangiectasia-mutated and Rad3-related (ATR) activation requires replication stress. J Biol Chem 279(11): 9677-9680.

Wilson, C.L., K.J. Heppner, P.A. Labosky, B.L. Hogan and L.M. Matrisian. 1997. Intestinal tumorigenesis is suppressed in mice lacking the metalloproteinase matrilysin. Proc Natl Acad Sci U S A 94(4): 1402-1407.

Wolf, F.I., A. Torsello, V. Covacci, S. Fasanella, M. Montanari, A. Boninsegna and A. Cittadini. 2002. Oxidative DNA damage as a marker of aging in WI-38 human fibroblasts. Exp Gerontol 37(5): 647-656.

Xue, W., L. Zender, C. Miething, R.A. Dickins, E. Hernando, V. Krizhanovsky, C. Cordon-Cardo and S.W. Lowe. 2007. Senescence and tumour clearance is triggered by p53 restoration in murine liver carcinomas. Nature 445(7128): 656-660.

Yang, N.C. and M.L. Hu. 2005. The limitations and validities of senescence associated-beta-galactosidase activity as an aging marker for human foreskin fibroblast Hs68 cells. Exp Gerontol 40(10): 813-819.

Zhao, W., Z.X. Lin and Z.Q. Zhang. 2004. Cisplatin-induced premature senescence with concomitant reduction of gap junctions in human fibroblasts. Cell Res 14(1): 60-66.

CHAPTER 8

MicroRNAs: From Regulators of Gene Expression to Cancer Biomarkers

Stamatina Giannouli[1], Panagiotis Maragozidis[1,2], Konstantinos I. Gourgoulianis[2] and Nikolaos A.A. Balatsos[1,*]

[1]Department of Biochemistry and Biotechnology, University of Thessaly, Ploutonos 26, 412 21 Larissa, Greece

[2]Respiratory Medicine Department, University of Thessaly Medical School, University Hospital of Larissa, Biopolis, 411 10 Larissa, Greece

ABSTRACT

The research on RNA during the last three decades has witnessed impressive discoveries with an ever-increasing impact on biomedical science. Undoubtedly, RNAi has revolutionized our understanding on how gene expression is regulated. Of the many small-noncoding RNAs, microRNAs are the best studied ones, so far. They are encoded in our genome and are produced through different pathways. MicroRNAs mediate their effects by base-pairing to target mRNAs reducing their stability and are involved in numerous processes, such as cell cycle regulation, differentiation, apoptosis. As regulators of gene expression, they mediate important roles in cancer; their deregulated expression is used to diagnose and categorize human cancers, also suggesting a correlation between prognosis and therapeutic outcome. To provide an overview of the significance of microRNAs as cancer biomarkers, we introduce the reader to important aspects of their biogenesis and mechanisms of action, and then we review the impact of their deregulated expression in the disease.

*Corresponding author: E-mail: balatsos@bio.uth.gr

INTRODUCTION

Analyses of human genome revealed that almost 98% of the genetic material does not code for proteins. Part of this non-coding genome is transcribed into a fascinating repertoire of functional non-protein-coding RNAs (ncRNAs) (Eddy 2001; Mattick 2009). In fact, most human transcriptional units are ncRNAs (Bertone et al. 2004; Birney et al. 2007; Consortium et al. 2007). Until three decades ago, transfer RNA (tRNA) and ribosomal RNA (rRNA), were the most prominent members of ncRNAs, both involved in the biosynthesis of proteins. The discovery that RNAs are able to catalyze biochemical reactions, such as cleavage and ligation of RNA molecules, or the formation of the peptide bond, expanded the repertoire of ncRNAs (Rossi 1992; Nissen et al. 2000). Apart their "processive" role, the discovery of the RNA interference mechanism (RNAi) revealed that specific ncRNAs are major posttranscriptional regulators of gene expression. The latter constitute a class of short (~19-28 nt in length) RNAs produced by double stranded RNA processing. The short length of such regulatory ncRNAs compared to other RNA types christened these as microRNAs (miRNAs), short interfering RNAs (siRNAs), small nucleolar RNAs (snoRNAs), small modulatory RNAs (smRNAs) and tiny non-coding RNAs (tncRNAs) (Lagos-Quintana et al. 2001; Lau et al. 2001; Lee and Ambros 2001; Kim 2005b). Apart the short ncRNAs, genome-wide surveys have recently revealed that eukaryotic genomes are extensively transcribed into thousands of transcripts much longer (>200 nt) than the "traditional" short, conserved ncRNAs, adding a new subclass, namely long non-coding RNAs (lncRNAs). lncRNAs are thus far implicated in gene-regulatory roles, such as transcription, splicing, translation and others (Guttman et al. 2009; Mattick 2009; Ponting et al. 2009; Wilusz et al. 2009).

Of the vast repertoire of ncRNAs, miRNAs have been extensively studied due to their involvement in biological pathways and diseases. More than one thousand miRNAs have been identified so far, and have been shown to regulate more than 60% of human protein-coding genes (Friedman et al. 2009; Sayed and Abdellatif 2011). The link of miRNAs to cancer is attributed to sequences transcribed to miRNAs that locate in genomic regions involved in cancer, as well as to their deregulated expression profiles in malignancies (Calin et al. 2004a; Calin et al. 2004b; Lu et al. 2005). Hence, miRNAs have emerged as important prognostic and diagnostic markers, as well as to therapeutic targets against cancer with increasing potential. Herein, we focus on the biogenesis, the mechanisms of miRNA-mediated gene silencing and the recent advances of miRNAs as cancer biomarkers.

MiRNA BIOGENESIS AND FUNCTION

Discovery and Nomenclature

Lin-4 gene was studied as a key factor for the development of the nematode *C. elegans*. Upon cloning of lin-4 though, no cognate protein was assigned to it. Instead, it generated two RNA products, the premature and the mature lin-4 forms. The latter is complementary to repeats of the 3′-untranslated region (UTR) of lin-14 through the translational repression of lin-14 (Lee et al. 1993), leading to the identification of the first miRNA. It was only seven years later when let-7, another small regulatory RNA in *C. elegans* with orthologs in other species, was discovered implying the existence of a conserved mechanism of post-transcriptional regulation (Reinhart et al. 2000). Apart from controlling development in *C. elegans*, miRNAs participate in almost all cellular functions, such as differentiation and organogenesis, growth control and programmed cell death in plants, worms, flies, fish, frogs and mammals. Systematic cloning of small RNAs from plants to humans revealed an increasing number of small RNAs, suggesting that a novel regulatory mechanism was potentially conserved during evolution (Lagos-Quintana et al. 2003; Lim et al. 2003; Lagos-Quintana et al. 2002; Lee et al. 2001; Lagos-Quintana et al. 2001). Using computational target predictions and genome-wide identification of miRNA targets, over 10,000 miRNAs have been identified in a wide range of species, and it is estimated that each animal miRNA regulates hundreds of different mRNAs. Further, at least 60% of the human transcriptome is subject to regulation by miRNAs (Rajewsky and Socci 2004; Rajewsky 2006; Mashima et al. 2008; Persson et al. 2009). However, due to the large quantity predicted, only a few of them have been experimentally validated (Y. Wang et al. 2009). Once a sequence has been verified to be a unique miRNA, it is assigned to the miRBase Registry according to existing nomenclature guidelines (Song et al. 2004; Y. Wang et al. 2009). The latest version of miRBase Registry (release April 2011) contains more than 1400 unique microRNAs found in the human genome. A mature miRNA is named as "miR" followed by a unique identifying number (miR-1, miR-2, etc.), whereas the precursor hairpin as "mir". A 3-letter prefix designates the origin of the miR ("hsa-miR-1" for Homo sapiens miR-1), though this labeling is less used in the literature. Herein, the "miRNA" term will be used in general, except for certain identified ones, where the "miR-number" labeling will be used.

Biogenesis of miRNAs

The biogenesis of miRNAs is diverse following several different pathways, reflecting control of the processes by different mechanisms and evolutionary flexibility. The pathways are generally divided into canonical and alternative. The canonical pathway refers to RNA polymerase II (RNAP II) transcripts processed by two RNase III[1] enzymes, Drosha and Dicer, generating the majority of miRNAs. The alternative pathways generate non-canonical miRNAs usually bypassing one of the RNase III processing steps. These pathways came into light following analyses of data from deep sequencing and high-throughput sequencing technologies, revealing miRNAs that only partially meet the classical criteria (Fig. 1a).

Canonical miRNA precursors are transcribed by RNAP II into long primary transcripts (pri-miRNAs) of several thousand nucleotides in length that acquire 5′ cap and 3′ poly(A) tails at their ends (Lee et al. 2004). RNAP III can also transcribe sequences to produce miRNAs as is the case for the human chromosome 19 miRNA cluster (C19MC) (Bohmert et al. 1998). Often, the primary transcripts are organized as polycistronic primary transcripts consisting a miRNA cluster (Peters and Meister 2007). These clusters are highly conserved implying important regulatory and functional roles (Elbashir et al. 2001b). Either in clusters or alone, the resulting transcripts are capped, spliced and polyadenylated (Cai et al. 2004; Lee et al. 2004). Subsequently, pri-miRNAs are cleaved in the nucleus by Drosha, an RNase III endonuclease, in complex with the double-stranded RNA-binding domain protein DGCR8 (DiGeorge syndrome critical region gene 8), also known as Pasha (Partner of Drosha) in *D. melanogaster* and Pash-1 in *C. elegans*, forming the microprocessor complex. DGCR8 interacts directly with the pri-miRNA guiding Drosha to cleave the 5′ and 3′ arms of the pri-miRNA hairpin producing a pre-miRNA with a 5′ phosphate, a 3′ hydroxyl and a 2-nt overhang at its 3′ end that serves as a recognition site for the second processing step (Denli et al. 2004; Gregory et al. 2004; Han et al. 2004; Landthaler et al. 2004; Miyoshi et al. 2010; Yang and Lai 2011). An unexpected origin of miRNAs was revealed when it was reported that introns might process to functional miRNAs, also revealing an alternative

---

[1]RNase III enzymes consist a family of specific endoribonucleases that catalyze the cleavage of double-stranded RNA producing a characteristic 2-nt 3' overhang in their products (Lamontagne et al. 2001; Ji 2008). They are divided into three classes based on their domain organization. Class I include RNase III enzymes found in bacteria and yeasts. Class II includes Drosha, with homologs found only in animals. Class III includes Dicer homologs that are conserved in *Schizosaccharomyces pombe*, plants and animals.

pathway for their biogenesis. These pre-miRNAs/introns, called mirtrons, can bypass Drosha cleavage given that the intron resulting from mRNA splicing has the appropriate size to form a hairpin resembling a precursor miRNA (pre-miRNA) (Okamura et al. 2007; Ruby et al. 2007). Thus far, 11 human mirtrons have been predicted (Berezikov et al. 2007). However, research on the biogenesis of mirtrons revealed a subset of predicted mirtrons that are splicing-independent, thus termed "simtrons" (splicing- independent mirtron-like miRNAs) (Havens et al. 2012). Simtrons do not follow the canonical miRNA pathway either, but they are Drosha-dependent. Either originated from the canonical or the mirtron pathway, the resulting pre-miRNA consists of ~ 60-75 nucleotides with a hairpin shape and a two-nucleotide overhang at its 3′ end (Lee et al. 2002; Bartel 2004; Cai et al. 2004; Han et al. 2004; Singh et al. 2008).

Another alternative biogenesis pathway and an exception to the Drosha-processing revealed upon studies on non-coding RNAs contained in vault particles known as vault RNAs (vRNAs or vtRNAs)[2]. It was shown that vRNAs generate small vault RNAs bypassing Drosha processing, while Dicer is involved in svRNA cleavage (Persson et al. 2009)[3]. When a pre-miRNA is correctly processed (i.e bearing a short stem loop and a 3′ overhang), it can be exported to the cytoplasm through successful binding to Exportin 5 in complex with Ran-GTP (Yi et al. 2003; Lund et al. 2004). Further processing in the cytoplasm by Dicer, a RNase III endonuclease, removes the loop of the pre-miRNA leaving another 3′ overhang to the other end (reviewed in (Kim 2005a). The previous processes result to a mature miRNA of 19–24 nucleotide duplex with two nucleotides protruding as overhangs at each 3′ end. Therefore, in an appropriate loop sequence in the nucleus, Drosha generates one end of pre-mature miRNA (a two-nt 3′ overhang), while in the cytoplasm Dicer measures ~22 nucleotides from the pre-existing end (generated by Drosha) producing the other end of the miRNA through appropriate cleavage of the loop.

[2]Vault particles or vault ribonucleoprotein complexes are conserved organelles associated in nuclear pore complexes, implicated in multidrug resistance and intracellular transport (Unwin and Milligan 1982; Persson et al. 2009). They contain three different proteins and up to 16 non-coding vault RNAs (Kedersha et al. 1991) transcribed by RNAP III.

[3]It should be noted here that among vtRNAs is pre-miR-886. The latter has been proposed to be renamed as vtRNA2, since it fulfils certain characteristics of human vtRNAs: a) they are 88 or 98 nt in length, they are RNA polymerase III (Pol III) transcripts, and c) they have a stem loop secondary structure (Nandy et al. 2009; Stadler et al. 2009). However, recent evidence showed that pre-miR-886 is neither a Drosha-product, nor matures into a functional miRNA, suggesting that it is not a vtRNA, nor a canonical RNA. Importantly, pre-miR-886 may be involved in tumor suppression by modulating the activity of the tumor suppressor PKR (Protein Kinase RNA-activated) (Lee et al. 2011).

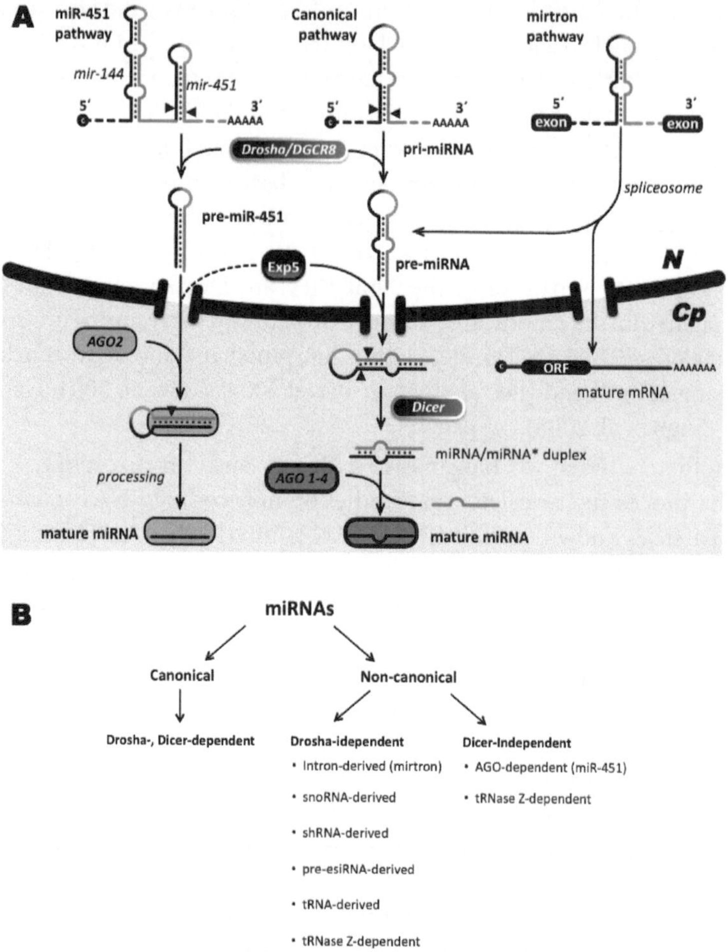

Fig. 1 Biogenesis and categories of miRNAs. A. Biogenesis of miRNAs. The illustration summarizes the main steps of the canonical pathway (middle) based on two major endoribonucleases, Drosha and Dicer. Shown are also two examples of non-canonical pathways, the Dicer-independent (left) and the mirtron (Drosha-independent). Upon miRISC assembly and base pairing of miRNA with the 3′UTR of target mRNA, silencing is mediated either via mRNA degradation and/or translational repression, or via endonucleolytic cleavage, as a consequence of partial or full complementarity of the single-stranded miRNA with the target mRNA, respectively. For simplicity reasons potential auxilliary/additional RISC subunits, currently uncharacterized, are not included. Arrowheads indicate approximate cleavage positions by Drosha and Dicer. N, nucleus; Ct, cytoplasm; Exp5, exportin 5; ORF, open reading frame; c, 5′cap; poly(A)-tail is shown as a stretch of As. Dashed line indicates hypothetical export of miR-451 by Exp5. B. Categories of miRNAs. MiRNAs may be categorized according to their biogenesis origin, as canonical, Drosha- and Dicer-independent.

The Dicer-mediated cleavage has been described in many organisms, including mammals, and is essential for miRNA processing. An exception to the Dicer-dependent pathway is the maturation of miR-451. Initially cleaved by the Drosha/DGCR8, pri-miR-451 generates a short stem (~18 bp), which has not an appropriate length for Dicer-cleavage, and it is loaded directly onto AGO2 (Cheloufi et al. 2010; Cifuentes et al. 2010; Yang et al. 2010).

Apart the above-mentioned alternative biogenesis pathways (mirtrons, vtRNAs, miR-451), several other sources contribute to the non-canonical miRNA pool. These include snoRNAs, endogenous short-hairpin RNAs (endo-shRNAs) and tRNA precursors, tRNase Z-derived and miRNAs derived from a terminal hairpin of an endo-siRNA precursor (Fig. 1B). Clearly, non-canonical miRNAs set more challenges for the future, such search for novel classes, regulatory mechanisms of canonical miRNAs and biological functions (Miyoshi et al. 2010).

In addition to the Drosha/Dicer processing, miRNA precursors are subject to editing modifications that regulate miRNA biogenesis. Editing of pri-miR-142 by specific enzymes (ADARs)[4] suppresses its processing by Drosha (Yang et al. 2006), while precursors of miR-151 may block cleavage by Dicer (Kawahara et al. 2007a). The significance of this modification on miRNA-mediated silencing revealed when demonstrated that the edited and unedited versions of miR-376a target different set of genes (Kawahara et al. 2007b).

Target Recognition

The main role of miRNAs is the regulation of eukaryotic gene expression. In metazoans miRNAs base pair with mRNAs, to which they are partially complementary, resulting to gene silencing via translational repression and poly(A) tail removal leading to rapid mRNA degradation (Kedersha et al. 1991; Giraldez et al. 2006; Wu et al. 2006; Mashima et al. 2008). Less often, when a miRNA is fully complementary to its target mRNA, degradation of the

[4]RNA editing refers to the alteration of RNA as a consequence of nucleotide modification, insertion or deletion. The most common type of RNA editing is the hydrolytic deamination of adenosine (A) to inosine (I), catalyzed by double-stranded RNA-specific adenosine deaminase or Adenosine deaminase that act on RNA (ADAR) family of enzymes (Hogg et al. 2011). Inosine is recognized as guanosine (G) by the splicing and translation machineries altering the coding information of an mRNA. Further, inosine preferentially pairs with cytidine (C). Thus, ADARs also either destabilize dsRNA by converting AU base pairs to IU mismatches, or increase stability by changing AC mismatches to IC. Moreover, introns and untranslated regions are the primary non-coding targets of ADARs (Hogg et al. 2011).

latter may occur by directing endonucleolytic cleavage at the miRNA-binding site (Bass 2002; Nandy et al. 2009; Tanaka et al. 2009a; Lee et al. 2011).

Animal miRNAs recognize complementary regions of target mRNAs, generally located in their 3′UTRs, by Watson-Crick base pairing (Fig . 1C). This is mediated by one of the two strands of miRNA, the functional guide strand, while the other is the opposite strand (* or passenger). The central region is flanked by the 5′ or 'seed', and the 3′ regions. The most important requirement in the miRNA-mRNA interaction is 5′ nucleotides 2-8 of the miRNA, also known as the "seed" region (Fig. 2A). With very few known exceptions, the full complementarity of the seed region is a prerequisite for effective silencing, although this is not always the case (Vella et al. 2004; Bartel 2009). The central region (nt 9/10-12) generally bulges out preventing endonucleolytic cleavage by Argonaute proteins (AGOs)[5] (Bartel 2009; Voinnet 2009) (see below). Though, in case of endonucleolytic cleavage of the target-mRNA, the central region determines the cleavage position of the target. The 3′ region is considered to be more tolerant to mismatches, and in rare cases of weak seed-match, complementarity may contribute to target selection (Brennecke et al. 2005; Grimson et al. 2007). Thus, miRNA targets may be classified as: a) canonical, b) 3′-supplementary and c) 3′-compensatory sites (Bartel 2009). Most of the canonical sites have full complementarity within the seed region. Sites with shorter, 6-nt seed complementarity are thought to have a limited functionality. These regions may have an additional supplementary pairing within 3′ part of miRNA (3′-supplementary sites) but usually have a less profound effect on target recognition (Grimson et al. 2007). Enhancement of the effectiveness of the miRNA-mRNA interaction requires pairing of 3-4 nucleotides in positions 13-16 of miRNA. Extended pairing in the 3′ region may counterbalance for possible mismatches in the seed (3′-compensatory sites) (Grimson et al. 2007; Bartel 2009; Witkos et al. 2011).

Two models have been proposed for target recognition. A 'two-state' model for target recognition has been proposed, a nucleation step of target recognition, that is the pairing of the 5′ nucleotide of the miRNA to an accessible site on the target 3′ UTR, followed by a propagation or "zippering" step, that is the disruption of local secondary structure, elongation, and formation of a stable duplex (Filipowicz 2005; Tomari and Zamore 2005;

[5]Argonaute (AGO) proteins were first identified in *Arabidopsis thaliana*. In humans there are four AGOs (AGO1-4); AGO2 gene is located on chromosome 8, while AGO1, AGO3 and AGO4 genes are clustered on chromosome 1 (Hock and Meister 2008). The term argonaute refers to the appearance of null AGO1 mutants in *Arabidopsis thaliana*; they looked like the tentacles of a squid (*Argonauta argo*) (Bohmert et al. 1998).

Yuan et al. 2005; Long et al. 2007). A 'fixed-end' model proposes that both ends of the guide strand remain anchored during both the nucleation and propagation steps of RNA recognition (Filipowicz 2005). Recently though, results based on suppression of miRNAs by synthetic inhibitors suggest that the 3′ region is equally important as the seed for target recognition (Robertson et al. 2010).

MiRNA-mediated Gene Silencing

The regulatory effects of miRNAs are mainly mediated by the protein components of miRNA-induced silencing complex (miRISC) (Eulalio et al. 2009; Ketting 2011). AGOs are the major protein component of miRISC consisting of a miRNA-loaded AGO protein and the AGO-interacting GW182[6] protein (Yang et al. 2006; Kawahara et al. 2007a; Kawahara et al. 2007b). AGO proteins are characterized by a bilobal architecture and contain three evolutionarily conserved domains, PAZ, MID, and PIWI[7]. PAZ binds the 3′ end of guide RNA. The MID-PIWI interface provides the binding pocket for binding the guide 5′ phosphate end of the miRNA (Boland et al. 2011). The PIWI domain adopts a RNase H fold and mediates the endonucleolytic cleavage of target mRNA occurring specifically 10 nt from the 5′- end of the miRNA, leaving the miRNA intact for another round of cleavage (Elbashir et al. 2001a; Elbashir et al. 2001b; Song et al. 2004; Peters and Meister 2007; Y. Wang et al. 2008; Y. Wang et al. 2009). The cleavage reaction requires the presence of Mg(II) ions and proceeds via hydrolysis and release of a 3'-hydroxyl and a 5'-phosphate terminus (Martinez and Tuschl 2004; Schwarz et al. 2004). AGO2 is the only of the four human AGOs that have endonucleolytic activity (often referred as "slicer") through its RNaseH-like PIWI domain, thus mediating mRNA cleavage (Liu et al. 2004). Following Dicer cleavage, the double-stranded duplex is separated by helicases into the functional guide strand and the opposite strand (Kai and Pasquinelli 2010). The miRNA strand with the less stable base pair at its 5′ end in the duplex is recruited by AGO to form the core of the miRISC along with GW182 protein, while the opposite strand is subsequently eliminated (Bernstein et al.

[6]GW182 consist a protein family. Its name reflects the presence of glycine (G)/tryptophan (W) repeats (GW, WG, or GWG motifs) at the N terminus and its molecular mass (182kDa) (Eystathioy et al. 2002). In humans there are three paralogs of GW182 also known as TNRC (TriNucleotide Repeat-Containing) proteins, TNRC6A/GW182, TNRC6B, and TNRC6C, (Behm-Ansmant et al. 2006). The GW repeats mediate the binding to AGOs (Behm-Ansmant et al. 2006; Till et al. 2007).

[7]PIWI, P-element induced wimpy testis; MID, middle; PAZ, Piwi-Argonaute-Zwilli.

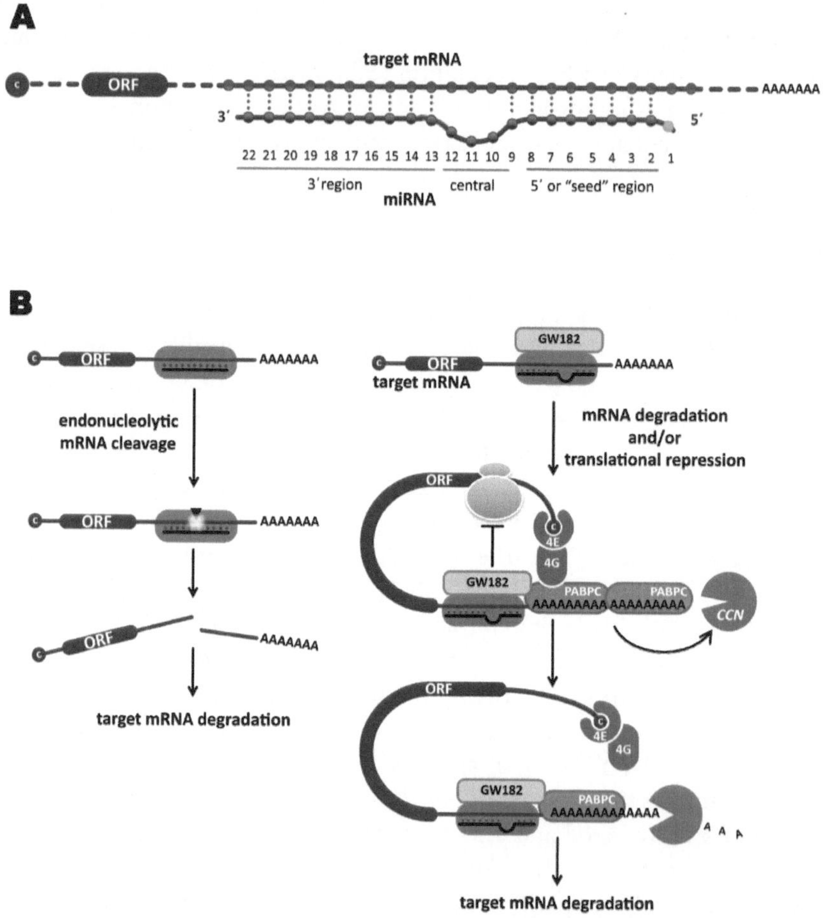

Fig. 2 Target recognition and pathways of miRNA-mediated silencing. A. Animal miRNAs (lower strand) recognize their target mRNAs (upper strand) through complementary regions, generally located in their 3′UTRs, by Watson-Crick base pairing. The 5′ or 'seed', and the 3′ regions flank the central region. Full complementarity to seed sequence (nt 2-7/8) is an almost absolute determinant for silencing. The central region (nt 9/10-12) generally bulges out preventing endonucleolytic cleavage by AGOs. In case of endonucleolytic cleavage, it determines the cleavage position of the target at nt 10 from the 5′ end of miRNA. The 5′ terminal nucleotide (nt 1, light grey sphere) is not available for pairing as it is bound by AGOs. Spheres indicate mRNA or miRNA nucleotides (nt). Numbers 1-22 indicate positions of miRNA nucleotides. Sphere with "c" indicates the 5′-cap. ORF, Open reading frame. B. The main pathways of miRNA-mediated silencing. In case of full complementarity between miRNA and target mRNA, the latter is degraded via direct endonucleolytic cleavage by AGO2 (left). Most often, in case of partial complementarity silencing is mediated by mRNA degradation by deadenylation and translational repression (right).

Color image of this figure appears in the color plate section at the end of the book.

2001; Grishok et al. 2001; Hammond et al. 2001; Khvorova et al. 2003; Zhu et al. 2008; Fabian et al. 2010) (Fig. 2B). Upon binding to AGO proteins, the half-lives of most miRNAs are greater than 14 hours, more stable than average mRNAs (Hock and Meister 2008).

Targeting Translation Initiation

As mentioned, miRNAs either inhibit translation of target mRNAs, or facilitate their deadenylation and subsequent degradation. Translational repression by miRNAs may occur at the initiation (Humphreys et al. 2005; Pillai et al. 2005), or at post-initiation steps (Maroney et al. 2006; Nottrott et al. 2006; Petersen et al. 2006). Since translation initiation is generally the rate-limiting step of protein biosynthesis, it is the main target subject to control mechanisms. Such mechanisms include (a) cap-dependent processes, where the 40S ribosomal subunit and a number of eukaryotic initiation factors (eIFs) bind close to the cap, and (b) cap-independent process, where a sequence known as internal ribosome entry site (IRES) bypass the cap guiding translation initiation.

Let-7 miRNA drive miRISC to target mRNAs triggering deadenylation, which in turn abolishes the synergy between 5′-cap and poly(A), and represses translation (Yuan et al. 2005). In line with the previous, let-7 repressed 5' cap-mediated translation initiation suggesting that this step is the earliest molecular event affected by miRNAs, while other mechanisms, such as mRNA degradation, may subsequently consolidate mRNA silencing (Robertson et al. 2010). Further, using chemically modified cap analogs it was shown that the 5′-m7GpppN structure is a primary target for miRNA-mediated translational inhibition and show that deadenylation is a rapid independent process that can contribute to repression (Ji 2008). Concerning the role of cap-associated factors, it was shown that miRISC may impair the function of the 5′-cap-binding protein eIF4E, or interfere upstream of eIF4E - eIF4G interaction during the initiation step of translation, most likely through the recognition of the cap by eIF4E (Tomari and Zamore 2005; Yuan et al. 2005). Contrary to these results, Wang and coworkers used cell-free reactions with rabbit reticulocyte lysates and provided evidence supporting a model that miRNAs repress translation initiation by preventing 60S ribosome subunit recruitment to target mRNAs, after cap-facilitated, 40S ribosome subunit recruitment (Du and Zamore 2005).

However, the role of 60S ribosomal subunit in miRNA-mediated inhibition of translation raised more questions. The joining of 60S to 40S is negatively regulated by eIF6 protein. Upon depletion of eIF6 in human cells or *C. elegans* the repressive effects of miRNAs on the target mRNA and protein levels were diminished, hence the assembly of the 80S complex was inhibited due to the miRISC-eIF6 association (Landthaler et al. 2004). This was not the case though in *Drosophila melanogaster* S2 cells, where depletion of eIF6 had no significant effect miRNA-mediated repression (Liu et al. 2004; Voinnet 2009). Moreover, in marked contrast it was reported that knockdown of eIF6 in *Caenorhabditis elegans* resulted in let-7 suppression, suggesting that eIF6 does not seem to facilitate miRNA function (Gregory et al. 2004).

More debate has been raised on the role of cap in miRNA-mediated silencing. It has been reported that human AGO2 binds weakly on m^7GTP beads through two important phenylalanines residing in a motif (MC) that shares sequence similarity with the eIF4E cap-binding motif. These findings support the idea that miRNAs inhibit translation at the cap-recognition step by displacing eIF4E from the cap structure (Long et al. 2007). However, in *D. melanogaster* cells it was found that mutation of these residues prevented AGO1 from interacting with GW182 and miRNAs, and their role is unrelated to cap binding (Voinnet 2009). Furthermore, a modeling study showed that AGO2 structure is incompatible to eIF4E-type binding of the cap (Lamontagne et al. 2001).

MiRNA-mediated Deadenylation

The link between miRISC and deadenylation is mediatied by GW182 protein, which serves as: a) a platform that recruits deadenylases and b) a deadenylase coactivator, as it interacts independently with the PAN2-PAN3 (PAN) and CCR4-NOT deadenylase complexes promoting processive poly(A) removal of target mRNAs (Kim et al. 1994; Miyoshi et al. 2010; Hogg et al. 2011; Yang and Lai 2011). Another important factor is poly(A)-binding protein (PABP) as it seems to mediate the association between PAN2 (the catalytic subunit of PAN complex) and GW182. In contrast, the CCR4-NOT complex is recruited by GW182 to miRISC in a PABP-independent manner (Miyoshi et al. 2010). It should be noted that according to a proposed model, the shortening of poly(A) tails in the cytoplasm is biphasic. In the first phase, a synchronous and stepwise shortening of the poly(A) tail to ~1/3 – 1/2 of its initial length is followed by the second phase, where the poly(A) lengths become highly heterogeneous. The transition to the second phase is crucial because it triggers the decay of the entire mRNA body. In this biphasic process, PAN2 catalyzes poly(A) shortening in the first phase and CCR4-NOT catalyzes the second one (Yamashita et al. 2005). The model proposes that the multiple PABPs associated with the poly(A) tail may (a) provide sufficient interactions with PAN complex to stimulate the shortening of the poly(A) tail by PAN2, and (b) block the activity of CCR4. Thus, as deadenylation proceeds, the shorter the poly(A) the fewer PABPs are associated with it, which in turn leads to diminish of PAN2 and release of CCR4 activity, and the mRNA enters the second phase of deadenylation leading to degradation (Yamashita et al. 2005). The CCR4–NOT complex is considered to be integral to miRNA-mediated deadenylation acting in concert with GW182 to displace PABP from the poly(A) tail, whereas the PAN complex is dispensable (Kim et al. 1994; Behm-Ansmant et al. 2006; Grimson et al. 2007). Keeping the previous in mind, it may be speculated that GW182 and PABP initially recruit and stabilize PAN on poly(A) to ignite the first phase of poly(A) shortening. As the number of the poly(A)-bound PABP molecules decreases as a consequence of the shortening, the activity of PAN is reduced [as PAN binds to poly(A) in a PABP-dependent manner]. In the second phase, GW182 acts in concert with the CCR4-NOT complex to displace the remaining PABPs from the poly(A) tail, allowing processive deadenylation by its nuclease subunits.

Degradation of miRNAs

MiRNAs tend to target multiple mRNA transcripts within common cellular response pathways (Elbashir et al. 2001a). Upon completion of their silencing effect, miRNAs are degraded, though few things are known on miRNA turnover (Krol et al. 2010). A family of exoribonucleases degrading miRNAs has been identified in *A. thaliana*, so far (Ramachandran and Chen 2008). It is assumed that miRNA turnover is needed for rapid changes in miRNA expression profiles. AGO proteins are thought to stabilize the RISC-incorporated guide strands, while the opposite one is destroyed. Also AGO may preferentially retain miRNAs with many targets over the ones with few or no targets, leading to their degradation (Kai and Pasquinelli 2010).

MiRNAs AND HUMAN CANCER

The link between miRNAs and cancer emerged from studies revealing that deregulation of miRNAs disrupts cellular differentiation and proliferation, while the expression of several miRNAs is altered in primary human tumors (Calin et al. 2002; Calin et al. 2004a; He et al. 2005a; Lu et al. 2005). MiRNAs seem to be functionally important in human cancers due to the fact that their levels are specifically deregulated according to the cancer type, providing an accurate fingerprint (Volinia et al. 2006; Lanza et al. 2007). In addition, they seem to be involved in processes crucial for cancer development and progression (Iorio et al. 2008), while manipulation of miRNA expression levels in cancer cell lines can partially lead to phenotype recovery (Gaur et al. 2007). The following part of this chapter will focus on the role of miRNAs in human cancers.

MiRNAs as Oncogenes and Tumor Suppressors

An increasing number of reports indicate that miRNAs can function as tumor suppressors and oncogenes. Such miRNAs are designated as oncogenic miRNAs, or "oncomiRs" (Esquela-Kerscher and Slack 2006; Slack and Weidhaas 2006; Krutovskikh and Herceg 2010). MiRNAs with decreased expression in cancerous cells are considered as tumor suppressors. They usually prevent tumor development by negatively inhibiting oncogenes and genes that control cell differentiation or apoptosis. Many miRNA genes have been found significantly overexpressed in different types of cancer. Comparative analysis of miRNA levels in solid tumors and normal tissues revealed a specific "miRNome" expression signature, consisting of both downregulated and upregulated miRNA levels (Volinia et al. 2006).

The first evidence for the involvement of miRNAs in cancers came from a molecular study characterizing the 13q14 deletion in human B cell chronic lymphocytic leukemia (B-CLL). This region contains miR-15α and miR-16-1 and it is deleted in more than half of cases (Calin et al. 2002). Later on it was found that both miR-15α and miR-16-1 potentiate the normal apoptotic response by targeting the anti-apoptotic gene BCL-2 (Cimmino et al. 2005). Let-7, the second miRNA identified, is considered as a tumor suppressor. Twelve let-7 homologs have been recognized in the human genome organized in eight clusters, four of which are within regions frequently deleted in diverse human malignancies (Calin et al. 2004a). Their potential tumor suppressor activity is further supported by the fact that let-7 family members negatively regulate the expression of specific oncogenes, such as RAS, MYC and high mobility group A2 (HMGA2) (Johnson et al. 2005; Mayr et al. 2007; He et al. 2010). Thus, loss of let-7 in lung tumors correlates with high expression of oncogenic RAS protein, providing one mechanism through which reduced let-7 levels promote tumorigenesis (Johnson et al. 2005). Tumor suppressor activity is also characteristic of miR-29 family members that can induce p53-dependent cell apoptosis as they target factors, such as p85 and CDC42, which negatively regulate p53 (Park et al. 2009b).

MiR-17-92 polycistron, also known as *oncomir-1*, is the one of the first reported miRNA acting as a mammalian oncogene. This cluster is embedded in 13q31 human genomic locus, a region that is amplified in several types of lymphoma and solid tumors (He et al. 2005b). The precursor transcript contains six tandem hairpin loop structures generating six mature miRNAs: miR-17, miR-18a, miR-19a, miR-20a, miR-19b-1, and miR-92-1 (Tanzer and Stadler 2004; Olive et al. 2010). The transcription factor E2F1 that promotes G1 to S phase progression in mammalian (Bracken et al. 2004) is regulated by miR-17–5p and miR-20a, members of the miR-17-92 cluster, which in turn is directly activated by c-Myc (O'Donnell et al. 2005). Two members of *miR-17-92* cluster possibly mediate the oncogenic activity of the entire cluster, as they target the tumor suppressor PTEN thus activating Akt-mTOR signaling which leads to cell survival (Mu et al. 2009; Olive et al. 2009). In parallel, miR-17-92 cluster target the pro-apoptotic protein Bim, a tumor suppressor that also regulates B-cell survival (Ventura et al. 2008; Xiao et al. 2008). However, these miRNAs may have tumor suppressor activity under other circumstances, as loss of heterozygosity of 13q31 also results in several types of tumors including hepatocellular carcinoma, pituitary adenoma, nasopharyngeal carcinoma, breast cancer and bladder cancer (Eiriksdottir et al. 1998; Richter et al. 1999; Tsang et al. 1999; Honda et al. 2003; Koo et al. 2003; Lin et al. 2003).

Another example of oncogenic miRNA is found at B-cell integration cluster (BIC) noncoding RNA. BIC integration site for the avian leukosis virus generates a noncoding RNA (Tam et al. 1997), but since this observation was reported before the emergence of miRNAs, the role of a transcript lacking an open reading frame was unclear. Later it was shown that this noncoding transcript harbored miR-155 located in the final exon of this transcript (van den Berg et al. 2003; Eis et al. 2005; Calin and Croce 2006a; M. Wang et al. 2008). Overexpression of *BIC*/mir-155 has been linked to several types of lymphomas such as Hodgkin and Burkitt lymphoma, breast, lung, colon and thyroid cancers (Eis et al. 2005; Kluiver et al. 2005; Volinia et al. 2006). Several miR-155 targets have been validated in normal B cells and macrophages as described above, such as Angiotensin II type I receptor, but it remains uncertain if the same targets are relevant in the lymphoma cell background, while the complete mechanism through which BIC/miR-155 promotes tumorigenesis remains unknown (Martin et al. 2006; Sayed and Abdellatif 2011).

Interestingly, different studies revealed that the same miRNA molecule could act as either an oncogene or a tumor suppressor depending on the tissue context and target genes. Such an example is the group of miR-221 and miR-222. Upregulation of these miRNAs promotes the progression of liver tumorgenesis and it has been proposed that miR-221 and miR-222 together with miR-21, miR-93, miR-96, and miR-106b should be considered as *bona fide* oncomiRs (Pineau et al. 2010). However, it was shown that miR-221 and miR-222 inhibit erythropoiesis by downregulating CD117 (or c-kit), which is the receptor of stem cell factor and a known proto-oncogene (McIntyre et al. 2005), thus acting as tumor suppressors (Felli et al. 2005). It should be noted here that in normal human erythroid progenitors, miR-155 levels were significantly reduced, and miR-221 and miR-223 expression showed a moderate decrease, while in marked contrast, the expression of miR-451 was significantly increased, suggesting that these miRNAs are key regulators for normal erythroid differentiation (Masaki et al. 2007).

Regulation of miRNA Deregulation in Cancer

More than half of the miRNAs that are deregulated in cancer reside in cancer hotspot chromosomal regions (Calin et al. 2004a). These regions exhibit loss of heterozygosity where tumor suppressor genes, amplification regions, common breakpoint regions and fragile sites are possibly located (Calin and Croce 2006b). The underlying cause of miRNA deregulation in cancer is far from clear, although some reports have shed light in this issue, as is the case for the reduced expression of miR-15 and miR-16 in CLL (Calin et al. 2002), and miR-218 in bladder cancer (Schwarz et al. 2004) which is associated with specific chromosomal deletions. Furthermore, Zhang and coworkers showed that in human cancers there is a high frequency of genomic alterations in miRNAs loci (Zhang et al. 2006), while the overexpression of miR-17-92 correlates with the amplification of its gene locus (He et al. 2005b). An indirect mechanism to regulate miRNA-induced silencing is the loss of miRNA binding sites in the target gene: chromosomal translocations in HMGA2, led to loss of the let-7 miRNA binding sites on the 3′UTR of its cognate mRNA. Although this mechanism does not alter the miRNA levels, the disruption of miRNA-induced gene silencing can lead to abnormal phenotypes enhancing oncogenic transformation (Lee et al. 2005; Mayr et al. 2007).

Apart from the chromosomal localization, miRNA deregulation has been shown to rely on transcriptional or epigenetic regulation (Lujambio et al. 2007), as promoter hypermethylation seems to be responsible for the epigenetic silencing of miR127 in cancer cells (Saito et al. 2006).

MiRNAs in Tumor Growth and Metastasis

The deregulated levels of miRNA in cancer seem to correlate not only with the tumor-initiating events, but also with cancer progression, invasion and metastasis. An important step in tumor growth and cancer progression is angiogenesis. Work on the role of miRNAs in this direction revealed that specific members of the miR-17-92 cluster and miR-378 play important role in tumorigenesis and angiogenesis (Dews et al. 2006; Lee et al. 2007a; Fang et al. 2011). Additionally, several miRNA expression signatures have been identified and confirmed in tumor angiogenesis (Wang and Olson 2009; Patella and Rainaldi 2011; Yang et al. 2011). Leite and colleagues reported significant changes in miRNA expression profiles during the progression of prostate cancer from localized carcinoma to metastatic (Leite et al. 2011). Similarly, using miRNA profiles it is possible to accurately distinguish between primary and metastatic brain tumors, gastric cancer and other tumor types (Fassan et al. 2011; Li et al. 2011). In breast cancer, the elevated expression levels of miR-10b seem to promote cell migration and invasion in cancer cells (Ma et al. 2007), while miR-31 was absent in metastatic breast tumors (Valastyan et al. 2009).

MiRNAs also have been shown to relate with crucial aspect of the metastatic process, such as the epithelial-mesenchymal transition (EMT)[8] (Thiery and Sleeman 2006; de Krijger et al. 2011). Several miRNAs are associated with EMT such as miR221/222 in breast cancer metastasis (Shah and Calin 2011), or through regulation of E-cadherin (Korpal et al. 2008). As mentioned above, miR-155 is considered to promote tumor formation, and it was shown recently that it remarkably promotes macroscopic tumor formation in the lung (Xiang et al. 2011). However, in the same report, miR-155 has been shown to reduce significantly the aggressiveness of tumor cell dissemination

[8]EMT is a process where epithelial cells acquire mesenchymal, fibroblast-like properties and show reduced intercellular adhesion and increased motility. It is regarded as a critical event, since it is involved in morphogenetic changes during embryonic development, wound healing, but also in malignant transformation and cancer metastasis. The reduced cell adhesion is reflected by the downregulation of E-cadherin, an adhesion molecule present in most epithelial cell, acting also as a tumor suppressor inhibiting invasion and metastasis. Together with decreased E-cadherin, in cancer cells, EMT is associated with increased expression of N-cadherin, as well as of mesenchymal proteins, including vimentin, fibronectin, periostin. The combination of loss of E-cadherin with the increase of N-cadherin leads to alterations in the epithelial cell shape and motility (Larue and Bellacosa 2005; Thiery and Sleeman 2006; Morra and Moch 2011).

by preventing EMT *in vivo* through the suppression of important EMT regulators. The results of this work supported another parameter in the regulation of miRNAs, the location of the tumor cells; miR-155 prevents tumor dissemination in mammary fat pads, whereas in the lung it contributes to the macroscopic formation of the tumor (Xiang et al. 2011).

Perhaps, the most promising conclusion of the studies on the role of miRNAs in cell migration and establishment of metastatic tumors (Baranwal and Alahari 2010) is that these miRNAs are highly tissue-specific, giving the opportunity to develop novel biomarkers for cancer detection and identification of the origins of metastasis (Ullah and Aatif 2009). The latter remains a challenge, as cancers of unknown origin are histologically confirmed metastatic tumors but their tissue of origin cannot be easily identified. As a result, these types of cancer are not treated properly, leading to insufficient therapy schemes and poor outcome. Consequently, the classification of tumors that have been poorly differentiated or of unknown origin would greatly help in cancer treatment.

Distinguishing Clinical Subtypes

To tackle with the problem of tumor classification, Lu and colleagues set up a classifier including few miRNAs that could be used for diagnosis of tumors of unknown origin providing more information than the traditional profiling of several thousands of mRNAs (Lu et al. 2005). Importantly, miRNA levels are more reliable to classify poorly differentiated tumors than mRNA profiles, which proved as highly inaccurate (Lu et al. 2005). In this direction, Rosenfeld and colleagues, succeeded to determine a classification system for unknown tumors based on 48 miRNAs (Rosenfeld et al. 2008). These miRNA fingerprints could track the origin of metastatic cancers with high accuracy and were more indicative of cancer type, stage and pathophysiological features than mRNA profiles. Two distinct but histologically similar subtypes of colorectal cancer, classified as microsatellite-stable and microsatellite-unstable cancers, can be differentiated as 14 miRNA present specific fingerprints between the two subtypes (Lanza et al. 2007; Schepeler et al. 2008). Concerning breast cancer, two research groups identified unique subsets of miRNAs that would accurately classify breast tumors as luminal A, luminal B, basal-like, $HER2^+$, $HER2^-$, ER^+, ER^- (Mattie et al. 2006; Blenkiron et al. 2007). Further, Lebanony and colleagues showed that miR-205 is a highly specific marker that can differentiate adenocarcinoma from squamous lung cancer with high sensitivity and specificity (Lebanony et al. 2009). Differential expression profiles of miRNAs are also observed in cytogenetic and molecular subtypes of pediatric acute myeloid leukemia (Nowsheen et al. 2011) and between acral and non-acral melanomas (Chan et al. 2011). The above-mentioned studies suggest that miRNAs may influence the pathogenesis of cancer subtypes, thus may function as effective diagnostic differentiating markers.

MiRNAs and Cancer Diagnosis

The establishment of new clinical tools for cancer prognosis and diagnosis is of primary importance, as the majority of the tumors are identified in late disease stage, thus impeding timely response and treatment. Therefore, current studies focus on defining novel markers with the potential to identify cancers at early stage or precursor lesions (Nowsheen et al. 2011). Naturally, the correlation between deregulation of miRNAs and cancer raised the question whether specific miRNA expression levels could be used as cancer biomarkers. As mentioned, Lu and colleagues reported that miRNAs exhibit specific profiles in different human cancers, allowing their accurate classification (Lu et al. 2005). Another large-scale miRNA profiling study on breast, lung, prostate, stomach, colon and pancreatic tumors, revealed differentiated miRNA profiles for each type of cancer (Volinia et al. 2006). A key observation from these studies was that in all types of cancer there were some similarly deregulated miRNAs, which probably participate in basal cancer pathways (Calin and Croce 2006a).

To embody miRNAs as biomarkers for cancer diagnosis in clinical practice they should meet specific requirements such as a) to feasibly identify a tumor at early stages, before metastasis, thus increasing the survivability, b) to monitor the molecular diversity of cancer and disease progression, and c) to serve as an indicator of patient's response to therapy. Indeed, miRNAs can be detected from a minute amount of the tissue sample needed for histological evaluation and the results are readily interpretable (Mattie et al. 2006). More significantly, miRNAs are stably expressed in various biological samples that are currently used for diagnosis, such as formalin-fixed tissues (Nelson et al. 2004; Xi et al. 2007; Iorio and Croce 2009). Further, miRNAs show some significant advantages over classical protein biomarkers or mRNAs. Unlike the currently used protein biomarkers, they can be isolated by available high affinity miRNA capture reagents and they present less complexity, as they do not bear post-processing modifications (K. Wang et al. 2009). Unlike mRNAs, miRNAs have the advantage of a longer lifetime *in vivo* (Lim et al. 2005) and are more stable *in vitro* (Tang et al. 2006; Waldman and Terzic 2007). Importantly, miRNAs fingerprints may be superior to mRNA expression profiles, since miRNAs are master regulators of protein expression, while only few mRNAs are regulatory molecules (Chen 2005; Lu et al. 2005).

Probably, the most important feature that favours the use of miRNAs in cancer diagnosis is the fact that they participate in fundamental pathways of cell development and differentiation. Therefore, miRNA levels can indicate abnormalities in the balance between cell surveillance, proliferation and apoptosis. Indeed, miR-221 and miR-222 levels are elevated in ovarian cancer (Wurz et al. 2010), hepatocellular cancer (Pineau et al. 2010), and glioblastomas (Zhang et al. 2010), indicating their implication in basal pathways correlated with cancer. In some cases only few miRNAs are consistently deregulated, thus serving as a potential feasible tool for cancer detection and evaluation. Analysis of formalin-fixed tissues revealed a significant miRNA signature in breast cancer, where miR-21, miR-155, miR-191, miR-196a were upregulated, while miR-125b and miR-221 downregulated (Hui et al. 2009). Other studies indicate alternative miRNA signatures including the above-mentioned miRNAs that could also be used for breast cancer tracking (Iorio et al. 2005; Iorio et al. 2008). Noteworthy, miR-21 elevated levels are associated with breast cancer severity, lymph node metastasis and poor prognosis (Yan et al. 2008). Many studies focused on the miRNA expression profile in breast cancer revealed deregulated levels in various miRNAs, a combination of which could provide an accurate diagnostic tool (Bachour and Bennett 2011).

Trying to define specific miRNA fingerprints through surveying the expression levels of various miRNAs, several groups found that miRNAs are both upregulated and downregulated in several cancer types. In gastric cancer, specific miRNAs levels constitute a signature, that can distinguish the malignant from normal tissue (Michael et al. 2003; Akao et al. 2006; Bandres et al. 2006; Schetter et al. 2008; Guo et al. 2009; Xiao et al. 2009). Pancreatic cancer is one of the most challenging ones due to limited specific biomarkers and prognosis remains extremely poor. MiR-21 reveals as a promising biomarker, as it is significantly overexpressed in pancreatic cancer and it may predict limited survival in patients with node-negative disease (Dillhoff et al. 2008). It may facilitate diagnosis at early stage, as characteristic miRNA profiles can distinguish pancreatic tumors from chronic pancreatitis and benign tissue (Bloomston et al. 2007; Lee et al. 2007a; Szafranska et al. 2007; Lee et al. 2011).

Other cases of miRNAs deregulated in specific cancer types are, for example, miR-122a, which is downregulated in hepatocellular carcinoma (Kutay et al. 2006), miR-204 and miR-211 that are upregulated in insulinomas (Roldo et al. 2006), let-7 that is downregulated in lung cancer (Takamizawa et al. 2004; Dillhoff et al. 2008; Khabar 2010), a combination of upregulated miRNAs and downregulated miR-199a-3p in osteosarcoma (Dillhoff et al. 2008; Khabar 2010), and many others that hold many promises for the clinical application of miRNA profiles. Table 1 summarizes important miRNAs with altered expression in cancer types.

Circulating miRNAs in Cancer Diagnosis

The quest for reliable protein-based biomarkers from body fluids, such as plasma, serum, and urine, face many challenges mainly due to a) the low abundance of protein-marker of interest in the fluid, b) the complex composition of the protein itself, and c) the posttranslational modifications, proteolysis and protein denaturation that add to the difficulty of developing high-affinity capture agents and easy assay methods (Ebert et al. 2006; Cowan and Vera 2008). Another challenge in cancer diagnosis is the application of non-invasive tools which has led to increased interest in the integration of circulating nucleic acids in clinical practice for cancer diagnosis (Tsang and Lo 2007; Chan et al. 2011). Recent studies suggest that circulating miRNAs may be considered as auxiliary non-invasive biomarkers for various physiological or pathological conditions (Gilad et al. 2008; Cortez and Calin 2009), as prenatal diagnosis (Chim et al. 2008; Luo et al. 2009), myocardial infraction (Adachi et al. 2010; Wang et al. 2010), drug induced liver injury (K. Wang et al. 2009), and inflammation (Vasilescu et al. 2009; Wang et al. 2010). Concerning cancer diagnosis, it was in 2008 when several studies showed that miRNAs could be isolated from human body fluids and used as biomarkers (Chen et al. 2008; Lawrie et al. 2008; Mitchell et al. 2008).

Thus, it was shown that miR-21 levels are elevated in the serum of patients suffering from diffuse large B-cell lymphoma and ovarian cancer (Lawrie et al. 2008; Resnick et al. 2009), and a close correlation between circulating miRNAs and tumors has been revealed (Chen et al. 2008). However, miR-21 was reduced in sera from chronic viral hepatitis B (HBV)

Table 1 miRNAs with altered expression levels in cancer.

Cancer type	*Upregulated miRNAs*	*Downregulated miRNAs*	*References*
Breast	miR-21, 29a/c, 29b-2, 10b, 22, 27a/b, 102, 125b, 191, 196a/b, 203, 210, 213, 221, 222, 328, 373, 516-3p, 520c,	miR-7, 9-1, 17, 17-5p, 20, 26a, 27b, 30-3p, 30e, 31, 93, 101, 106, 125a/b, 140, 143, 145, 146, 155, 205, 206, 335,	(Iorio et al. 2005; Hossain et al. 2006; Mattie et al. 2006; Ma et al. 2007; Mertens-Talcott et al. 2007; Negrini et al. 2007; Scott et al. 2007; Yu et al. 2007; Zhu et al. 2007; Bhaumik et al. 2008; Camps et al. 2008; Gregory et al. 2008; Huang et al. 2008; Kondo et al. 2008; Kong et al. 2008; Lehmann et al. 2008; Lowery et al. 2008; Miller et al. 2008; Negrini and Calin 2008; Zhu et al. 2008; Pandey and Picard 2009; Valastyan et al. 2009; Webster et al. 2009; Cho 2010; Bueno and Malumbres 2011)
Colon	miR-21, 223, 17, 106, 34s, 17-92 cluster, 31, 96, 183, 92, 10a, 20a, 24-1, 29b-2	miR-34a/b/c, 143, 145, 195, 130a, 331, 129, 127, 342, 9, 137 let-7 family	(Lu et al. 2005; Bandres et al. 2006; Cummins et al. 2006; Saito et al. 2006; Volinia et al. 2006; Slaby et al. 2007; Tazawa et al. 2007; Toyota et al. 2007; Asangani et al. 2008; Grady et al. 2008; Schetter et al. 2008; Cho 2010; Bueno and Malumbres 2011)
Hepatocellular	miR-17-92 cluster, 18, 21, 34a, 34s, 106, 143, 221, 222, 224, lin-28b	miR-17-19b cluster, 1, 19, 26a/b, 101,122, 125a, 141, 142, 145, 181a/c, 199, 200a, 301, let-7 family	(Guo et al. 2006; Murakami et al. 2006; Gramantieri et al. 2007; Meng et al. 2007; Connolly et al. 2008; Datta et al. 2008; Jiang et al. 2008; Y. Wang et al. 2008; Ji et al. 2009; Ji and Wang 2009; Su et al. 2009; Zhang et al. 2009; Barshack et al. 2010; Gao et al. 2011; Wu et al. 2011)
Leukemia, Chronic Lymphocytic (CLL)	miR-21, 92, 155	miR-15, 16, 29a/b/c, 34a, 92, 143, 145, 150, 181b, 223, 34 family	(Calin et al. 2002; Calin et al. 2005; Pekarsky et al. 2006; Akao et al. 2007; Fulci et al. 2007; Dijkstra et al. 2009; Stamatopoulos et al. 2009; Cho 2010; Bueno and Malumbres 2011)

Table 1 Contd. ...

Table 1 Contd. ...

Lung	miR-21, 17-3p, 17-92 cluster, 24-2, 93, 106a, 106b, 146, 150, 155, 189, 191, 192, 200b, 212	miR-148a/b, 375, 15 family let-7 family	(Takamizawa et al. 2004; Eder and Scherr 2005; Lu et al. 2005; Volinia et al. 2006; Yanaihara et al. 2006; Fabbri et al. 2007; Lowery et al. 2008; Nasser et al. 2008; Bandi et al. 2009; Spizzo et al. 2009; Webster et al. 2009; George and Mittal 2010; Bueno and Malumbres 2011)
Ovarian	miR-141, 200a/b/c, 214, 221 lin-28b	miR-26a, 125a/b, 129, 140, 145, 199a, 199a/b, 204, 424, let-7 family	(Corney et al. 2007; Iorio et al. 2007; Nam et al. 2008; Taylor and Gercel-Taylor 2008; Yang et al. 2008; Heneghan et al. 2009; Hu et al. 2009; Lu et al. 2009)
Pancreatic	miR-21, 24-2, 100, 103, 103-1, 107, 125b-1 155, 181a, 196a/b, 203, 210, 221, 222, 301, 376a	miR-148a/b, 375, 15 family, let-7 family	(Lu et al. 2005; Roldo et al. 2006; Volinia et al. 2006; Bloomston et al. 2007; Lee et al. 2007b; Szafranska et al. 2007; Lowery et al. 2008; George and Mittal 2010; Bueno and Malumbres 2011)
Prostate	miR-20a, 21, 125b, 195, 203, 221, 222 let-7d	miR-101, 127, 126, 128a, 143, 145, 146, 205, 330, 449a/b, 15 family	(Lu et al. 2005; Mattie et al. 2006; Saito et al. 2006; Volinia et al. 2006; Galardi et al. 2007; Ambs et al. 2008; Bonci et al. 2008; Varambally et al. 2008; Noonan et al. 2009; Cho 2010; Bueno and Malumbres 2011; George et al. 2011)
Thyroid	miR-21, 24-2, 102, 125b-1, 125b-2, 146, 155, 181a/b/c, 197, 202, 220, 221, 222, 224, 346	miR-15b, 26a, 140, 142, 219-1, let-7f	(He et al. 2005a; Pallante et al. 2006; Volinia et al. 2006; Weber et al. 2006; Negrini et al. 2007; Visone et al. 2007; Nikiforova et al. 2008; Bueno and Malumbres 2011; Marini et al. 2011; Vriens et al. 2011)

patients with or without hepatocellular carcinoma[9], indicating that miR-21 may behave either as an oncogene or as a tumor suppressor, depending on the targets and the factors that deregulate its expression (Qi et al. 2011). Further studies showed that the levels of miRNA found in sera of patients with lung adenocarcinoma are in agreement with the corresponding levels in cancer cells (Rabinowits et al. 2009). Accordingly, lung cancer-specific miRNA signatures have been detected in blood serum, which could enable the disease diagnosis in asymptomatic or symptomatic patients (Bianchi et al. 2011; Chen et al. 2011; Roth et al. 2011). Other studies described cancer-specific miRNA expression pattern in serum. This was the case of colon cancer study, where elevated levels of five miRNAs were found in plasma of patients compared to healthy individuals and two members of miR-17-92 cluster, miR-92 and miR-17-3p, were confirmed as effective diagnostic markers (Ng et al. 2009; Schetter and Harris 2009). Noteworthy, miRNA profiles in the blood stream may even detect colon cancer at early stages, while it is clinically asymptomatic (Huang et al. 2010).

Despite RNA is generally considered as a very vulnerable molecule, it was shown that miRNA levels in circulating fluids may be determined reliably, as they remain stable under manipulations, such as boiling, extreme pH changes, long-term storage and multiple freeze-thaw cycles, and resistant to RNase treatment (Chen et al. 2008; Gilad et al. 2008; Mitchell et al. 2008; Taylor and Gercel-Taylor 2008; Ho et al. 2010). Concerning the origin of circulating miRNAs, they are supposed to be released from proliferating, dying or lysed tumor cells, or from cancer cells ongoing invasion. The miRNAs are released in blood stream through microvesicles, also called exosomes[10]. The mechanism of miRNA release through exosomes mediates a ceramide-dependent secretory pathway, while, importantly, the secreted miRNAs are transferable and functional in the recipient cells (Kosaka et al. 2010).

Table 2 summarizes current expression alterations of specific miRNAs that alter their expression in biological fluids from several cancers. The studies in this table underline the importance of the expression levels of serum miRNAs and the expectations towards the improvement of current diagnostic tools.

[9]Hepatocellular carcinoma (HCC) accounts for 90% of primary liver cancers. Among the major risk factors for the development of HCC are infections by hepatitis B and C viruses (Zemel et al. 2011).

[10]Exosomes are small membrane particles excreted from the cell plasma membrane into the extracellular space, that contain both mRNA and microRNA, called "exosomal shuttle RNA" (esRNA) (Valadi et al. 2007). Their ribonucleic acid content can be delivered to another cell, and may be functional in their new host, thus playing an important role in cell-cell communication (van Niel et al. 2006; Valadi et al. 2007).

Table 2 miRNAs in biological fluids.

Cancer type	*Fluid source*	*Upregulated*	*Downregulated*	*References*
Bladder	Plasma	miR-126, 152, 182		(Hanke et al. 2010)
Breast	Serum	miR-195	miR-195, let-7a*	(Heneghan et al. 2010)
Colorectal	Plasma	miR-92a, 92b	miR-17-3p, 92*	(Ng et al. 2009; Huang et al. 2010)
Diffuse large B-cell lymphoma	Serum	miR-21, 155, 210		(Lawrie et al. 2008)
Gastric	Serum	miR-17-5p, 21, 106a, 106b	let-7a	(Tsujiura et al. 2010)
	Plasma		miRs-17-5p, 21, 106a, 106b*	(Tsujiura et al. 2010)
Hepatocellular carcinoma (HCC)	Serum	miR-500	miR-21	(Yamamoto et al. 2009; Qi et al. 2011)
Leukemia, Acute	Plasma		miR-92a	(Tanaka et al. 2009b)
Oral squamous-cell	Plasma	miR-31	miR-31*	(Liu et al. 2010)
	Saliva	miR-31	miR-31*, miRs--125a, -200a	(Park et al. 2009a; Liu et al. 2010)
Ovarian		miR-21, -29a, -92, -93, -126		(Resnick et al. 2009)
Pancreatic	Plasma	miR-210		(Ho et al. 2010)
Pancreatic ductal adenocarcinoma	Plasma	miR-21, 155, 196a, 210		(J. Wang et al. 2009)
Squamous cell carcinoma of tongue	Serum	miR-184		(Wong et al. 2008)

* Decreased levels postoperatively.

MiRNAs and Cancer Prognosis

Evidence supporting the active participation of miRNAs in human oncogenesis suggested their contribution to cancer prognosis. Specific miRNA signatures have been shown to facilitate disease surveillance in various cancer types. Calin and coworkers reported that two prognostic factors of CLL, the expression of ZAP-70 and the mutation status of IgV_H, correlate with the expression profiles of certain miRNAs. Specifically, downregulated miR-15a and miR-16-1 are found in patients with good prognosis, while an expression pattern of downregulated miR-29b and miR-181 and upregulated miR-155 and miR-21 associates with patients with poor prognosis (Calin et al. 2005). Moreover, high expression levels of miR-155 are also found in diffuse large B-cell lymphoma patients with poor prognosis (Eis et al. 2005; Kluiver et al. 2005).

MiRNAs seem to be further involved in leukemia prognosis. Acute myeloid leukemia (AML) is a cytogenetically and molecularly heterogeneous disorder, where the majority of AML patients represent intermediate-and poor-risk cytogenetic profile, while for almost half of them no cytogenetic abnormality is detectable in leukemic cells[11] (Lowenberg et al. 1999; Garzon et al. 2008; Marcucci et al. 2008). As chemotherapy most often fails to cure the patient, and transplantation is frequently the treatment of choice but not always the option, in order to understand the biology of the disease and develop novel therapeutic approaches research has turned to miRNAs. In this direction, it has been shown that the prognosis for low-risk acute myeloid leukemia (AML) is poor for patients with high expression of miR-191 and miR-199a; patients with high expression of these miRNAs had significantly worse overall and event-free survival than AML patients with low expression (Garzon et al. 2008). In this work also, the authors identified miRNAs associated with survival despite poor prognosis, and also showed that the outcome signature of AML consisting of up-regulated miRNAs (including miR-20, miR-25, miR-199a, and miR-191) in common with signatures of six solid tumors, (breast, colon, lung, pancreatic, prostate and stomach) (Volinia et al. 2006; Garzon et al. 2008). The prognostic miRNA expression profile differs in the case of high-risk AML patients. Elevated levels of specific miR-181a and miR-181b family members and reduced expression of seven different miRNAs were shown to be associated with the risk of an event, meaning relapse or death (Marcucci et al. 2008). Further work showed that the expression miR-181a alone associates with the clinical outcome of patients with cytogenetically normal AML and may refine their molecular risk classification (Schwind et al. 2010). Thus, a treatment leading to the increase of endogenous miR-181a levels might represent a novel therapeutic strategy against AML.

MiRNA signatures with prognostic value have been also reported for various types of solid tumors. In lung cancer for example, let-7 expression was reduced

[11]Cytogenetic analysis revealed abnormalities in 50–60% of newly diagnosed AML of adult patients, while the remaining 40 to 50% do not have clonal chromosomal aberrations and is characterized as cytogenetically normal AML (Mrozek et al. 2004). Cytogenetics is the most important tool to classify AML patients into low-, intermediate- and high-risk groups. Cytogenetically normal AML are currently categorized in the intermediate-risk group (Mrozek et al. 2007). The classification usually dictates the best post-remission therapy for single patient; high-risk patients are candidates for an allogeneic hematopoietic stem cells transplantation, whereas low risk cytogenetics patients will receive only standard chemotherapy (Blum 2008). Cytogenetics is also the most important prognostic factor to predict the outcome of AML patients (Mrozek and Bloomfield 2006; Morra et al. 2009).

and this was associated to shortened postoperative survival (Takamizawa et al. 2004). Noteworthy, low levels of let-7 have prognostic impact only for non-small-cell cancer and no other subtypes of lung cancer (Lu et al. 2007). Later, it was reported that a unique miRNA expression pattern, including reduced levels of let-7 and high levels of miR-155, could serve as a predictor of non-small-cell lung cancer outcome (Yanaihara et al. 2006). Further work supported that specific miRNAs signatures can predict cancer relapse and survival of lung cancer patients (Yu et al. 2008), while the decreased levels of miR-451 in non-small cell lung cancer patients is associated with poor prognosis (Wang et al. 2011). In colon cancer patients, upregulated miR-21 correlates with poor survival and therapeutic outcome (Schetter et al. 2008). Li et al. in 2010 identified an expression profile consisting of seven miRNAs that could sufficiently enough predict overall survival and relapse-free survival of gastric cancer patients. In pancreatic tumors, elevated levels of a subset of miRNAs is linked to poor survival (Greither et al. 2010). Independent studies focusing on prognostic factors for breast cancers revealed that overexpressed miR-21 (Yan et al. 2008) and miR-210 (Camps et al. 2008) exhibit inverse correlation with disease-free and overall survival. The prognostic value of miRNAs profile is sometimes gender specific, as was shown for miR-210 deregulation that associates with poor prognosis of soft-tissue sarcomas only in female patients (Greither et al. 2011).

FUTURE PERSPECTIVES

The regulation of protein expression by miRNAs was critical for the association of miRNAs to biological pathways and processes. The involvement also in pathways responsible for cell proliferation, apoptosis and differentiation revealed their role in cancer. Perhaps, the most promising feature is the tumor-specific miRNA fingerprints present in both tissue and circulating fluid samples that may serve for cancer diagnosis and prognosis. However, the bench results are still far away from the clinical practice, as there are many critical issues that should be encountered. Firstly, it is of primary importance to understand miRNA function in full detail as well as their cancer-related target and pathways. In this context, genome-wide analyses would reveal quantitative and qualitative alterations in miRNAs in various human cancers, leading to identification of putative oncogenes and tumor suppressors. Secondly, the tumor-specific fingerprints should be further analysed, as there are studies linking the same miRNA to different cancer types, especially to those of common origin. These are expected to lead to well-characterised miRNA expression profiles that could reveal the exact tumor type, progression stage, patient outcome and potential recurrence. Third, current published studies analyze relative small sample sizes and there is need for long-term outcome data. Larger validation studies representing genders, all ethnicities and age groups are needed to verify the preliminary results. The previous issues also suggest that in order to obtain reliable and reproducible results, common techniques from specimen preparation to miRNA purification and detection, and widely accepted standard genes as controls, should be adopted unanimously. Finally, perhaps the most challenging issue for therapeutic development remains the effective *in vivo* delivery of miRNAs, in order to harness and take full advantage of their exciting potential. Clearly, miRNAs represent one of the most promising partners in the efforts for cancer prognosis, diagnosis and treatment. The advances of molecular biochemistry and the global interest of public and private research on the therapeutic use of miRNAs may bring individualized therapy a step closer.

REFERENCES

Adachi, T., M. Nakanishi, Y. Otsuka, K. Nishimura, G. Hirokawa, Y. Goto, H. Nonogi and N. Iwai. 2010. Plasma microRNA 499 as a biomarker of acute myocardial infarction. Clin. Chem. 56: 1183-1185.

Akao, Y., Y. Nakagawa, Y. Kitade, T. Kinoshita and T. Naoe. 2007. Downregulation of microRNAs-143 and -145 in B-cell malignancies. Cancer Sci. 98: 1914-1920.

Akao, Y., Y. Nakagawa and T. Naoe. 2006. let-7 microRNA functions as a potential growth suppressor in human colon cancer cells. Biol. Pharm. Bull. 29: 903-906.

Ambs, S., R.L. Prueitt, M. Yi, R.S. Hudson, T.M. Howe, F. Petrocca, T.A. Wallace, C.G. Liu, S. Volinia, G.A. Calin, H.G. Yfantis, R.M. Stephens and C.M. Croce. 2008. Genomic profiling of microRNA and messenger RNA reveals deregulated microRNA expression in prostate cancer. Cancer Res. 68: 6162-6170.

Asangani, I.A., S.A. Rasheed, D.A. Nikolova, J.H. Leupold, N.H. Colburn, S. Post and H. Allgayer. 2008. MicroRNA-21 (miR-21) post-transcriptionally downregulates tumor suppressor Pdcd4 and stimulates invasion, intravasation and metastasis in colorectal cancer. Oncogene 27: 2128-2136.

Bachour, T. and K. Bennett. 2011. The Role of MicroRNAs in Breast Cancer. J. Assoc. Genet. Technol. 37: 21-28.

Bandi, N., S. Zbinden, M. Gugger, M. Arnold, V. Kocher, L. Hasan, A. Kappeler, T. Brunner and E. Vassella. 2009. miR-15a and miR-16 are implicated in cell cycle regulation in a Rb-dependent manner and are frequently deleted or down-regulated in non-small cell lung cancer. Cancer Res. 69: 5553-5559.

Bandres, E., E. Cubedo, X. Agirre, R. Malumbres, R. Zarate, N. Ramirez, A. Abajo, A. Navarro, I. Moreno, M. Monzo and J. Garcia-Foncillas. 2006. Identification by Real-time PCR of 13 mature microRNAs differentially expressed in colorectal cancer and non-tumoral tissues. Mol. Cancer 5: 29.

Baranwal, S. and S.K. Alahari. 2010. miRNA control of tumor cell invasion and metastasis. Int. J. Cancer 126: 1283-1290.

Barshack, I., E. Meiri, et al. 2010. Differential diagnosis of hepatocellular carcinoma from metastatic tumors in the liver using microRNA expression. Int. J. Biochem. Cell Biol. 42: 1355-1362.

Bartel, D.P. 2004. MicroRNAs: genomics, biogenesis, mechanism, and function. Cell 116: 281-297.

Bartel, D.P. 2009. MicroRNAs: target recognition and regulatory functions. Cell 136: 215-233.

Bass, B.L. 2002. RNA editing by adenosine deaminases that act on RNA. Annu. Rev. Biochem. 71: 817-846.

Behm-Ansmant, I., J. Rehwinkel, T. Doerks, A. Stark, P. Bork and E. Izaurralde. 2006. mRNA degradation by miRNAs and GW182 requires both CCR4:NOT deadenylase and DCP1:DCP2 decapping complexes. Genes Dev. 20: 1885-1898.

Berezikov, E., W.J. Chung, J. Willis, E. Cuppen and E.C. Lai. 2007. Mammalian mirtron genes. Mol. Cell 28: 328-336.

Bernstein, E., A.A. Caudy, S.M. Hammond and G.J. Hannon. 2001. Role for a bidentate ribonuclease in the initiation step of RNA interference. Nature 409: 363-366.

Bertone, P., V. Stolc, T.E. Royce, J.S. Rozowsky, A.E. Urban, X. Zhu, J.L. Rinn, W. Tongprasit, M. Samanta, S. Weissman, M. Gerstein and M. Snyder. 2004. Global identification of human transcribed sequences with genome tiling arrays. Science 306: 2242-2246.

Bhaumik, D., G.K. Scott, S. Schokrpur, C.K. Patil, J. Campisi and C.C. Benz. 2008. Expression of microRNA-146 suppresses NF-kappaB activity with reduction of metastatic potential in breast cancer cells. Oncogene 27: 5643-5647.

Bianchi, F., F. Nicassio, M. Marzi, E. Belloni, V. Dall'olio, L. Bernard, G. Pelosi, P. Maisonneuve, G. Veronesi and P.P. Di Fiore. 2011. A serum circulating miRNA diagnostic test to identify asymptomatic high-risk individuals with early stage lung cancer. EMBO Mol. Med. 3: 495-503.

Birney, E., J.A. Stamatoyannopoulos, et al. 2007. Identification and analysis of functional elements in 1% of the human genome by the ENCODE pilot project. Nature 447: 799-816.

Blenkiron, C., L.D. Goldstein, N.P. Thorne, I. Spiteri, S.F. Chin, M.J. Dunning, N.L. Barbosa-Morais, A.E. Teschendorff, A.R. Green, I.O. Ellis, S. Tavare, C. Caldas and E.A. Miska. 2007. MicroRNA expression profiling of human breast cancer identifies new markers of tumor subtype. Genome Biol. 8: R214.

Bloomston, M., W.L. Frankel, F. Petrocca, S. Volinia, H. Alder, J.P. Hagan, C.G. Liu, D. Bhatt, C. Taccioli and C.M. Croce. 2007. MicroRNA expression patterns to differentiate pancreatic adenocarcinoma from normal pancreas and chronic pancreatitis. JAMA 297: 1901-1908.

Blum, W. 2008. Post-remission therapy in acute myeloid leukemia: what should I do now? Haematologica 93: 801-805.

Bohmert, K., I. Camus, C. Bellini, D. Bouchez, M. Caboche and C. Benning. 1998. AGO1 defines a novel locus of Arabidopsis controlling leaf development. Embo J. 17: 170-180.

Boland, A., E. Huntzinger, S. Schmidt, E. Izaurralde and O. Weichenrieder. 2011. Crystal structure of the MID-PIWI lobe of a eukaryotic Argonaute protein. Proc. Natl. Acad. Sci. U S A 108: 10466-10471.

Bonci, D., V. Coppola, et al. 2008. The miR-15a-miR-16-1 cluster controls prostate cancer by targeting multiple oncogenic activities. Nat. Med. 14: 1271-1277.

Bracken, A.P., M. Ciro, A. Cocito and K. Helin. 2004. E2F target genes: unraveling the biology. Trends Biochem. Sci. 29: 409-417.

Brennecke, J., A. Stark, R.B. Russell and S.M. Cohen. 2005. Principles of microRNA-target recognition. PLoS Biol. 3: e85.

Bueno, M.J. and M. Malumbres. 2011. MicroRNAs and the cell cycle. Biochim Biophys. Acta 1812: 592-601.

Cai, X., C.H. Hagedorn and B.R. Cullen. 2004. Human microRNAs are processed from capped, polyadenylated transcripts that can also function as mRNAs. RNA 10: 1957-1966.

Calin, G.A. and C.M. Croce. 2006a. MicroRNA signatures in human cancers. Nat. Rev. Cancer 6: 857-866.

Calin, G.A. and C.M. Croce. 2006b. MicroRNAs and chromosomal abnormalities in cancer cells. Oncogene 25: 6202-6210.

Calin, G.A., C.D. Dumitru, et al. 2002. Frequent deletions and down-regulation of micro- RNA genes miR15 and miR16 at 13q14 in chronic lymphocytic leukemia. Proc. Natl. Acad. Sci. U S A 99: 15524-15529.

Calin, G.A., M. Ferracin, et al. 2005. A MicroRNA signature associated with prognosis and progression in chronic lymphocytic leukemia. N. Engl. J. Med. 353: 1793-1801.

Calin, G.A., C.G. Liu, et al. 2004a. MicroRNA profiling reveals distinct signatures in B cell chronic lymphocytic leukemias. Proc Natl Acad Sci U S A 101: 11755-11760.

Calin, G.A., C. Sevignani, C.D. Dumitru, T. Hyslop, E. Noch, S. Yendamuri, M. Shimizu, S. Rattan, F. Bullrich, M. Negrini and C.M. Croce. 2004b. Human microRNA genes are frequently located at fragile sites and genomic regions involved in cancers. Proc. Natl. Acad. Sci. U S A 101: 2999-3004.

Camps, C., F.M. Buffa, S. Colella, J. Moore, C. Sotiriou, H. Sheldon, A.L. Harris, J.M. Gleadle and J. Ragoussis. 2008. hsa-miR-210 Is induced by hypoxia and is an independent prognostic factor in breast cancer. Clin. Cancer Res. 14: 1340-1348.

Chan, E., R. Patel, et al. 2011. MicroRNA signatures differentiate melanoma subtypes. Cell Cycle 10: 1845-1852.

Cheloufi, S., C.O. Dos Santos, M.M. Chong and G.J. Hannon. 2010. A dicer-independent miRNA biogenesis pathway that requires Ago catalysis. Nature 465: 584-589.

Chen, C.Z. 2005. MicroRNAs as oncogenes and tumor suppressors. N. Engl. J. Med. 353: 1768-1771.

Chen, X., Y. Ba, et al. 2008. Characterization of microRNAs in serum: a novel class of biomarkers for diagnosis of cancer and other diseases. Cell Res. 18: 997-1006.

Chen, X., Z. Hu, et al. 2012. Identification of ten serum microRNAs from a genome-wide serum microRNA expression profile as novel noninvasive biomarkers for nonsmall cell lung cancer diagnosis. Int. J. Cancer 130: 1620-1628.

Chim, S.S., T.K. Shing, E.C. Hung, T.Y. Leung, T.K. Lau, R.W. Chiu and Y.M. Lo. 2008. Detection and characterization of placental microRNAs in maternal plasma. Clin. Chem. 54: 482-490.

Cho, W.C. 2010. MicroRNAs: potential biomarkers for cancer diagnosis, prognosis and targets for therapy. Int. J. Biochem. Cell Biol. 42: 1273-1281.

Cifuentes, D., H. Xue, D.W. Taylor, H. Patnode, Y. Mishima, S. Cheloufi, E. Ma, S. Mane, G.J. Hannon, N.D. Lawson, S.A. Wolfe and A.J. Giraldez. 2010. A novel miRNA processing pathway independent of Dicer requires Argonaute2 catalytic activity. Science 328: 1694-1698.

Cimmino, A., G.A. Calin, et al. 2005. miR-15 and miR-16 induce apoptosis by targeting BCL2. Proc. Natl. Acad. Sci. U S A 102: 13944-13949.

Connolly, E., M. Melegari, P. Landgraf, T. Tchaikovskaya, B.C. Tennant, B.L. Slagle, L.E. Rogler, M. Zavolan, T. Tuschl and C.E. Rogler. 2008. Elevated expression of the miR-17-92 polycistron and miR-21 in hepadnavirus-associated hepatocellular carcinoma contributes to the malignant phenotype. Am. J. Pathol. 173: 856-864.

Consortium, E.P., E. Birney, et al. 2007. Identification and analysis of functional elements in 1% of the human genome by the ENCODE pilot project. Nature 447: 799-816.

Corney, D.C., A. Flesken-Nikitin, A.K. Godwin, W. Wang and A.Y. Nikitin. 2007. MicroRNA-34b and MicroRNA-34c are targets of p53 and cooperate in control of cell proliferation and adhesion-independent growth. Cancer Res. 67: 8433-8438.

Cortez, M.A. and G.A. Calin. 2009. MicroRNA identification in plasma and serum: a new tool to diagnose and monitor diseases. Expert. Opin. Biol. Ther. 9: 703-711.

Cowan, M.L. and J. Vera. 2008. Proteomics: advances in biomarker discovery. Expert. Rev. Proteomics 5: 21-23.

Cummins, J.M., Y. He, et al. 2006. The colorectal microRNAome. Proc. Natl. Acad. Sci. U S A 103: 3687-3692.

Datta, J., H. Kutay, M.W. Nasser, G.J. Nuovo, B. Wang, S. Majumder, C.G. Liu, S. Volinia, C.M. Croce, T.D. Schmittgen, K. Ghoshal and S.T. Jacob. 2008. Methylation mediated silencing of MicroRNA-1 gene and its role in hepatocellular carcinogenesis. Cancer Res. 68: 5049-5058.

de Krijger, I., L.J. Mekenkamp, C.J. Punt and I.D. Nagtegaal. 2011. MicroRNAs in colorectal cancer metastasis. J. Pathol. 224: 438-447.

Denli, A.M., B.B. Tops, R.H. Plasterk, R.F. Ketting and G.J. Hannon. 2004. Processing of primary microRNAs by the Microprocessor complex. Nature 432: 231-235.

Dews, M., A. Homayouni, D. Yu, D. Murphy, C. Sevignani, E. Wentzel, E.E. Furth, W.M. Lee, G.H. Enders, J.T. Mendell and A. Thomas-Tikhonenko. 2006. Augmentation of tumor angiogenesis by a Myc-activated microRNA cluster. Nat. Genet. 38: 1060-1065.

Dijkstra, M.K., K. van Lom, D. Tielemans, F. Elstrodt, A.W. Langerak, M.B. van 't Veer and M. Jongen-Lavrencic. 2009. 17p13/TP53 deletion in B-CLL patients is associated with microRNA-34a downregulation. Leukemia 23: 625-627.

Dillhoff, M., J. Liu, W. Frankel, C. Croce and M. Bloomston. 2008. MicroRNA-21 is overexpressed in pancreatic cancer and a potential predictor of survival. J. Gastrointest. Surg. 12: 2171-2176.

Du, T. and P.D. Zamore. 2005. microPrimer: the biogenesis and function of microRNA. Development 132: 4645-4652.

Ebert, M.P., M. Korc, P. Malfertheiner and C. Rocken. 2006. Advances, challenges, and limitations in serum-proteome-based cancer diagnosis. J. Proteome Res. 5: 19-25.

Eddy, S.R. 2001. Non-coding RNA genes and the modern RNA world. Nat. Rev. Genet. 2: 919-929.

Eder, M. and M. Scherr. 2005. MicroRNA and lung cancer. N. Engl. J. Med. 352: 2446-2448.

Eiriksdottir, G., G. Johannesdottir, S. Ingvarsson, I.B. Bjornsdottir, J.G. Jonasson, B.A. Agnarsson, J. Hallgrimsson, J. Gudmundsson, V. Egilsson, H. Sigurdsson and R.B. Barkardottir. 1998. Mapping loss of heterozygosity at chromosome 13q: loss at 13q12-q13 is associated with breast tumour progression and poor prognosis. Eur. J. Cancer 34: 2076-2081.

Eis, P.S., W. Tam, L. Sun, A. Chadburn, Z. Li, M.F. Gomez, E. Lund and J.E. Dahlberg. 2005. Accumulation of miR-155 and BIC RNA in human B cell lymphomas. Proc. Natl. Acad. Sci. U S A 102: 3627-3632.

Elbashir, S.M., W. Lendeckel and T. Tuschl. 2001a. RNA interference is mediated by 21- and 22-nucleotide RNAs. Genes Dev. 15: 188-200.

Elbashir, S.M., J. Martinez, A. Patkaniowska, W. Lendeckel and T. Tuschl. 2001b. Functional anatomy of siRNAs for mediating efficient RNAi in Drosophila melanogaster embryo lysate. EMBO J. 20: 6877-6888.

Esquela-Kerscher, A. and F.J. Slack. 2006. Oncomirs - microRNAs with a role in cancer. Nat. Rev. Cancer 6: 259-269.

Eulalio, A., F. Tritschler and E. Izaurralde. 2009. The GW182 protein family in animal cells: new insights into domains required for miRNA-mediated gene silencing. Rna 15: 1433-1442.

Eystathioy, T., E.K. Chan, S.A. Tenenbaum, J.D. Keene, K. Griffith and M.J. Fritzler. 2002. A phosphorylated cytoplasmic autoantigen, GW182, associates with a unique population of human mRNAs within novel cytoplasmic speckles. Mol. Biol. Cell 13: 1338-1351.

Fabbri, M., R. Garzon, et al. 2007. MicroRNA-29 family reverts aberrant methylation in lung cancer by targeting DNA methyltransferases 3A and 3B. Proc. Natl. Acad. Sci. U S A 104: 15805-15810.

Fabian, M.R., N. Sonenberg and W. Filipowicz. 2010. Regulation of mRNA translation and stability by microRNAs. Annu. Rev. Biochem. 79: 351-379.

Fang, L., Z. Deng, T. Shatseva, J. Yang, C. Peng, W.W. Du, A.J. Yee, L.C. Ang, C. He, S.W. Shan and B.B. Yang. 2011. MicroRNA miR-93 promotes tumor growth and angiogenesis by targeting integrin-beta8. Oncogene 30: 806-821.

Fassan, M., K. Sachsenmeir, M. Rugge and R. Baffa. 2011. Role of miRNA in distinguishing primary brain tumors from secondary tumors metastatic to the brain. Front. Biosci. (Schol Ed) 3: 970-979.

Felli, N., L. Fontana, et al. 2005. MicroRNAs 221 and 222 inhibit normal erythropoiesis and erythroleukemic cell growth via kit receptor down-modulation. Proc. Natl. Acad. Sci. U S A 102: 18081-18086.

Filipowicz, W. 2005. RNAi: the nuts and bolts of the RISC machine. Cell 122: 17-20.

Friedman, R.C., K.K. Farh, C.B. Burge and D.P. Bartel. 2009. Most mammalian mRNAs are conserved targets of microRNAs. Genome Res. 19: 92-105.

Fulci, V., S. Chiaretti, et al. 2007. Quantitative technologies establish a novel microRNA profile of chronic lymphocytic leukemia. Blood 109: 4944-4951.

Galardi, S., N. Mercatelli, E. Giorda, S. Massalini, G.V. Frajese, S.A. Ciafre and M.G. Farace. 2007. miR-221 and miR-222 expression affects the proliferation potential of human prostate carcinoma cell lines by targeting p27Kip1. J. Biol. Chem. 282: 23716-23724.

Gao, P., C.C. Wong, E.K. Tung, J.M. Lee, C.M. Wong and I.O. Ng. 2011. Deregulation of microRNA expression occurs early and accumulates in early stages of HBV-associated multistep hepatocarcinogenesis. J. Hepatol. 54: 1177-1184.

Garzon, R., S. Volinia, et al. 2008. MicroRNA signatures associated with cytogenetics and prognosis in acute myeloid leukemia. Blood 111: 3183-3189.

Gaur, A., D.A. Jewell, Y. Liang, D. Ridzon, J.H. Moore, C. Chen, V.R. Ambros and M.A. Israel. 2007. Characterization of microRNA expression levels and their biological correlates in human cancer cell lines. Cancer Res. 67: 2456-2468.

George, G.P., R. Gangwar, R.K. Mandal, S.N. Sankhwar and R.D. Mittal. 2011. Genetic variation in microRNA genes and prostate cancer risk in North Indian population. Mol. Biol. Rep. 38: 1609-1615.

George, G.P. and R.D. Mittal. 2010. Caspase 8 gene variants in healthy North Indian population and comparison with worldwide ethnic group variations. Indian J. Hum. Genet. 16: 144-148.

Gilad, S., E. Meiri, et al. 2008. Serum microRNAs are promising novel biomarkers. PLoS One 3: e3148.

Giraldez, A.J., Y. Mishima, J. Rihel, R.J. Grocock, S. Van Dongen, K. Inoue, A.J. Enright and A.F. Schier. 2006. Zebrafish MiR-430 promotes deadenylation and clearance of maternal mRNAs. Science 312: 75-79.

Grady, W.M., R.K. Parkin, et al. 2008. Epigenetic silencing of the intronic microRNA hsa-miR-342 and its host gene EVL in colorectal cancer. Oncogene 27: 3880-3888.

Gramantieri, L., M. Ferracin, F. Fornari, A. Veronese, S. Sabbioni, C.G. Liu, G.A. Calin, C. Giovannini, E. Ferrazzi, G.L. Grazi, C.M. Croce, L. Bolondi and M. Negrini. 2007. Cyclin G1 is a target of miR-122a, a microRNA frequently down-regulated in human hepatocellular carcinoma. Cancer Res. 67: 6092-6099.

Gregory, P.A., A.G. Bert, E.L. Paterson, S.C. Barry, A. Tsykin, G. Farshid, M.A. Vadas, Y. Khew-Goodall and G.J. Goodall. 2008. The miR-200 family and miR-205 regulate epithelial to mesenchymal transition by targeting ZEB1 and SIP1. Nat. Cell. Biol. 10: 593-601.

Gregory, R.I., K.P. Yan, G. Amuthan, T. Chendrimada, B. Doratotaj, N. Cooch and R. Shiekhattar. 2004. The Microprocessor complex mediates the genesis of microRNAs. Nature 432: 235-240.

Greither, T., L.F. Grochola, A. Udelnow, C. Lautenschlager, P. Wurl and H. Taubert. 2010. Elevated expression of microRNAs 155, 203, 210 and 222 in pancreatic tumors is associated with poorer survival. Int. J. Cancer 126: 73-80.

Greither, T., P. Wurl, L. Grochola, G. Bond, M. Bache, M. Kappler, C. Lautenschlager, H.J. Holzhausen, S. Wach, A.W. Eckert and H. Taubert. 2012. Expression of microRNA 210 associates with poor survival and age of tumor onset of soft-tissue sarcoma patients. Int. J. Cancer 30: 1230-1235.

Grimson, A., K.K. Farh, W.K. Johnston, P. Garrett-Engele, L.P. Lim and D.P. Bartel. 2007. MicroRNA targeting specificity in mammals: determinants beyond seed pairing. Mol. Cell 27: 91-105.

Grishok, A., A.E. Pasquinelli, D. Conte, N. Li, S. Parrish, I. Ha, D.L. Baillie, A. Fire, G. Ruvkun and C.C. Mello. 2001. Genes and mechanisms related to RNA interference regulate expression of the small temporal RNAs that control C. elegans developmental timing. Cell 106: 23-34.

Guo, J., Y. Miao, B. Xiao, R. Huan, Z. Jiang, D. Meng and Y. Wang. 2009. Differential expression of microRNA species in human gastric cancer versus non-tumorous tissues. J. Gastroenterol. Hepatol. 24: 652-657.

Guo, Y., Y. Chen, H. Ito, A. Watanabe, X. Ge, T. Kodama and H. Aburatani. 2006. Identification and characterization of lin-28 homolog B (LIN28B) in human hepatocellular carcinoma. Gene 384: 51-61.

Guttman, M., I. Amit, et al. 2009. Chromatin signature reveals over a thousand highly conserved large non-coding RNAs in mammals. Nature 458: 223-227.

Hammond, S.M., A.A. Caudy and G.J. Hannon. 2001. Post-transcriptional gene silencing by double-stranded RNA. Nat. Rev. Genet. 2: 110-119.

Han, J., Y. Lee, K.H. Yeom, Y.K. Kim, H. Jin and V.N. Kim. 2004. The Drosha-DGCR8 complex in primary microRNA processing. Genes Dev. 18: 3016-3027.

Hanke, M., K. Hoefig, H. Merz, A.C. Feller, I. Kausch, D. Jocham, J.M. Warnecke and G. Sczakiel. 2010. A robust methodology to study urine microRNA as tumor marker: microRNA-126 and microRNA-182 are related to urinary bladder cancer. Urol. Oncol. 28: 655-661.

Havens, M.A., A.A. Reich, D.M. Duelli and M.L. Hastings. 2012. Biogenesis of mammalian microRNAs by a non-canonical processing pathway. Nucleic Acids Res. 40: 4626-4640.

He, H., K. Jazdzewski, W. Li, S. Liyanarachchi, R. Nagy, S. Volinia, G.A. Calin, C.G. Liu, K. Franssila, S. Suster, R.T. Kloos, C.M. Croce and A. de la Chapelle. 2005a. The role of microRNA genes in papillary thyroid carcinoma. Proc. Natl. Acad. Sci. U S A 102: 19075-19080.

He, L., J.M. Thomson, M.T. Hemann, E. Hernando-Monge, D. Mu, S. Goodson, S. Powers, C. Cordon-Cardo, S.W. Lowe, G.J. Hannon and S.M. Hammond. 2005b. A microRNA polycistron as a potential human oncogene. Nature 435: 828-833.

He, X.Y., J.X. Chen, Z. Zhang, C.L. Li, Q.L. Peng and H.M. Peng. 2010. The let-7a microRNA protects from growth of lung carcinoma by suppression of k-Ras and c-Myc in nude mice. J. Cancer Res. Clin. Oncol. 136: 1023-1028.

Heneghan, H.M., N. Miller, A.J. Lowery, K.J. Sweeney and M.J. Kerin. 2009. MicroRNAs as Novel Biomarkers for Breast Cancer. J Oncol 2009: 950201.

Heneghan, H.M., N. Miller, A.J. Lowery, K.J. Sweeney, J. Newell and M.J. Kerin. 2010. Circulating microRNAs as novel minimally invasive biomarkers for breast cancer. Ann. Surg. 251: 499-505.

Ho, A.S., X. Huang, H. Cao, C. Christman-Skieller, K. Bennewith, Q.T. Le and A.C. Koong. 2010. Circulating miR-210 as a Novel Hypoxia Marker in Pancreatic Cancer. Transl. Oncol. 3: 109-113.

Hock, J. and G. Meister. 2008. The Argonaute protein family. Genome Biol 9: 210.

Hogg, M., S. Paro, L.P. Keegan and M.A. O'Connell. 2011. RNA editing by mammalian ADARs. Adv. Genet. 73: 87-120.

Honda, S., C. Tanaka-Kosugi, S. Yamada, T. Sano, T. Matsumoto, M. Itakura and K. Yoshimoto. 2003. Human pituitary adenomas infrequently contain inactivation of retinoblastoma 1 gene and activation of cyclin dependent kinase 4 gene. Endocr. J. 50: 309-318.

Hossain, A., M.T. Kuo and G.F. Saunders. 2006. Mir-17-5p regulates breast cancer cell proliferation by inhibiting translation of AIB1 mRNA. Mol. Cell. Biol. 26: 8191-8201.

Hu, X., D.M. Macdonald, P.C. Huettner, Z. Feng, I.M. El Naqa, J.K. Schwarz, D.G. Mutch, P.W. Grigsby, S.N. Powell and X. Wang. 2009. A miR-200 microRNA cluster as prognostic marker in advanced ovarian cancer. Gynecol. Oncol. 114: 457-464.

Huang, Q., K. Gumireddy, et al. 2008. The microRNAs miR-373 and miR-520c promote tumour invasion and metastasis. Nat. Cell Biol. 10: 202-210.

Huang, Z., D. Huang, S. Ni, Z. Peng, W. Sheng and X. Du. 2010. Plasma microRNAs are promising novel biomarkers for early detection of colorectal cancer. Int. J. Cancer 127: 118-126.

Hui, A.B., W. Shi, P.C. Boutros, N. Miller, M. Pintilie, T. Fyles, D. McCready, D. Wong, K. Gerster, L. Waldron, I. Jurisica, L.Z. Penn and F.F. Liu. 2009. Robust global micro-RNA profiling with formalin-fixed paraffin-embedded breast cancer tissues. Lab. Invest. 89: 597-606.

Humphreys, D.T., B.J. Westman, D.I. Martin and T. Preiss. 2005. MicroRNAs control translation initiation by inhibiting eukaryotic initiation factor 4E/cap and poly(A) tail function. Proc. Natl. Acad. Sci. U S A 102: 16961-16966.

Iorio, M.V., P. Casalini, E. Tagliabue, S. Menard and C.M. Croce. 2008. MicroRNA profiling as a tool to understand prognosis, therapy response and resistance in breast cancer. Eur. J. Cancer 44: 2753-2759.

Iorio, M.V. and C.M. Croce. 2009. MicroRNAs in cancer: small molecules with a huge impact. J. Clin. Oncol 27: 5848-5856.

Iorio, M.V., M. Ferracin, et al. 2005. MicroRNA gene expression deregulation in human breast cancer. Cancer Res. 65: 7065-7070.

Iorio, M.V., R. Visone, G. Di Leva, V. Donati, F. Petrocca, P. Casalini, C. Taccioli, S. Volinia, C.G. Liu, H. Alder, G.A. Calin, S. Menard and C.M. Croce. 2007. MicroRNA signatures in human ovarian cancer. Cancer Res. 67: 8699-8707.

Ji, J., J. Shi, et al. 2009. MicroRNA expression, survival, and response to interferon in liver cancer. N. Engl. J. Med. 361: 1437-1447.

Ji, J. and X.W. Wang. 2009. New kids on the block: diagnostic and prognostic microRNAs in hepatocellular carcinoma. Cancer Biol. Ther. 8: 1686-1693.

Ji, X. 2008. The mechanism of RNase III action: how dicer dices. Curr. Top. Microbiol. Immunol. 320: 99-116.

Jiang, J., Y. Gusev, I. Aderca, T.A. Mettler, D.M. Nagorney, D.J. Brackett, L.R. Roberts and T.D. Schmittgen. 2008. Association of MicroRNA expression in hepatocellular carcinomas with hepatitis infection, cirrhosis, and patient survival. Clin Cancer Res. 14: 419-427.

Johnson, S.M., H. Grosshans, J. Shingara, M. Byrom, R. Jarvis, A. Cheng, E. Labourier, K.L. Reinert, D. Brown and F.J. Slack. 2005. RAS is regulated by the let-7 microRNA family. Cell 120: 635-647.

Kai, Z.S. and A.E. Pasquinelli. 2010. MicroRNA assassins: factors that regulate the disappearance of miRNAs. Nat. Struct. Mol. Biol. 17: 5-10.

Kawahara, Y., B. Zinshteyn, T.P. Chendrimada, R. Shiekhattar and K. Nishikura. 2007a. RNA editing of the microRNA-151 precursor blocks cleavage by the Dicer-TRBP complex. EMBO Rep. 8: 763-769.

Kawahara, Y., B. Zinshteyn, P. Sethupathy, H. Iizasa, A.G. Hatzigeorgiou and K. Nishikura. 2007b. Redirection of silencing targets by adenosine-to-inosine editing of miRNAs. Science 315: 1137-1140.

Kedersha, N.L., J.E. Heuser, D.C. Chugani and L.H. Rome. 1991. Vaults. III. Vault ribonucleoprotein particles open into flower-like structures with octagonal symmetry. J. Cell Biol. 112: 225-235.

Ketting, R.F. 2011. microRNA Biogenesis and Function : An overview. Adv. Exp. Med. Biol. 700: 1-14.

Khabar, K.S. 2010. Post-transcriptional control during chronic inflammation and cancer: a focus on AU-rich elements. Cell Mol. Life Sci. 67: 2937-2955.

Khvorova, A., A. Reynolds and S.D. Jayasena. 2003. Functional siRNAs and miRNAs exhibit strand bias. Cell 115: 209-216.

Kim, U., Y. Wang, T. Sanford, Y. Zeng and K. Nishikura. 1994. Molecular cloning of cDNA for double-stranded RNA adenosine deaminase, a candidate enzyme for nuclear RNA editing. Proc. Natl. Acad. Sci. U S A 91: 11457-11461.

Kim, V.N. 2005a. MicroRNA biogenesis: coordinated cropping and dicing. Nat. Rev. Mol. Cell Biol. 6: 376-385.

Kim, V.N. 2005b. Small RNAs: classification, biogenesis, and function. Mol. Cells 19: 1-15.

Kluiver, J., S. Poppema, D. de Jong, T. Blokzijl, G. Harms, S. Jacobs, B.J. Kroesen and A. van den Berg. 2005. BIC and miR-155 are highly expressed in Hodgkin, primary mediastinal and diffuse large B cell lymphomas. J. Pathol. 207: 243-249.

Kondo, N., T. Toyama, H. Sugiura, Y. Fujii and H. Yamashita. 2008. miR-206 Expression is down-regulated in estrogen receptor alpha-positive human breast cancer. Cancer Res. 68: 5004-5008.

Kong, W., H. Yang, L. He, J.J. Zhao, D. Coppola, W.S. Dalton and J.Q. Cheng. 2008. MicroRNA-155 is regulated by the transforming growth factor beta/Smad pathway and contributes to epithelial cell plasticity by targeting RhoA. Mol. Cell Biol. 28: 6773-6784.

Koo, S.H., C.H. Ihm, K.C. Kwon, J.S. Lee, J.W. Park and J.W. Kim. 2003. Microsatellite alterations in hepatocellular carcinoma and intrahepatic cholangiocarcinoma. Cancer Genet. Cytogenet. 146: 139-144.

Korpal, M., E.S. Lee, G. Hu and Y. Kang. 2008. The miR-200 family inhibits epithelial-mesenchymal transition and cancer cell migration by direct targeting of E-cadherin transcriptional repressors ZEB1 and ZEB2. J. Biol. Chem. 283: 14910-14914.

Kosaka, N., H. Iguchi, Y. Yoshioka, F. Takeshita, Y. Matsuki and T. Ochiya. 2010. Secretory mechanisms and intercellular transfer of microRNAs in living cells. J. Biol. Chem. 285: 17442-17452.

Krol, J., I. Loedige and W. Filipowicz. 2010. The widespread regulation of microRNA biogenesis, function and decay. Nat. Rev. Genet. 11: 597-610.

Krutovskikh, V.A. and Z. Herceg. 2010. Oncogenic microRNAs (OncomiRs) as a new class of cancer biomarkers. Bioessays 32: 894-904.

Kutay, H., S. Bai, J. Datta, T. Motiwala, I. Pogribny, W. Frankel, S.T. Jacob and K. Ghoshal. 2006. Downregulation of miR-122 in the rodent and human hepatocellular carcinomas. J. Cell. Biochem. 99: 671-678.

Lagos-Quintana, M., R. Rauhut, W. Lendeckel and T. Tuschl. 2001. Identification of novel genes coding for small expressed RNAs. Science 294: 853-858.

Lamontagne, B., S. Larose, J. Boulanger and S.A. Elela. 2001. The RNase III family: a conserved structure and expanding functions in eukaryotic dsRNA metabolism. Curr. Issues Mol. Biol. 3: 71-78.

Landthaler, M., A. Yalcin and T. Tuschl. 2004. The human DiGeorge syndrome critical region gene 8 and Its D. melanogaster homolog are required for miRNA biogenesis. Curr. Biol. 14: 2162-2167.

Lanza, G., M. Ferracin, R. Gafa, A. Veronese, R. Spizzo, F. Pichiorri, C.G. Liu, G.A. Calin, C.M. Croce and M. Negrini. 2007. mRNA/microRNA gene expression profile in microsatellite unstable colorectal cancer. Mol. Cancer 6: 54.

Larue, L. and A. Bellacosa. 2005. Epithelial-mesenchymal transition in development and cancer: role of phosphatidylinositol 3' kinase/AKT pathways. Oncogene 24: 7443-7454.

Lau, N.C., L.P. Lim, E.G. Weinstein and D.P. Bartel. 2001. An abundant class of tiny RNAs with probable regulatory roles in Caenorhabditis elegans. Science 294: 858-862.

Lawrie, C.H., S. Gal, H.M. Dunlop, B. Pushkaran, A.P. Liggins, K. Pulford, A.H. Banham, F. Pezzella, J. Boultwood, J.S. Wainscoat, C.S. Hatton and A.L. Harris. 2008. Detection of elevated levels of tumour-associated microRNAs in serum of patients with diffuse large B-cell lymphoma. Br. J. Haematol. 141: 672-675.

Lebanony, D., H. Benjamin, et al. 2009. Diagnostic assay based on hsa-miR-205 expression distinguishes squamous from nonsquamous non-small-cell lung carcinoma. J. Clin. Oncol. 27: 2030-2037.

Lee, D.Y., Z. Deng, C.H. Wang and B.B. Yang. 2007a. MicroRNA-378 promotes cell survival, tumor growth, and angiogenesis by targeting SuFu and Fus-1 expression. Proc. Natl. Acad. Sci. U S A 104: 20350-20355.

Lee, E.J., Y. Gusev, J. Jiang, G.J. Nuovo, M.R. Lerner, W.L. Frankel, D.L. Morgan, R.G. Postier, D.J. Brackett and T.D. Schmittgen. 2007b. Expression profiling identifies microRNA signature in pancreatic cancer. Int. J. Cancer 120: 1046-1054.

Lee, K., N. Kunkeaw, S.H. Jeon, I. Lee, B.H. Johnson, G.Y. Kang, J.Y. Bang, H.S. Park, C. Leelayuwat and Y.S. Lee. 2011. Precursor miR-886, a novel noncoding RNA repressed in cancer, associates with PKR and modulates its activity. RNA 17: 1076-1089.

Lee, R.C. and V. Ambros. 2001. An extensive class of small RNAs in Caenorhabditis elegans. Science 294: 862-864.

Lee, R.C., R.L. Feinbaum and V. Ambros. 1993. The C. elegans heterochronic gene lin-4 encodes small RNAs with antisense complementarity to lin-14. Cell 75: 843-854.

Lee, Y., K. Jeon, J.T. Lee, S. Kim and V.N. Kim. 2002. MicroRNA maturation: stepwise processing and subcellular localization. EMBO J. 21: 4663-4670.

Lee, Y., M. Kim, J. Han, K.H. Yeom, S. Lee, S.H. Baek and V.N. Kim. 2004. MicroRNA genes are transcribed by RNA polymerase II. EMBO J. 23: 4051-4060.

Lee, Y.S., H.K. Kim, S. Chung, K.S. Kim and A. Dutta. 2005. Depletion of human micro-RNA miR-125b reveals that it is critical for the proliferation of differentiated cells but not for the down-regulation of putative targets during differentiation. J. Biol. Chem. 280: 16635-16641.

Lehmann, U., B. Hasemeier, M. Christgen, M. Muller, D. Romermann, F. Langer and H. Kreipe. 2008. Epigenetic inactivation of microRNA gene hsa-mir-9-1 in human breast cancer. J. Pathol. 214: 17-24.

Leite, K.R., A. Tomiyama, S.T. Reis, J.M. Sousa-Canavez, A. Sanudo, L.H. Camara-Lopes and M. Srougi. 2011. MicroRNA expression profiles in the progression of prostate cancer-from high-grade prostate intraepithelial neoplasia to metastasis. Urol. Oncol. PMID: 21880514

Li, X., F. Luo, Q. Li, M. Xu, D. Feng, G. Zhang and W. Wu. 2011. Identification of new aberrantly expressed miRNAs in intestinal-type. Oncol. Rep. 26: 1431-1439.

Lim, L.P., N.C. Lau, P. Garrett-Engele, A. Grimson, J.M. Schelter, J. Castle, D.P. Bartel, P.S. Linsley and J.M. Johnson. 2005. Microarray analysis shows that some microRNAs downregulate large numbers of target mRNAs. Nature 433: 769-773.

Lin, J.K., S.C. Chang, Y.C. Yang and A.F. Li. 2003. Loss of heterozygosity and DNA aneuploidy in colorectal adenocarcinoma. Ann. Surg. Oncol. 10: 1086-1094.

Liu, C.J., S.Y. Kao, H.F. Tu, M.M. Tsai, K.W. Chang and S.C. Lin. 2010. Increase of microRNA miR-31 level in plasma could be a potential marker of oral cancer. Oral Dis. 16: 360-364.

Liu, J., M.A. Carmell, F.V. Rivas, C.G. Marsden, J.M. Thomson, J.J. Song, S.M. Hammond, L. Joshua-Tor and G.J. Hannon. 2004. Argonaute2 is the catalytic engine of mammalian RNAi. Science 305: 1437-1441.

Long, D., R. Lee, P. Williams, C.Y. Chan, V. Ambros and Y. Ding. 2007. Potent effect of target structure on microRNA function. Nat Struct Mol Biol 14: 287-294.

Lowenberg, B., J.R. Downing and A. Burnett. 1999. Acute myeloid leukemia. N. Engl. J. Med. 341: 1051-1062.

Lowery, A.J., N. Miller, R.E. McNeill and M.J. Kerin. 2008. MicroRNAs as prognostic indicators and therapeutic targets: potential effect on breast cancer management. Clin. Cancer Res. 14: 360-365.

Lu, J., G. Getz, et al. 2005. MicroRNA expression profiles classify human cancers. Nature 435: 834-838.

Lu, L., D. Katsaros, I.A. de la Longrais, O. Sochirca and H. Yu. 2007. Hypermethylation of let-7a-3 in epithelial ovarian cancer is associated with low insulin-like growth factor-II expression and favorable prognosis. Cancer Res. 67: 10117-10122.

Lu, L., D. Katsaros, K. Shaverdashvili, B. Qian, Y. Wu, I.A. de la Longrais, M. Preti, G. Menato and H. Yu. 2009. Pluripotent factor lin-28 and its homologue lin-28b in epithelial ovarian cancer and their associations with disease outcomes and expression of let-7a and IGF-II. Eur. J. Cancer 45: 2212-2218.

Lujambio, A., S. Ropero, et al. 2007. Genetic unmasking of an epigenetically silenced microRNA in human cancer cells. Cancer Res. 67: 1424-1429.

Lund, E., S. Guttinger, A. Calado, J.E. Dahlberg and U. Kutay. 2004. Nuclear export of microRNA precursors. Science 303: 95-98.

Luo, S.S., O. Ishibashi, G. Ishikawa, T. Ishikawa, A. Katayama, T. Mishima, T. Takizawa, T. Shigihara, T. Goto, A. Izumi, A. Ohkuchi, S. Matsubara and T. Takeshita. 2009. Human villous trophoblasts express and secrete placenta-specific microRNAs into maternal circulation via exosomes. Biol. Reprod. 81: 717-729.

Ma, L., J. Teruya-Feldstein and R.A. Weinberg. 2007. Tumour invasion and metastasis initiated by microRNA-10b in breast cancer. Nature 449: 682-688.

Marcucci, G., M.D. Radmacher, et al. 2008. MicroRNA expression in cytogenetically normal acute myeloid leukemia. N. Engl. J. Med. 358: 1919-1928.

Marini, F., E. Luzi and M.L. Brandi. 2011. MicroRNA Role in Thyroid Cancer Development. J. Thyroid Res. 2011: 407123.

Maroney, P.A., Y. Yu and T.W. Nilsen. 2006. MicroRNAs, mRNAs, and translation. Cold Spring Harb. Symp. Quant. Biol. 71: 531-535.

Martin, M.M., E.J. Lee, J.A. Buckenberger, T.D. Schmittgen and T.S. Elton. 2006. MicroRNA-155 regulates human angiotensin II type 1 receptor expression in fibroblasts. J. Biol. Chem. 281: 18277-18284.

Martinez, J. and T. Tuschl. 2004. RISC is a 5' phosphomonoester-producing RNA endonuclease. Genes Dev. 18: 975-980.

Masaki, S., R. Ohtsuka, Y. Abe, K. Muta and T. Umemura. 2007. Expression patterns of microRNAs 155 and 451 during normal human erythropoiesis. Biochem. Biophys. Res. Commun. 364: 509-514.

Mashima, T., M. Kudo, Y. Takada, A. Matsugami, S.C. Gopinath, P.K. Kumar and M. Katahira. 2008. Interactions between antitumor drugs and vault RNA. Nucleic Acids Symp. Ser. (Oxf) 217-218.

Mattick, J.S. 2009. The genetic signatures of noncoding RNAs. PLoS Genet. 5: e1000459.

Mattie, M.D., C.C. Benz, J. Bowers, K. Sensinger, L. Wong, G.K. Scott, V. Fedele, D. Ginzinger, R. Getts and C. Haqq. 2006. Optimized high-throughput microRNA expression profiling provides novel biomarker assessment of clinical prostate and breast cancer biopsies. Mol. Cancer 5: 24.

Mayr, C., M.T. Hemann and D.P. Bartel. 2007. Disrupting the pairing between let-7 and Hmga2 enhances oncogenic transformation. Science 315: 1576-1579.

McIntyre, A., B. Summersgill, et al. 2005. Amplification and overexpression of the KIT gene is associated with progression in the seminoma subtype of testicular germ cell tumors of adolescents and adults. Cancer Res. 65: 8085-8089.

Meng, F., R. Henson, H. Wehbe-Janek, K. Ghoshal, S.T. Jacob and T. Patel. 2007. MicroRNA-21 regulates expression of the PTEN tumor suppressor gene in human hepatocellular cancer. Gastroenterology 133: 647-658.

Mertens-Talcott, S.U., S. Chintharlapalli, X. Li and S. Safe. 2007. The oncogenic microRNA-27a targets genes that regulate specificity protein transcription factors and the G2-M checkpoint in MDA-MB-231 breast cancer cells. Cancer Res. 67: 11001-11011.

Michael, M.Z., S.M. O'Connor, N.G. van Holst Pellekaan, G.P. Young and R.J. James. 2003. Reduced accumulation of specific microRNAs in colorectal neoplasia. Mol. Cancer Res. 1: 882-891.

Miller, T.E., K. Ghoshal, B. Ramaswamy, S. Roy, J. Datta, C.L. Shapiro, S. Jacob and S. Majumder. 2008. MicroRNA-221/222 confers tamoxifen resistance in breast cancer by targeting p27Kip1. J. Biol. Chem. 283: 29897-29903.

Mitchell, P.S., R.K. Parkin, et al. 2008. Circulating microRNAs as stable blood-based markers for cancer detection. Proc. Natl. Acad. Sci. U S A 105: 10513-10518.

Miyoshi, K., T. Miyoshi and H. Siomi. 2010. Many ways to generate microRNA-like small RNAs: non-canonical pathways for microRNA production. Mol. Genet. Genomics 284: 95-103.

Morra, E., G. Barosi, A. Bosi, F. Ferrara, F. Locatelli, M. Marchetti, G. Martinelli, C. Mecucci, M. Vignetti and S. Tura. 2009. Clinical management of primary non-acute promyelocytic leukemia acute myeloid leukemia: Practice Guidelines by the Italian Society of Hematology, the Italian Society of Experimental Hematology, and the Italian Group for Bone Marrow Transplantation. Haematologica 94: 102-112.

Morra, L. and H. Moch. 2011. Periostin expression and epithelial-mesenchymal transition in cancer: a review and an update. Virchows Arch. 459: 465-475.

Mrozek, K. and C.D. Bloomfield. 2006. Chromosome aberrations, gene mutations and expression changes, and prognosis in adult acute myeloid leukemia. Hematology Am. Soc. Hematol. Educ. Program 169-177.

Mrozek, K., H. Dohner and C.D. Bloomfield. 2007. Influence of new molecular prognostic markers in patients with karyotypically normal acute myeloid leukemia: recent advances. Curr. Opin. Hematol. 14: 106-114.

Mrozek, K., N.A. Heerema and C.D. Bloomfield. 2004. Cytogenetics in acute leukemia. Blood Rev. 18: 115-136.

Mu, P., Y.C. Han, D. Betel, E. Yao, M. Squatrito, P. Ogrodowski, E. de Stanchina, A. D'Andrea, C. Sander and A. Ventura. 2009. Genetic dissection of the miR-17~92 cluster of microRNAs in Myc-induced B-cell lymphomas. Genes Dev. 23: 2806-2811.

Murakami, Y., T. Yasuda, K. Saigo, T. Urashima, H. Toyoda, T. Okanoue and K. Shimotohno. 2006. Comprehensive analysis of microRNA expression patterns in hepatocellular carcinoma and non-tumorous tissues. Oncogene 25: 2537-2545.

Nam, E.J., H. Yoon, S.W. Kim, H. Kim, Y.T. Kim, J.H. Kim, J.W. Kim and S. Kim. 2008. MicroRNA expression profiles in serous ovarian carcinoma. Clin. Cancer Res. 14: 2690-2695.

Nandy, C., J. Mrazek, H. Stoiber, F.A. Grasser, A. Huttenhofer and N. Polacek. 2009. Epstein-barr virus-induced expression of a novel human vault RNA. J. Mol. Biol. 388: 776-784.

Nasser, M.W., J. Datta, G. Nuovo, H. Kutay, T. Motiwala, S. Majumder, B. Wang, S. Suster, S.T. Jacob and K. Ghoshal. 2008. Down-regulation of micro-RNA-1 (miR-1) in lung cancer. Suppression of tumorigenic property of lung cancer cells and their sensitization to doxorubicin-induced apoptosis by miR-1. J. Biol. Chem. 283: 33394-33405.

Negrini, M. and G.A. Calin. 2008. Breast cancer metastasis: a microRNA story. Breast Cancer. Res. 10: 203.

Negrini, M., M. Ferracin, S. Sabbioni and C.M. Croce. 2007. MicroRNAs in human cancer: from research to therapy. J. Cell Sci. 120: 1833-1840.

Nelson, P.T., D.A. Baldwin, L.M. Scearce, J.C. Oberholtzer, J.W. Tobias and Z. Mourelatos. 2004. Microarray-based, high-throughput gene expression profiling of microRNAs. Nat. Methods 1: 155-161.

Ng, E.K., W.W. Chong, H. Jin, E.K. Lam, V.Y. Shin, J. Yu, T.C. Poon, S.S. Ng and J.J. Sung. 2009. Differential expression of microRNAs in plasma of patients with colorectal cancer: a potential marker for colorectal cancer screening. Gut 58: 1375-1381.

Nikiforova, M.N., G.C. Tseng, D. Steward, D. Diorio and Y.E. Nikiforov. 2008. MicroRNA expression profiling of thyroid tumors: biological significance and diagnostic utility. J. Clin. Endocrinol. Metab. 93: 1600-1608.

Nissen, P., J. Hansen, N. Ban, P.B. Moore and T.A. Steitz. 2000. The structural basis of ribosome activity in peptide bond synthesis. Science 289: 920-930.

Noonan, E.J., R.F. Place, D. Pookot, S. Basak, J.M. Whitson, H. Hirata, C. Giardina and R. Dahiya. 2009. miR-449a targets HDAC-1 and induces growth arrest in prostate cancer. Oncogene 28: 1714-1724.

Nottrott, S., M.J. Simard and J.D. Richter. 2006. Human let-7a miRNA blocks protein production on actively translating polyribosomes. Nat Struct Mol Biol 13: 1108-1114.

Nowsheen, S., K. Aziz, M.I. Panayiotidis and A.G. Georgakilas. 2011. Molecular markers for cancer prognosis and treatment: Have we struck gold? Cancer Lett. PMID: 22120674

O'Donnell, K.A., E.A. Wentzel, K.I. Zeller, C.V. Dang and J.T. Mendell. 2005. c-Myc-regulated microRNAs modulate E2F1 expression. Nature 435: 839-843.

Okamura, K., J.W. Hagen, H. Duan, D.M. Tyler and E.C. Lai. 2007. The mirtron pathway generates microRNA-class regulatory RNAs in Drosophila. Cell 130: 89-100.

Olive, V., M.J. Bennett, J.C. Walker, C. Ma, I. Jiang, C. Cordon-Cardo, Q.J. Li, S.W. Lowe, G.J. Hannon and L. He. 2009. miR-19 is a key oncogenic component of mir-17-92. Genes Dev. 23: 2839-2849.

Olive, V., I. Jiang and L. He. 2010. mir-17-92, a cluster of miRNAs in the midst of the cancer network. Int. J. Biochem. Cell Biol. 42: 1348-1354.

Pallante, P., R. Visone, M. Ferracin, A. Ferraro, M.T. Berlingieri, G. Troncone, G. Chiappetta, C.G. Liu, M. Santoro, M. Negrini, C.M. Croce and A. Fusco. 2006. MicroRNA deregulation in human thyroid papillary carcinomas. Endocr. Relat. Cancer 13: 497-508.

Pandey, D.P. and D. Picard. 2009. miR-22 inhibits estrogen signaling by directly targeting the estrogen receptor alpha mRNA. Mol. Cell Biol. 29: 3783-3790.

Park, N.J., H. Zhou, D. Elashoff, B.S. Henson, D.A. Kastratovic, E. Abemayor and D.T. Wong. 2009a. Salivary microRNA: discovery, characterization, and clinical utility for oral cancer detection. Clin. Cancer Res. 15: 5473-5477.

Park, S.Y., J.H. Lee, M. Ha, J.W. Nam and V.N. Kim. 2009b. miR-29 miRNAs activate p53 by targeting p85 alpha and CDC42. Nat. Struct. Mol. Biol. 16: 23-29.

Patella, F. and G. Rainaldi. 2011. MicroRNAs mediate metabolic stresses and angiogenesis. Cell Mol. Life Sci. 69: 1049-1065.

Pekarsky, Y., U. Santanam, et al. 2006. Tcl1 expression in chronic lymphocytic leukemia is regulated by miR-29 and miR-181. Cancer Res. 66: 11590-11593.

Persson, H., A. Kvist, J. Vallon-Christersson, P. Medstrand, A. Borg and C. Rovira. 2009. The non-coding RNA of the multidrug resistance-linked vault particle encodes multiple regulatory small RNAs. Nat. Cell Biol. 11: 1268-1271.

Peters, L. and G. Meister. 2007. Argonaute proteins: mediators of RNA silencing. Mol. Cell 26: 611-623.

Petersen, C.P., M.E. Bordeleau, J. Pelletier and P.A. Sharp. 2006. Short RNAs repress translation after initiation in mammalian cells. Mol. Cell 21: 533-542.

Pillai, R.S., S.N. Bhattacharyya, C.G. Artus, T. Zoller, N. Cougot, E. Basyuk, E. Bertrand and W. Filipowicz. 2005. Inhibition of translational initiation by Let-7 MicroRNA in human cells. Science 309: 1573-1576.

Pineau, P., S. Volinia, K. McJunkin, A. Marchio, C. Battiston, B. Terris, V. Mazzaferro, S.W. Lowe, C.M. Croce and A. Dejean. 2010. miR-221 overexpression contributes to liver tumorigenesis. Proc. Natl. Acad. Sci. U S A 107: 264-269.

Ponting, C.P., P.L. Oliver and W. Reik. 2009. Evolution and functions of long noncoding RNAs. Cell 136: 629-641.

Qi, P., S.Q. Cheng, H. Wang, N. Li, Y.F. Chen and C.F. Gao. 2011. Serum MicroRNAs as Biomarkers for Hepatocellular Carcinoma in Chinese Patients with Chronic Hepatitis B Virus Infection. PLoS One 6: e28486.

Rabinowits, G., C. Gercel-Taylor, J.M. Day, D.D. Taylor and G.H. Kloecker. 2009. Exosomal microRNA: a diagnostic marker for lung cancer. Clin. Lung Cancer 10: 42-46.

Rajewsky, N. 2006. microRNA target predictions in animals. Nat. Genet. 38 Suppl: S8-13.

Rajewsky, N. and N.D. Socci. 2004. Computational identification of microRNA targets. Dev. Biol. 267: 529-535.

Ramachandran, V. and X. Chen. 2008. Degradation of microRNAs by a family of exoribonucleases in Arabidopsis. Science 321: 1490-1492.

Reinhart, B.J., F.J. Slack, M. Basson, A.E. Pasquinelli, J.C. Bettinger, A.E. Rougvie, H.R. Horvitz and G. Ruvkun. 2000. The 21-nucleotide let-7 RNA regulates developmental timing in Caenorhabditis elegans. Nature 403: 901-906.

Resnick, K.E., H. Alder, J.P. Hagan, D.L. Richardson, C.M. Croce and D.E. Cohn. 2009. The detection of differentially expressed microRNAs from the serum of ovarian cancer patients using a novel real-time PCR platform. Gynecol. Oncol. 112: 55-59.

Richter, J., U. Wagner, P. Schraml, R. Maurer, G. Alund, H. Knonagel, H. Moch, M.J. Mihatsch, T.C. Gasser and G. Sauter. 1999. Chromosomal imbalances are associated with a high risk of progression in early invasive (pT1) urinary bladder cancer. Cancer Res. 59: 5687-5691.

Robertson, B., A.B. Dalby, J. Karpilow, A. Khvorova, D. Leake and A. Vermeulen. 2010. Specificity and functionality of microRNA inhibitors. Silence 1: 10.

Roldo, C., E. Missiaglia, J.P. Hagan, M. Falconi, P. Capelli, S. Bersani, G.A. Calin, S. Volinia, C.G. Liu, A. Scarpa and C.M. Croce. 2006. MicroRNA expression abnormalities in pancreatic endocrine and acinar tumors are associated with distinctive pathologic features and clinical behavior. J. Clin. Oncol. 24: 4677-4684.

Rosenfeld, N., R. Aharonov, et al. 2008. MicroRNAs accurately identify cancer tissue origin. Nat. Biotechnol. 26: 462-469.

Rossi, J.J. 1992. Ribozymes. Curr. Opin. Biotechnol. 3: 3-7.

Roth, C., S. Kasimir-Bauer, K. Pantel and H. Schwarzenbach. 2011. Screening for circulating nucleic acids and caspase activity in the peripheral blood as potential diagnostic tools in lung cancer. Mol. Oncol. 5: 281-291.

Ruby, J.G., C.H. Jan and D.P. Bartel. 2007. Intronic microRNA precursors that bypass Drosha processing. Nature 448: 83-86.

Saito, Y., G. Liang, G. Egger, J.M. Friedman, J.C. Chuang, G.A. Coetzee and P.A. Jones. 2006. Specific activation of microRNA-127 with downregulation of the proto-oncogene BCL6 by chromatin-modifying drugs in human cancer cells. Cancer Cell 9: 435-443.

Sayed, D. and M. Abdellatif. 2011. MicroRNAs in development and disease. Physiol. Rev. 91: 827-887.

Schepeler, T., J.T. Reinert, M.S. Ostenfeld, L.L. Christensen, A.N. Silahtaroglu, L. Dyrskjot, C. Wiuf, F.J. Sorensen, M. Kruhoffer, S. Laurberg, S. Kauppinen, T.F. Orntoft and C.L. Andersen. 2008. Diagnostic and prognostic microRNAs in stage II colon cancer. Cancer Res. 68: 6416-6424.

Schetter, A.J. and C.C. Harris. 2009. Plasma microRNAs: a potential biomarker for colorectal cancer? Gut 58: 1318-1319.

Schetter, A.J., S.Y. Leung, et al. 2008. MicroRNA expression profiles associated with prognosis and therapeutic outcome in colon adenocarcinoma. JAMA 299: 425-436.

Schwarz, D.S., Y. Tomari and P.D. Zamore. 2004. The RNA-induced silencing complex is a Mg2+-dependent endonuclease. Curr. Biol. 14: 787-791.

Schwind, S., K. Maharry, et al. 2010. Prognostic significance of expression of a single microRNA, miR-181a, in cytogenetically normal acute myeloid leukemia: a Cancer and Leukemia Group B study. J. Clin. Oncol. 28: 5257-5264.

Scott, G.K., A. Goga, D. Bhaumik, C.E. Berger, C.S. Sullivan and C.C. Benz. 2007. Coordinate suppression of ERBB2 and ERBB3 by enforced expression of micro-RNA miR-125a or miR-125b. J. Biol. Chem. 282: 1479-1486.

Shah, M.Y. and G.A. Calin. 2011. MicroRNAs miR-221 and miR-222: a new level of regulation in aggressive breast cancer. Genome Med 3: 56.

Singh, S.K., M. Pal Bhadra, H.J. Girschick and U. Bhadra. 2008. MicroRNAs--micro in size but macro in function. FEBS J. 275: 4929-4944.

Slaby, O., M. Svoboda, P. Fabian, T. Smerdova, D. Knoflickova, M. Bednarikova, R. Nenutil and R. Vyzula. 2007. Altered expression of miR-21, miR-31, miR-143 and miR-145 is related to clinicopathologic features of colorectal cancer. Oncology 72: 397-402.

Slack, F.J. and J.B. Weidhaas. 2006. MicroRNAs as a potential magic bullet in cancer. Future Oncol. 2: 73-82.

Song, J.J., S.K. Smith, G.J. Hannon and L. Joshua-Tor. 2004. Crystal structure of Argonaute and its implications for RISC slicer activity. Science 305: 1434-1437.

Spizzo, R., M.S. Nicoloso, C.M. Croce and G.A. Calin. 2009. SnapShot: MicroRNAs in Cancer. Cell 137: 586-586.e1.

Stadler, P.F., J.J. Chen, J. Hackermuller, S. Hoffmann, F. Horn, P. Khaitovich, A.K. Kretzschmar, A. Mosig, S.J. Prohaska, X. Qi, K. Schutt and K. Ullmann. 2009. Evolution of vault RNAs. Mol. Biol. Evol. 26: 1975-1991.

Stamatopoulos, B., N. Meuleman, B. Haibe-Kains, P. Saussoy, E. Van Den Neste, L. Michaux, P. Heimann, P. Martiat, D. Bron and L. Lagneaux. 2009. microRNA-29c and microRNA-223 down-regulation has in vivo significance in chronic lymphocytic leukemia and improves disease risk stratification. Blood 113: 5237-5245.

Su, H., J.R. Yang, T. Xu, J. Huang, L. Xu, Y. Yuan and S.M. Zhuang. 2009. MicroRNA-101, down-regulated in hepatocellular carcinoma, promotes apoptosis and suppresses tumorigenicity. Cancer Res. 69: 1135-1142.

Szafranska, A.E., T.S. Davison, J. John, T. Cannon, B. Sipos, A. Maghnouj, E. Labourier and S.A. Hahn. 2007. MicroRNA expression alterations are linked to tumorigenesis and non-neoplastic processes in pancreatic ductal adenocarcinoma. Oncogene 26: 4442-4452.

Takamizawa, J., H. Konishi, K. Yanagisawa, S. Tomida, H. Osada, H. Endoh, T. Harano, Y. Yatabe, M. Nagino, Y. Nimura, T. Mitsudomi and T. Takahashi. 2004. Reduced expression of the let-7 microRNAs in human lung cancers in association with shortened postoperative survival. Cancer Res. 64: 3753-3756.

Tam, W., D. Ben-Yehuda and W.S. Hayward. 1997. bic, a novel gene activated by proviral insertions in avian leukosis virus-induced lymphomas, is likely to function through its noncoding RNA. Mol. Cell Biol. 17: 1490-1502.

Tanaka, H., K. Kato, E. Yamashita, T. Sumizawa, Y. Zhou, M. Yao, K. Iwasaki, M. Yoshimura and T. Tsukihara. 2009a. The structure of rat liver vault at 3.5 angstrom resolution. Science 323: 384-388.

Tanaka, M., K. Oikawa, M. Takanashi, M. Kudo, J. Ohyashiki, K. Ohyashiki and M. Kuroda. 2009b. Down-regulation of miR-92 in human plasma is a novel marker for acute leukemia patients. PLoS One 4: e5532.

Tang, F., P. Hajkova, S.C. Barton, K. Lao and M.A. Surani. 2006. MicroRNA expression profiling of single whole embryonic stem cells. Nucleic Acids Res. 34: e9.

Tanzer, A. and P.F. Stadler. 2004. Molecular evolution of a microRNA cluster. J. Mol. Biol. 339: 327-335.

Taylor, D.D. and C. Gercel-Taylor. 2008. MicroRNA signatures of tumor-derived exosomes as diagnostic biomarkers of ovarian cancer. Gynecol. Oncol. 110: 13-21.

Tazawa, H., N. Tsuchiya, M. Izumiya and H. Nakagama. 2007. Tumor-suppressive miR-34a induces senescence-like growth arrest through modulation of the E2F pathway in human colon cancer cells. Proc. Natl. Acad. Sci. U S A 104: 15472-15477.

Thiery, J.P. and J.P. Sleeman. 2006. Complex networks orchestrate epithelial-mesenchymal transitions. Nat. Rev. Mol. Cell Biol. 7: 131-142.

Till, S., E. Lejeune, R. Thermann, M. Bortfeld, M. Hothorn, D. Enderle, C. Heinrich, M.W. Hentze and A.G. Ladurner. 2007. A conserved motif in Argonaute-interacting proteins mediates functional interactions through the Argonaute PIWI domain. Nat. Struct. Mol. Biol. 14: 897-903.

Tomari, Y. and P.D. Zamore. 2005. Perspective: machines for RNAi. Genes Dev 19: 517-529.

Toyota, M., H. Suzuki, E. Yamamoto, K. Imai and Y. Shinomura. 2007. Gene mutation and DNA methylation in non-hereditary colon neoplasms. Nihon Naika Gakkai Zasshi 96: 226-230.

Tsang, J.C. and Y.M. Lo. 2007. Circulating nucleic acids in plasma/serum. Pathology 39: 197-207.

Tsang, Y.S., K.W. Lo, S.F. Leung, P.H. Choi, Y. Fong, J.C. Lee and D.P. Huang. 1999. Two distinct regions of deletion on chromosome 13q in primary nasopharyngeal carcinoma. Int. J. Cancer 83: 305-308.

Tsujiura, M., D. Ichikawa, S. Komatsu, A. Shiozaki, H. Takeshita, T. Kosuga, H. Konishi, R. Morimura, K. Deguchi, H. Fujiwara, K. Okamoto and E. Otsuji. 2010. Circulating microRNAs in plasma of patients with gastric cancers. Br. J. Cancer 102: 1174-1179.

Ullah, M.F. and M. Aatif. 2009. The footprints of cancer development: Cancer biomarkers. Cancer Treat Rev 35: 193-200.

Unwin, P.N. and R.A. Milligan. 1982. A large particle associated with the perimeter of the nuclear pore complex. J. Cell Biol. 93: 63-75.

Valadi, H., K. Ekstrom, A. Bossios, M. Sjostrand, J.J. Lee and J.O. Lotvall. 2007. Exosome-mediated transfer of mRNAs and microRNAs is a novel mechanism of genetic exchange between cells. Nat. Cell Biol. 9: 654-659.

Valastyan, S., F. Reinhardt, N. Benaich, D. Calogrias, A.M. Szasz, Z.C. Wang, J.E. Brock, A.L. Richardson and R.A. Weinberg. 2009. A pleiotropically acting microRNA, miR-31, inhibits breast cancer metastasis. Cell 137: 1032-1046.

van den Berg, A., B.J. Kroesen, K. Kooistra, D. de Jong, J. Briggs, T. Blokzijl, S. Jacobs, J. Kluiver, A. Diepstra, E. Maggio and S. Poppema. 2003. High expression of B-cell receptor inducible gene BIC in all subtypes of Hodgkin lymphoma. Genes Chromosomes Cancer 37: 20-28.

van Niel, G., I. Porto-Carreiro, S. Simoes and G. Raposo. 2006. Exosomes: a common pathway for a specialized function. J. Biochem. 140: 13-21.

Varambally, S., Q. Cao, et al. 2008. Genomic loss of microRNA-101 leads to overexpression of histone methyltransferase EZH2 in cancer. Science 322: 1695-1699.

Vasilescu, C., S. Rossi, et al. 2009. MicroRNA fingerprints identify miR-150 as a plasma prognostic marker in patients with sepsis. PLoS One 4: e7405.

Vella, M.C., E.Y. Choi, S.Y. Lin, K. Reinert and F.J. Slack. 2004. The C. elegans microRNA let-7 binds to imperfect let-7 complementary sites from the lin-41 3'UTR. Genes Dev. 18: 132-137.

Ventura, A., A.G. Young, M.M. Winslow, L. Lintault, A. Meissner, S.J. Erkeland, J. Newman, R.T. Bronson, D. Crowley, J.R. Stone, R. Jaenisch, P.A. Sharp and T. Jacks. 2008. Targeted deletion reveals essential and overlapping functions of the miR-17 through 92 family of miRNA clusters. Cell 132: 875-886.

Visone, R., P. Pallante, et al. 2007. Specific microRNAs are downregulated in human thyroid anaplastic carcinomas. Oncogene 26: 7590-7595.

Voinnet, O. 2009. Origin, biogenesis, and activity of plant microRNAs. Cell 136: 669-687.

Volinia, S., G.A. Calin, et al. 2006. A microRNA expression signature of human solid tumors defines cancer gene targets. Proc. Natl. Acad. Sci. U S A 103: 2257-2261.

Vriens, M.R., J. Weng, I. Suh, N. Huynh, M.A. Guerrero, W.T. Shen, Q.Y. Duh, O.H. Clark and E. Kebebew. 2011. MicroRNA expression profiling is a potential diagnostic tool for thyroid cancer. Cancer PMID: 22006248 doi: 10.1002/cncr.26587.

Waldman, S.A. and A. Terzic. 2007. Translating MicroRNA discovery into clinical biomarkers in cancer. JAMA 297: 1923-1925.

Wang, J., J. Chen, P. Chang, A. LeBlanc, D. Li, J.L. Abbruzzesse, M.L. Frazier, A.M. Killary and S. Sen. 2009. MicroRNAs in plasma of pancreatic ductal adenocarcinoma patients as novel blood-based biomarkers of disease. Cancer Prev. Res. (Phila) 2: 807-813.

Wang, J.F., M.L. Yu, G. Yu, J.J. Bian, X.M. Deng, X.J. Wan and K.M. Zhu. 2010. Serum miR-146a and miR-223 as potential new biomarkers for sepsis. Biochem Biophys Res Commun 394: 184-188.

Wang, K., S. Zhang, B. Marzolf, P. Troisch, A. Brightman, Z. Hu, L.E. Hood and D.J. Galas. 2009. Circulating microRNAs, potential biomarkers for drug-induced liver injury. Proc. Natl. Acad. Sci. U S A 106: 4402-4407.

Wang, M., L.P. Tan, et al. 2008. miRNA analysis in B-cell chronic lymphocytic leukaemia: proliferation centres characterized by low miR-150 and high BIC/miR-155 expression. J. Pathol. 215: 13-20.

Wang, R., Z.X. Wang, J.S. Yang, X. Pan, W. De and L.B. Chen. 2011. MicroRNA-451 functions as a tumor suppressor in human non-small cell lung cancer by targeting ras-related protein 14 (RAB14). Oncogene 30: 2644-2658.

Wang, S. and E.N. Olson. 2009. AngiomiRs--key regulators of angiogenesis. Curr. Opin. Genet. Dev. 19: 205-211.

Wang, Y., S. Juranek, H. Li, G. Sheng, G.S. Wardle, T. Tuschl and D.J. Patel. 2009. Nucleation, propagation and cleavage of target RNAs in Ago silencing complexes. Nature 461: 754-761.

Wang, Y., A.T. Lee, J.Z. Ma, J. Wang, J. Ren, Y. Yang, E. Tantoso, K.B. Li, L.L. Ooi, P. Tan and C.G. Lee. 2008. Profiling microRNA expression in hepatocellular carcinoma reveals microRNA-224 up-regulation and apoptosis inhibitor-5 as a microRNA-224-specific target. J. Biol. Chem. 283: 13205-13215.

Wang, Y., G. Sheng, S. Juranek, T. Tuschl and D.J. Patel. 2008. Structure of the guide-strand-containing argonaute silencing complex. Nature 456: 209-213.

Weber, F., R.E. Teresi, C.E. Broelsch, A. Frilling and C. Eng. 2006. A limited set of human MicroRNA is deregulated in follicular thyroid carcinoma. J. Clin. Endocrinol. Metab. 91: 3584-3591.

Webster, R.J., K.M. Giles, K.J. Price, P.M. Zhang, J.S. Mattick and P.J. Leedman. 2009. Regulation of epidermal growth factor receptor signaling in human cancer cells by microRNA-7. J. Biol. Chem. 284: 5731-5741.

Wilusz, J.E., H. Sunwoo and D.L. Spector. 2009. Long noncoding RNAs: functional surprises from the RNA world. Genes Dev. 23: 1494-1504.

Witkos, T.M., E. Koscianska and W.J. Krzyzosiak. 2011. Practical Aspects of microRNA Target Prediction. Curr. Mol. Med. 11: 93-109.

Wong, T.S., X.B. Liu, B.Y. Wong, R.W. Ng, A.P. Yuen and W.I. Wei. 2008. Mature miR-184 as Potential Oncogenic microRNA of Squamous Cell Carcinoma of Tongue. Clin. Cancer Res. 14: 2588-2592.

Wu, L., C. Cai, X. Wang, M. Liu, X. Li and H. Tang. 2011. MicroRNA-142-3p, a new regulator of RAC1, suppresses the migration and invasion of hepatocellular carcinoma cells. FEBS Lett. 585: 1322-1330.

Wu, L., J. Fan and J.G. Belasco. 2006. MicroRNAs direct rapid deadenylation of mRNA. Proc. Natl. Acad. Sci. U S A 103: 4034-4039.

Wurz, K., R.L. Garcia, B.A. Goff, P.S. Mitchell, J.H. Lee, M. Tewari and E.M. Swisher. 2010. MiR-221 and MiR-222 alterations in sporadic ovarian carcinoma: Relationship to CDKN1B, CDKNIC and overall survival. Genes Chromosomes Cancer 49: 577-584.

Xi, Y., J.R. Edwards and J. Ju. 2007. Investigation of miRNA biology by bioinformatic tools and impact of miRNAs in colorectal cancer--regulatory relationship of c-Myc and p53 with miRNAs. Cancer Inform. 3: 245-253.

Xiang, X., X. Zhuang, S. Ju, S. Zhang, H. Jiang, J. Mu, L. Zhang, D. Miller, W. Grizzle and H.G. Zhang. 2011. miR-155 promotes macroscopic tumor formation yet inhibits tumor dissemination from mammary fat pads to the lung by preventing EMT. Oncogene 30: 3440-3453.

Xiao, B., J. Guo, Y. Miao, Z. Jiang, R. Huan, Y. Zhang, D. Li and J. Zhong. 2009. Detection of miR-106a in gastric carcinoma and its clinical significance. Clin. Chim. Acta 400: 97-102.

Xiao, C., L. Srinivasan, D.P. Calado, H.C. Patterson, B. Zhang, J. Wang, J.M. Henderson, J.L. Kutok and K. Rajewsky. 2008. Lymphoproliferative disease and autoimmunity in mice with increased miR-17-92 expression in lymphocytes. Nat. Immunol. 9: 405-414.

Yamamoto, Y., N. Kosaka, M. Tanaka, F. Koizumi, Y. Kanai, T. Mizutani, Y. Murakami, M. Kuroda, A. Miyajima, T. Kato and T. Ochiya. 2009. MicroRNA-500 as a potential diagnostic marker for hepatocellular carcinoma. Biomarkers 14: 529-538.

Yamashita, A., T.C. Chang, Y. Yamashita, W. Zhu, Z. Zhong, C.Y. Chen and A.B. Shyu. 2005. Concerted action of poly(A) nucleases and decapping enzyme in mammalian mRNA turnover. Nat. Struct. Mol. Biol. 12: 1054-1063.

Yan, L.X., X.F. Huang, Q. Shao, M.Y. Huang, L. Deng, Q.L. Wu, Y.X. Zeng and J.Y. Shao. 2008. MicroRNA miR-21 overexpression in human breast cancer is associated with advanced clinical stage, lymph node metastasis and patient poor prognosis. RNA 14: 2348-2360.

Yanaihara, N., N. Caplen, et al. 2006. Unique microRNA molecular profiles in lung cancer diagnosis and prognosis. Cancer Cell 9: 189-198.

Yang, H., W. Kong, L. He, J.J. Zhao, J.D. O'Donnell, J. Wang, R.M. Wenham, D. Coppola, P.A. Kruk, S.V. Nicosia and J.Q. Cheng. 2008. MicroRNA expression profiling in human ovarian cancer: miR-214 induces cell survival and cisplatin resistance by targeting PTEN. Cancer Res. 68: 425-433.

Yang, J.S. and E.C. Lai. 2011. Alternative miRNA biogenesis pathways and the interpretation of core miRNA pathway mutants. Mol. Cell 43: 892-903.

Yang, J.S., T. Maurin, N. Robine, K.D. Rasmussen, K.L. Jeffrey, R. Chandwani, E.P. Papapetrou, M. Sadelain, D. O'Carroll and E.C. Lai. 2010. Conserved vertebrate mir-451 provides a platform for Dicer-independent, Ago2-mediated microRNA biogenesis. Proc. Natl. Acad. Sci. U S A 107: 15163-15168.

Yang, W., T.P. Chendrimada, Q. Wang, M. Higuchi, P.H. Seeburg, R. Shiekhattar and K. Nishikura. 2006. Modulation of microRNA processing and expression through RNA editing by ADAR deaminases. Nat. Struct. Mol. Biol. 13: 13-21.

Yang, W., D.Y. Lee and Y. Ben-David. 2011. The roles of microRNAs in tumorigenesis and angiogenesis. Int. J. Physiol. Pathophysiol. Pharmacol. 3: 140-155.

Yi, R., Y. Qin, I.G. Macara and B.R. Cullen. 2003. Exportin-5 mediates the nuclear export of pre-microRNAs and short hairpin RNAs. Genes Dev. 17: 3011-3016.

Yu, F., H. Yao, P. Zhu, X. Zhang, Q. Pan, C. Gong, Y. Huang, X. Hu, F. Su, J. Lieberman and E. Song. 2007. let-7 regulates self renewal and tumorigenicity of breast cancer cells. Cell 131: 1109-1123.

Yu, S.L., H.Y. Chen, et al. 2008. MicroRNA signature predicts survival and relapse in lung cancer. Cancer Cell 13: 48-57.

Yuan, Y.R., Y. Pei, J.B. Ma, V. Kuryavyi, M. Zhadina, G. Meister, H.Y. Chen, Z. Dauter, T. Tuschl and D.J. Patel. 2005. Crystal structure of A. aeolicus argonaute, a site-specific DNA-guided endoribonuclease, provides insights into RISC-mediated mRNA cleavage. Mol. Cell 19: 405-419.

Zemel, R., A. Issachar and R. Tur-Kaspa. 2011. The role of oncogenic viruses in the pathogenesis of hepatocellular carcinoma. Clin. Liver Dis. 15: 261-279, vii-x.

Zhang, C.Z., J.X. Zhang, A.L. Zhang, Z.D. Shi, L. Han, Z.F. Jia, W.D. Yang, G.X. Wang, T. Jiang, Y.P. You, P.Y. Pu, J.Q. Cheng and C.S. Kang. 2010. MiR-221 and miR-222 target PUMA to induce cell survival in glioblastoma. Mol. Cancer 9: 229.

Zhang, L., J. Huang, et al. 2006. microRNAs exhibit high frequency genomic alterations in human cancer. Proc. Natl. Acad. Sci. U S A 103: 9136-9141.

Zhang, X., S. Liu, T. Hu, Y. He and S. Sun. 2009. Up-regulated microRNA-143 transcribed by nuclear factor kappa B enhances hepatocarcinoma metastasis by repressing fibronectin expression. Hepatology 50: 490-499.

Zhu, S., M.L. Si, H. Wu and Y.Y. Mo. 2007. MicroRNA-21 targets the tumor suppressor gene tropomyosin 1 (TPM1). J. Biol. Chem. 282: 14328-14336.

Zhu, S., H. Wu, F. Wu, D. Nie, S. Sheng and Y.Y. Mo. 2008. MicroRNA-21 targets tumor suppressor genes in invasion and metastasis. Cell Res. 18: 350-359.

CHAPTER 9

Targeting PARP-1 in Breast Cancer: The Paradigm of a DNA Repair-based Therapeutic Strategy

Francisco Javier Oliver*, Jara Majuelos-Melguizo, José Manuel Rodriguez-Vargas, Ariannys González-Flores, A. Peralta-Leal, Laura López and María Isabel Rodríguez*

Instituto López Neyra de Parasitología y Biomedicina, Avda. Conocimiento, 18100 Armilla, Granada, Spain

ABSTRACT

Poly(ADP-ribose) polymerases (PARPs) are defined as cell signalling enzymes that catalyze the transfer of ADP-ribose units from NAD(+) to a number of acceptor proteins. PARP-1, the best characterized member of the PARP family, that presently includes eighteen members, is an abundant nuclear enzyme implicated in cellular responses to DNA injury provoked by genotoxic stress. PARP is involved in DNA repair and transcriptional regulation, and is now recognized as a key regulator of cell survival and cell death as well as a master component of a number of transcription factors involved in tumor development and inflammation. PARP-1 is essential to the repair of DNA single-strand breaks via the base excision repair pathway. Inhibitors of PARP-1 have been shown to enhance the cytotoxic effects of ionizing radiation and DNA damaging chemotherapy agents, such as the methylating agents and topoisomerase I inhibitors. Recent in vitro and in vivo evidence suggests that PARP inhibitors could be used not only as chemo/radiotherapy sensitizers, but as single agents to selectively kill cancers defective in DNA repair, specifically cancers with mutations in the breast cancer associated (BRCA) 1 and 2 genes and other

Corresponding authors*: Current address of **Dr. Francisco Javier Oliver and **Dr. María Isabel Rodríguez**: Dept. de Bioquímica y Biología Molecular I, Facultad de Ciencias, Universidad de Granada, Avda. Fuentenueva s/n, 18071 Granada, Spain. E-mails: joliver@ipb.csic.es; mirlara@ipb.csic.es mutations affecting the cell's efficiency to cope with homologous recombination. The aim of the present chapter is to overview the emerging data in the literature on the mechanistic insights of poly(ADP-ribose) polymerase-1 (PARP-1) in the pathway of homologous recombination and the consequences for the therapeutic intervention in familiar and refractory breast cancer.

Keywords: PARP, poly(ADP-ribose), BRCA1/2, DNA repair, homologous recombination, angiogenesis, anticancer therapy, breast cancer.

INTRODUCTION

PAR polymerases (PARPs) enzymes are an important family of proteins that regulates the integrity and correctness of transcription of DNA in both vegetal and animal cells, through the transfer of units of ADP-ribose to protein acceptors. The addition of polymers of ADP-ribose is an important post-translational modification, because the highly negative charge of ADP-monomer, changes the activities and function of these nuclear acceptors. PARPs family is defined as cell signaling enzymes that are key sensors of DNA damage (Kim et al. 2005). PARP-1 (poly(ADP-ribose) polymerase-1) is the most important member of this family, which includes others 16 members of different proteins, related with DNA metabolism: PARP-2, PARP-3, PARP-4 (Vault-PARP), PARP-5 (Tankyrases 1 and 2), PARP-6, PARP-7 (tiPARP), PARP-8, PARP-9 (BAL-1), PARP-10, PARP-11, PARP-12, PARP-13 (ZAP), PARP-14 (CoaSt6), PARP-15 and PARP-16. All these proteins have different structural domains and functions in the acid nucleic metabolism (Gagne et al. 2008).

PARP-1 play a key role as DNA damage sensor and in DNA repair provoked by genotoxic stress and maintenance of chromatin integrity (Burkle 2001; Kim et al. 2005). Moreover, PARP-1 is a chromatin-associated enzyme with important functions in regulation of transcription, cell cycle, tumorigenesis and other important human diseases through ROS/NF-kB/PARP-1 pathway including type I diabetes and associated endothelial dysfunction, cardiovascular injury and heart failure, sepsis, rheumatoid arthritis and stroke (4-6).

STRUCTURE AND METABOLISM OF PARP-1

PARP-1 is an abundant nuclear enzyme with 113 KDa; it may interact with DNA and different DNA repair molecules (Burkle 2001). This protein has three different structural/functional domains (Fig. 1): 1) A DNA-binding region or DBD domain, able to recognize DNA-strand breaks, through two zinc fingers

(FI/Zn1 and FII/Zn2), that mediates in the interaction with DNA. Recently one group has discovered a new zinc finger or FIII, which is very important for DNA-dependent PARP-1 activation (Langelier et al. 2010). The DBD domain contains a nuclear localization sequence (NLS) and a caspase-3 cleavage motif (DEVD); 2) A central automodification region rich in glutamic acid and containing a breast cancer associated protein C-terminal (BRCT) motif. This region allows PARPs enzymes to modify their activity by binding fragments of ADP-ribose polymer, synthesized by the PARP itself, which is able to increase their activity; and 3) A NAD-binding region with all the catalytic activities of the full-length enzyme. This region is the most conserved domain in PARPs family. The WGR motif has recently been described as a conserved amino acid sequence (Trp, Gly, Arg) but the function is still unknown (8, 9).

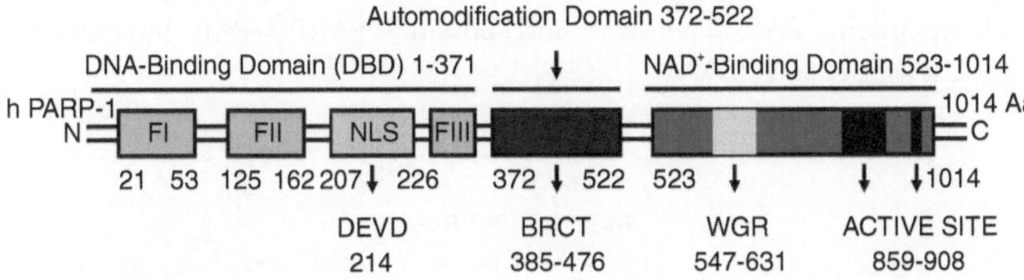

Fig. 1 Structural Organization of PARP-1: Schematic representation of human PARP-1 with the different domains and active regions.

Color image of this figure appears in the color plate section at the end of the book.

PARP-1 is activated following recognition of the damaged DNA and synthesizes a branched polymer of residues formed monomers of ADP-ribose (PAR), from NAD+ (nicotinamide adenine dinucleotide) as substrate (Fig. 2) (10, 11). The residues of ADP-ribose are covalently linked between them forming the various branches of the polymer, which will be covalently linked to glutamic residues of different nuclear acceptors related to the recognition and repair of damaged DNA, including nucleosomal core histones, topoisomerases I and II, High Mobility Group (HMGB) proteins or p53 (2, 12). PARP-1 can be hyperactivated after an important DNA damage induced by free radicals, reactive species of oxygen (ROS) or peroxynitrite, which produce oxidative stress and single strand-breaks in DNA and different alkylating agents that induce double strand-breaks (Fig. 3). In this situation, PARP-1 first links this polymer to itself resulting in an increased activity; then the branched PAR polymer is linked to different protein acceptors, which are recruited to the DNA lesion. Both the polymer is synthesized under normal conditions, for example during DNA replication, such as DNA stress, should be degraded and recycled; PARP-1 becomes inactive activation to avoid energy collapse, which would have resulted in a significant consumption of NAD+ and ATP. The only protein capable of degrading the polymer is the poly(ADP-ribose) glicohydrolase (PARG). PARG is a protein that has different isoforms, each of which locates in a different organelle (Whatcott et al. 2009). When the complex PARP-1/PAR/proteins have carried out their function and the genetic material has been repaired, PARG will be responsible to hydrolyze the polymer that, first is joined PARP-1, decreasing its activity, and then the polymer is attached to the base excision machinery, DNA repair and finally the free polymer that can play a role as a signaling molecule in nucleus. Recent studies have demonstrated that there are different mechanisms of recruitment of PARG to the area of damaged DNA, where there is a greater amount of polymer. These authors suggest different models of recruitment: PARP-1/PAR-dependent, PAR-dependent and possibly PARP-1/PAR independent (Mortusewicz et al. 2007).

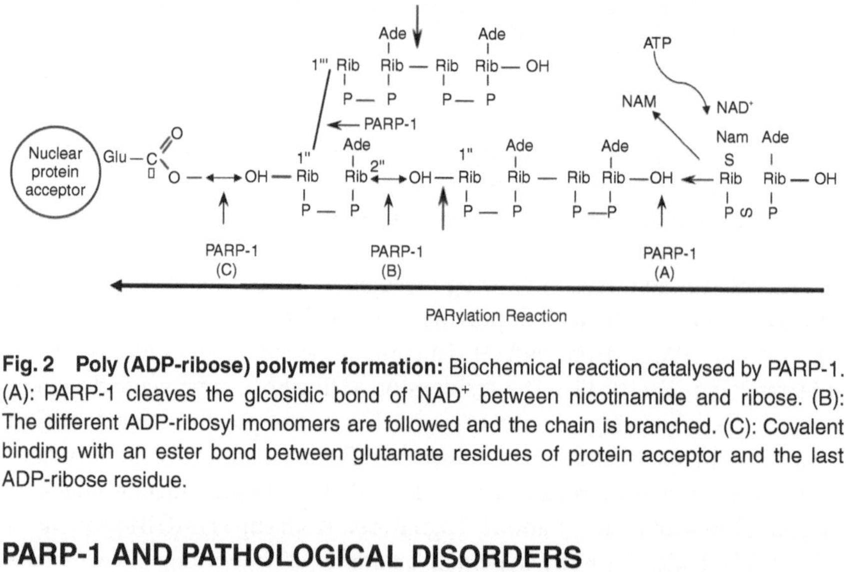

Fig. 2 Poly (ADP-ribose) polymer formation: Biochemical reaction catalysed by PARP-1. (A): PARP-1 cleaves the glcosidic bond of NAD^+ between nicotinamide and ribose. (B): The different ADP-ribosyl monomers are followed and the chain is branched. (C): Covalent binding with an ester bond between glutamate residues of protein acceptor and the last ADP-ribose residue.

PARP-1 AND PATHOLOGICAL DISORDERS

PARP is known to have been rapidly activated in various pathophysiological conditions, and its activation is prolonged and continued. For example, direct detection of PAR accumulation has demonstrated the activation of PARP in stroke induced by middle cerebral artery occlusion and reperfusion (Endres et al. 1998), and in heart after myocardial infarction and heart

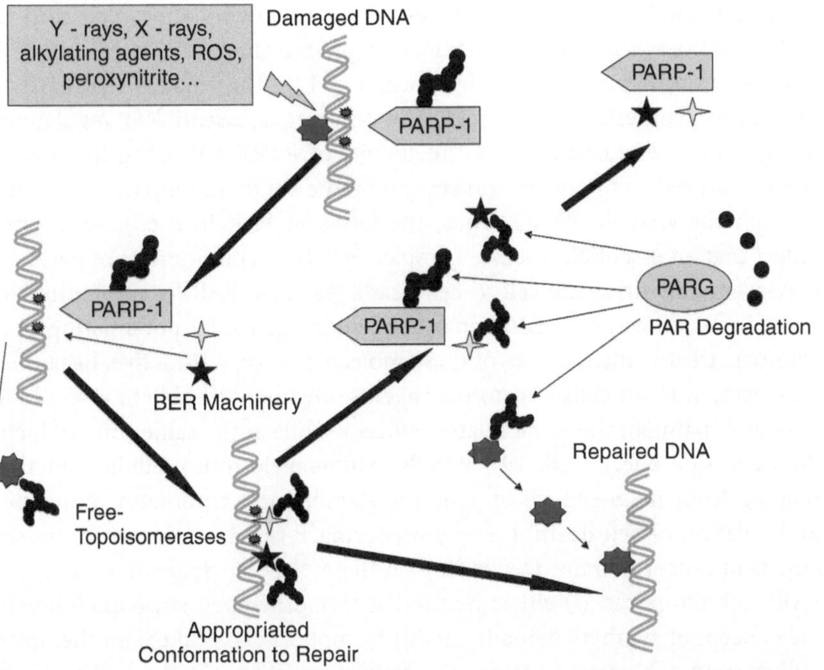

Fig. 3 Metabolism of Poly(ADP-ribosyl)ation: During DNA damage, PARP-1 units PAR polymer to himself and glutamate residues of different DNA-binding proteins and protein of BER system to repair the DNA. When the genetic material is repaired the polymer attached to PARP-1 and other proteins is degraded by PARG and the system returns to the initial situation.

transplantion (Faro et al. 2002). Also, PARP activation has been demonstrated in hemorrhagic and septic shock (Soriano et al. 2002), in the lung of mice subjected to a model of acute respiratory distress syndrome and well as in diabetic animals (Pacher et al. 2002).

POLY(ADP-RIBOSYL)ATION IN CANCER BIOLOGY

Poly(ADP-ribosyl)ation or PARylation is defined as the covalent modification of proteins with a negatively charged and highly branched polymer of ADP-ribose residues. This reaction occurs in a cell in many essential and physiological cellular processes such as replication of genetic material, where the mutation rate to duplicate the DNA is high, maintaining the integrity of the chromatin by interaction with different protein core histones, telomere integrity or modulation of the localisation of different transcriptional factor as p53 o KLF8, so PARP-1 and PAR are essential in a number of physiological and patho-physiological cellular processes, as programmed cell death, ischemia, inflammation, hypoxia, aging or cancer (15-17). The balance that exists in the cell to synthesize and degrade the polymer is essential in regulating a biological process, first because the activity of PARP-1 is going to consume energy and reducing power, and second by the ability to recycle the polymer through the enzyme PARG. Thus, the levels of PAR in a precise moment reflect that an essential process, as repair or DNA replication, take place and may eventually drive the cell to cell death. As such, PARP-1 inhibitors have been discussed as potential chemotherapeutic agents, by potentiating other cytotoxic treatments. The use of these molecules is very attractive, because by sensitizing tumour cells to cytotoxic agents one might be able to give a lower dose and maintain the same relative efficacy while at the same time reducing the toxic side effects (18, 19). PARPs exhibit pleiotropic cellular functions ranging from maintenance of genomic stability and chromatin remodeling to regulation of cell death, thereby rendering PARP homologues promising targets in cancer therapy. Depending on the molecular status of a cancer cell, PARP inhibitors can (i) either be used as monotherapeutic agents following the concept of synthetic lethality or (ii) to support classical chemotherapy or radiotherapy. The rationales are the following: (i) in cancers with selective defects in homologous recombination repair, inactivation of PARPs directly causes cell death. In cancer treatment, this phenomenon can be employed to specifically target tumor cells while sparing nonmalignant tissue. (ii) PARP inhibitors can also be used to sensitize cells to cytotoxic DNA-damaging treatments, as some PARPs actively participate in genomic maintenance. Apart from that, PARP inhibitors possess antiangiogenic functions, thus opening up a further option to inhibit tumor growth. In view of the above, a number of high-potency PARP inhibitors have been developed during the last decade and are currently evaluated as cancer therapeutics in clinical trials by several leading pharmaceutical companies. In this chapter we will analyze the role of PARP-1 in double strands break repair by homologous recombination and the potential of PARP inhibition induced synthetic lethality in breast cancer.

PARP inhibitors have been used in vitro (Tentori et al. 2007; Pyriochou et al. 2008) to inhibit vascular endothelial growth factor (VEGF)-induced proliferation, migration and tube formation in human umbilical vein endothelial cells (HUVECs) and in tumor models (Martin-Oliva et al. 2006).

Recently we have characterized a novel regulatory mechanism controlling Snail1 protein expression through poly(ADP-ribosyl)ation during epithelial to mesenchymal transition (Bryant et al. 2009). EMT iPARP inhibition have effects on EMT phenotype of different tumor cells, including mesenchymal markers downregulation and epithelial markers upregulation (Rodriguez et al. 2011). Also PARP inhibition is related to autophagy. Autophagy is a lysosome-dependent degradative pathway frequently activated in tumor cells treated with chemotherapy or radiation. PARP-1 deficient or inhibited cells display a decreased autophagic response (Munoz-Gamez et al. 2009). To have a more global overview, we will also review in short some aspects of the implications of PARP-1 in different non-neoplastic pathologies.

PARP in Neuronal Pathology

PARP-1 plays a significant role in necrotic cell death upon neuronal injury such as excitotoxicity, ischemia-reperfusion and traumatic brain injury (Virag and Szabo 2002). Traumatic brain injury (TBI), also known as intracranial injury, occurs when an external force traumatically injures the brain. TBI is a major cause of death and disability worldwide, especially in children and young adults. There is an involvement of nitric oxide (NO), oxidative and nitrosative stresses and PARP in the pathophysiology of TBI. In pathological conditions, an imbalance between the production and the elimination of reactive oxygen- and reactive nitrogen-containing intermediates (ROS and RNS, respectively) has been described and referred to as oxidative/nitrosative stress (Pacher et al. 2007). Many of these species (e.g., peroxynitrite and hydroxyl radical) can cause DNA damage, which results in the activation of PARP (Szabo and Dawson 1998; Pacher and Szabo 2008). PARP is markedly activated as early as 30 min after TBI and its activation persists for 72 h after TBI (LaPlaca et al. 1999; Besson et al. 2003; Satchell et al. 2003). This pattern of PARP activation is likely related to the continuing presence of peroxynitrite in the lesioned brain tissue. Chemical PARP inhibition and genetic depletion of PARP-1 gene block these neurotoxic effects. Therefore, INH2B (5-iodo-6-amino-1, 2-benzopyrone) (Satchell et al. 2003) and INO-1001 (Clark et al. 2007), prevent NAD depletion and improve post-TBI deficits. 3-aminobensamide (3-AB) and other derivatives induce neuroprotective effect on the neurological deficit and the brain lesion after closed head injury in mice (Mesenge et al. 1996) and after TBI induced by fluid percussion (Besson et al. 2003). PJ-34 [N-(6-oxo-5,6-dihydrophenanthridin-2-yl)-2-(N, N-dimethylamino)acetamide, HCl] and INO-1001 induce neurological recovery-promoting effects in a model of TBI caused by fluid percussion in rat (Besson et al. 2005).

Ischemic reperfused rats pretreated with PARP inhibitor PJ34 showed a near complete inhibition of microglia, macrophage activation and an 84% reduction in CA1 neuronal survival (Hamby et al. 2007). Furthermore, recent studies demonstrated PARP is activated during learning and memory (Sung and Ambron 2004).

PARP-1 in Cardiovascular Diseases

PARP-1 has been reported in various models of heart failure, cardiomyopathies, and myocardial hypertrophy. Importantly, recent studies have also demonstrated overexpression of PARP-1 or increased activity in biopsies from human subjects with heart failure (Molnar et al. 2006). Increased poly(ADP-ribosyl)ation, ROS formation and nitrosative stress were also reported in cardiomyocytes and endothelial cells in chronic heart failure associated with advanced aging or hypertension in rats and/or mice (Pacher et al. 2007). PARP inhibition with PJ-34 or INO-1001 was associated with improved heart failure-associated decreased endothelial function and decreased myocardial hypertrophy and adverse remodeling (Pacher and Szabo 2006). Endothelial dysfunction associated with diabetes, hypertension, heart failure and atherosclerosis is related to the local formation of reactive oxygen and nitrogen species in the vicinity of the vascular endothelium (Pacher et al. 2007). The use of PARP inhibitor PJ-34 during 1 week after induction of diabetes ameliorated vascular PARP accumulation and restored normal vascular function, without altering systemic glucose levels, plasma glycated haemoglobin levels or pancreatic insulin content (Akiyama et al. 2001).

PJ-34 and INO-1001 blocked hyperglycemia and also ameliorated diabetes-induced albumin excretion and mesangial expansion. There some evidences that suggest that NAD a weak PARP inhibitor reverses neurological and neurovascular deficits in streptozotocin diabetic mice (Stevens et al. 2007).

Poly (ADP-ribosylation) and Inflammation

The relationship between inflammation and PARP activity has been widely investigated. NF-κB is a key transcription factor in the regulation of the inflammatory response and PARP has been shown to act as a co-activator in the NF-κB-mediated transcription (Oliver et al. 1999). The administration of PARP inhibitors as 3-AB or new generation compounds to animals result in the attenuation of inflammation and rescued animals form many pathological signs was observed. The therapeutic effects of PARP inhibitors in various forms of heart failure, cardiomyopathies, circulatory shock, cardiovascular diseases and atherosclerosis have been described (Pacher et al. 2007).

PARP-1 in HIV Infection

An increasing body of evidence suggests the involvement of PARP in HIV infection. Prevention of PARP-1 nuclear localization may have greater consequences for HIV infection (Muthumani et al. 2006) and PARP is required for nicking of host cell DNA in order to incorporate the DNA of HIV-1 virus. Independent studies have shown that benzopyrone derivatives and nicotinamide possessed potent antiviral effects in HIV-infected cells. Another study using three different PARP inhibitory approach reported similar results (Kameoka et al. 2005). At present, it seems that PARP regulates HIV infection at two levels: integration and transcription but data with pharmacological inhibitors are still inconclusive.

PARP-1 in Parkinson

PARP activation has been implicated in the pathogenesis of Parkinson's disease, a degenerative disorder of the central nervous system. It results from the death of dopamine-containing cells in the substantia nigra, a region of the midbrain. ROS and RNS have been implicated in the pathogenesis of MPTP neurotoxicity (Blum et al. 2001). Direct evidence for the involvement of MPTP comes from a mouse model of Parkinson's disease. Simultaneous treatment with five different inhibitors of PARP ameliorates the catecholamine depletion induced by MPTP (Infante et al. 2007; Outeiro et al. 2007).

PARP-1 in Aging

Aging is a multi-factorial process defined as time-dependent general decline in physiological function, which is associated with a progressively increasing risk of frailty, morbidity and mortality (Murphy and Partridge 2008). A role for PARPs on longevity has been already postulated almost two decades ago (Grube and Burkle 1992). The oxidative stress theory of aging postulates that ROS induce a variety of macromolecular oxidative modification and accumulation of such oxidative damage, is a primary causal factor in the aging process (Csiszar et al. 2008). Interestingly, the enzymatic activity of PARP-1 can be strongly activated by treatment of cell with ROS such hydrogen peroxide promoting cellular demise and transcription of proinflammatory genes such as NF-κB, TNF-α, interleukin, cyclooxygenase-2 and adhesion molecules which may lead to the outcomes of aging (Chung et al. 2006). Recent results demonstrate that the relationship between PARP and SIRT (Sirtuin family of NAD dependent class III histone deacetylases) could affect not only metabolic control but also has an impact on aging (Canto and Auwerx). In fact PARP inhibition has proven beneficial in many cell culture and animal model system of acute and chronic inflammation and age-related diseases (Altmeyer and Hottiger 2009).

HOMOLOGOUS RECOMBINATION

A typical diploid human cell needs to maintain about 6×10^9 base pairs in the correct sequence and chromosomal organization, a task that is usually performed nearly perfectly from one somatic cell generation to the next (Drake et al. 1998). Homologous recombination (HR) (Fig. 4) is a key pathway to maintain genomic integrity between generations. Recombination is required for the repair or tolerance of DNA damage and the recovery of stalled or broken replication forks (Li and Heyer 2008). However, recombination is also potentially dangerous as it can lead to chromosomal rearrangements and potentially lethal intermediates (Kolodner et al. 2002). Defects in HR and associated processes define a number of human cancer predisposition syndromes associated with genome instability. HR repair of DSBs involves repair using preferentially the sister chromatid, which results in error-free repair (Alon 2003). HR is only active in the S and G2 phases of the cell cycle since the sister chromatid is present only after DNA replication (Aylon et al. 2003). It is possible that HR could use the homologous chromosome for DSB repair when cells are in the G1 phase of the cell cycle. However, this would invariably lead to loss of heterozygosity and may also result in translocations or other gene rearrangements (23, 25-30). To avoid such a scenario, HR is not only controlled by DNA damage response signaling pathways but also by the cell cycle and relies on a high cyclin-dependent kinase activity, present only in the S, G2 and M phases of the cell cycle. HR is suppressed in mitosis, for instance by inactivating phosphorylations on BRCA2 (25). Emerging data also suggest that a subset of DSBs present in the more condensed heterochromatin act as a substrate for HR in the G2 phase of the cell cycle (26).

PARP-1 in Homologous Recombination

For long it has been documented that PARP-1 has a key role on DNA repair of Single Strand Breaks (SSBs). However, the involvement of PARP-1 in

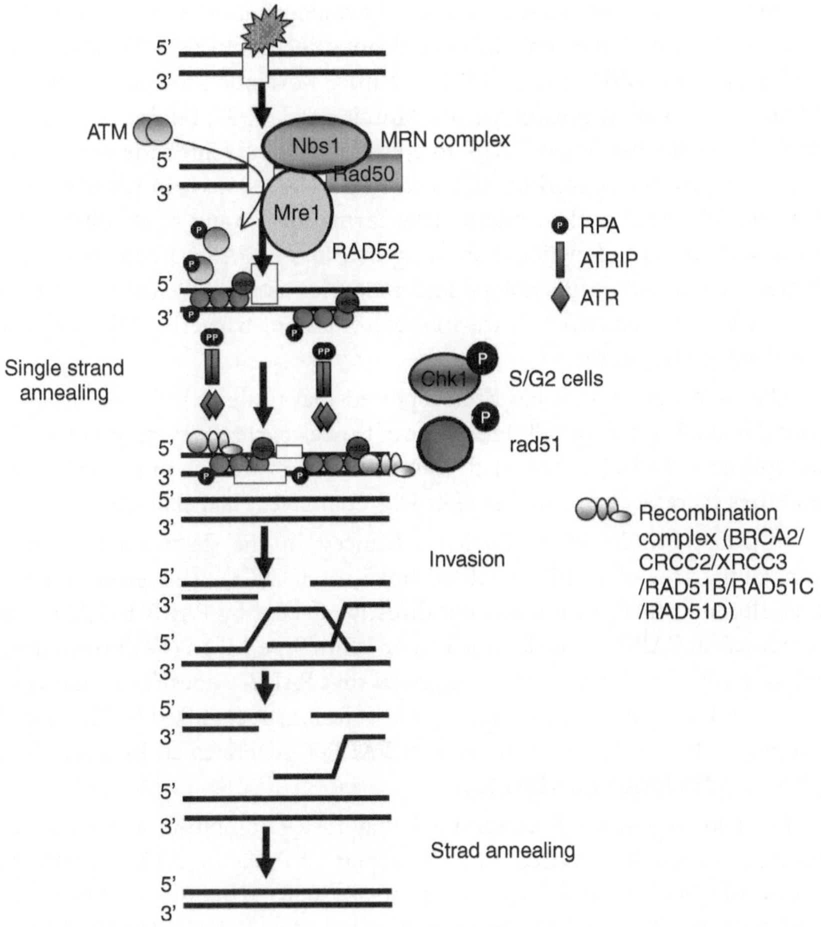

Fig. 4 Homologous recombination pathway. DSBs or those occurring in heterochromatic (packed) DNA activate the ATM kinase that inturn activates the full DNA damage response, allowing recruitment of endprocessing factors at the DNA ends to facilitate end joining (C). (D) Cells in the S, G2 or M phases have a high cyclin-dependent kinase activity, which trigger end resection. The exposed single stranded DNA (ssDNA) is coated with the ssDNA-binding protein RPA, which in turn triggers an ATR response that activates HR repair.

Color image of this figure appears in the color plate section at the end of the book.

DNA repair of Double Strand Breaks (DSBs) is still being investigated. DSBs can be repaired by two different processes: the "error-free" mechanism of Homologous Recombination (HR) and the "error-prone" mechanism of Non-Homologous-End-Joining (NHEJ). The current research shows that PARP-1 has a role regulating both DNA DSBs mechanism of repair. Nevertheless, this chapter will focus on the role of PARP in Homologous

Recombination. There have been several evidences about the role of PARP in HR. PARP inhibition stimulated Homologous Recombination (HR) (Waldman and Waldman 1991). The PARP-1 KO mice developed normally but had a high recombination phenotype (de Murcia et al. 1997; Simbulan-Rosenthal et al. 1999); the binding of PARP to a DNA break prevented the recombining machinery from recognizing and processing DNA lesions (Lindahl et al. 1995); the lack of PARP lead to micronuclei formation (Wang et al. 1997). There were several lines of evidence showing that loss of PARP-1 activity caused a hyper-recombination phenotype and genomic instability. However, it wasn't clear if PARP-1 affected only the number of lesions triggering HR or if it was involved in HR per se.

On the other hand, it has been reported that proteins involved in HR are found in nuclear foci in cells in S phase of the cell cycle, suggesting a role of HR at replication forks (Tashiro et al. 2000). These foci are rich in RAD51 protein and they increase their number after DNA damage (Haaf et al. 1995).

In 2003, Schultz et al. further advanced in the determination of the controlling role of PARP in HR (Schultz et al. 2003). This group reported that although this process was not directly affected by PARP-1, RAD51 foci increased in PARP-1 knockout MEFs or Wild Type MEFs (WT) treated with PARP inhibitors. These results suggested that PARP-1 doesn't execute HR as such, but has a negative controlling role of lesions recognized by HR (Fig. 5), supported by the fact that the lack of PARP-1 generates an increase of HR phenotype in form of RAD51 foci.

In 2006, Bryan et al. determined that PARP inhibition activated ATM protein, which is required for HR repair. A role of ATM in HR had previously been reported (Morrison et al. 2000). This group showed that although the PARP-1 KO mice was not lethal, the PARP-1/ATM KO and the PARP-1/ku80 KO mouse were not viable, suggesting that NHEJ has a role in the survival of PARP-inhibited cells due to the involvement of ku80 in this process and second, ATM is important for cellular survival after the inhibition of PARP-1 and this survival is caused by the involvement of ATM in PARP-inhibition –induced HR repair.

The inhibition of PARP generated an increase in unsolved SSBs at the DNA replication fork that turned into DSBs generating a stalled or even a collapsed replication fork. Since it has been reported that ATM is activated by DSBs (Khanna et al. 2001), it was determined that the collapsed replication fork activates this protein. Furthermore, it has been reported that ATM phosphorylates Artemis, which has a role in NHEJ (Riballo et al. 2004) via activation of the NHEJ protein DNA-PK.

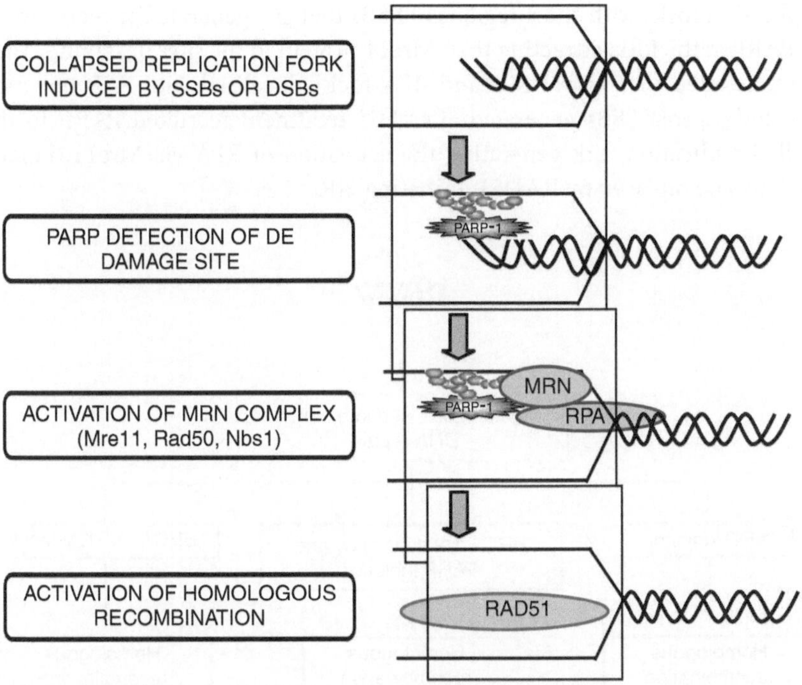

Fig. 5 Role of PARP-1 in the regulation of homologous recombination pathway.

Thus, the increased number of SSBs following PARP inhibition collapse into DSBs at the DNA replication forks. Such collapsed replication forks need HR for their repair, which is activated via ATM. Furthermore, an alternative pathway for the repair of the DSBs at the replication fork via NHEJ is also suggested after PARP inhibition through the ATM activation of Artemis protein.

Aguilar-Quesada et al. 2007 further advanced in analyzing the relationship between PARP-1 and ATM, showing that PARP-1 interacts with ATM and ATM is modified by poly-ADP-ribosylation and, secondly, the dual effect of PARP inhibition on ATM activation. While poly-ADP-ribosylation of ATM is probably needed for optimal ATM activation, long term exposure to PARP inhibitors result in the generation of DSBs and ATM activation. More recently, it has been shown that PARP is activated at stalled replication forks to mediate Mre11-NBS1 dependent replication restart and HR (Haince et al. 2008), (Bryant et al. 2009) (Fig. 6). Mre11/RAD50/Nbs1 complex had previously been reported to have a role in the restart of damaged replication forks (Trenz et al. 2006). Bryant et al. showed that HU treatment generates stalled replication forks with short regions of SSBs that can generate the recruitment of PARP at the fork, attracting then Mre11 protein to the sites of stalling. They demonstrated that Mre11 foci and RPA foci colocalised with PAR polymer, concluding that SSBs generated after HU treatment recruited PARP to the stalled replication fork generating the activation of RPA via Mre11. Finally RPA will be replaced by RAD51, initiating HR.

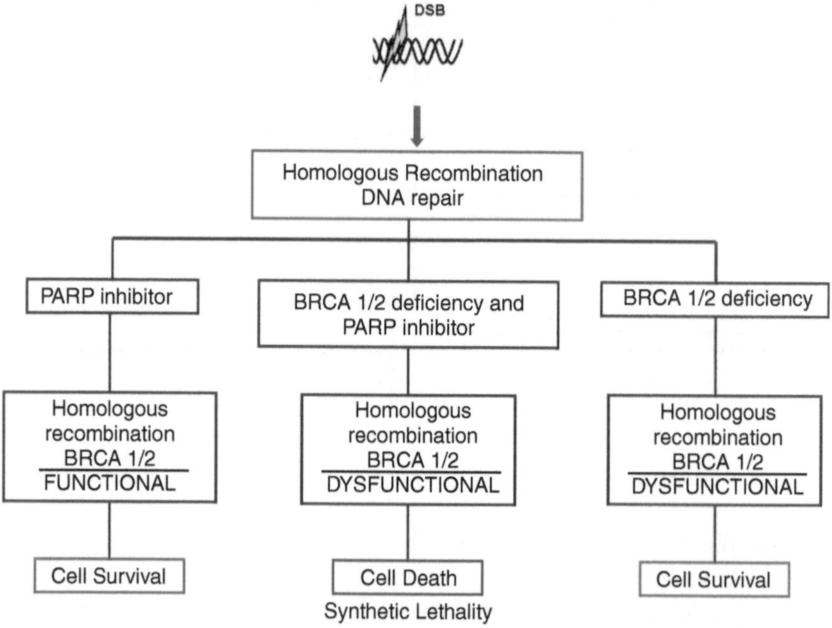

Fig. 6 Inhibition of PARP in HR deficient tumor cells selectively target mutant (tumoral cells).

Furthermore, Haince et al. 2008 have documented that the recruitment of MRE11 made by PARP, is due to their physical interaction. The fact of the presence of PARP-1 in the MRE11 complex suggested the formation of a common complex involving both proteins; on the other hand, the recruitment of MRE11 by PARP can also be generated by DSBs induced by laser microirradiation, which was determined by co-inmmunostaining of PARP, ATM and γH2AX upon multiphoton laser microirradiation.

In conclusion, PARP-1 has a dual role on the regulation of HR: (i) PARP inhibition provokes genomic instability producing the accumulation of non-repaired SSBs at the replication fork that turn into DSBs activating ATM and HR; (ii) the induction of DNA damage of both simple (HU) or double (microirradiation) strand drives to the recruitment of PARP-1 to the damage site, which forms a complex with MRE11 activating RPA and HR.

Homologous Recombination and Breast Cancer

Homologous recombination has a dual role in eukaryotic organisms. Firstly, it is responsible for the creation of genetic variability during meiosis by directing the formation of reciprocal crossover that result in random combination of alleles and traits. Secondly, in mitotic cells, it maintains the integrity of the genome by promoting the faithful repair of DNA double-strand breaks (DSBs). In vertebrates, it therefore plays a key role in tumor avoidance (Thorslund and West 2007). Genetic polymorphisms in homologous recombination repair genes that can lead to protein haploinsufficiency and germline mutations in many of the genes that are involved in HR-mediated DNA double-strand break repair, are generally associated with various genetic disorder and increased cancer risk (Romanowicz-Makowska et al. 2011; Somyajit et al. 2010).

Breast cancer is one of the major killers among malignant conditions worldwide, affecting one out of 10 women in industrialized countries, and being the leading cause of cancer-related morbidity and mortality in women. Genomic rearrangements such as translocations, deletions and duplications are extremely frequent in cancer cells and particularly, in breast cancer cells. These rearrangements are believed to result form an aberrant repair of DNA double-strand breaks (Davidson et al. 2000; Forozan et al. 2000; Loveday et al. 2000). Normally, these breaks are repaired by means of the homologous recombination double-stranded DNA repair pathway, key components of which are the tumor-suppressor proteins BRCA1 and BRCA2 (Gudmundsdottir and Ashworth 2006; Fong et al. 2009).

BRCA1 and *BRCA2* mutation carriers have a cumulative lifetime tumor risk of 84% and 60-80%, respectively, for breast cancer, and 40-50% and 10-20%, respectively, for ovarian cancer. Tumors that lack *BRCA1* or *BRCA2* have an insufficiency in the repair of DNA double-strand breaks by homologous recombination and, a resultant dependence on alternative mechanisms of DNA repair, potentially error-prone. Reliance on upregulated, lower fidelity and potentially mutagenic double-strand breaks DNA repair pathway confers cancer susceptibility through genomic instability (Sandhu et al. 2010). Among *BRCA1/2* gene mutations carriers, a large number of mutations have been described. While *BRCA1* operates in the maintenance of genomic integrity, chromatin remodelling, transcription regulation and cell cycle checkpoint control, the function of *BRCA2* is linked to DNA recombination and repair process (Amir et al. 2010).

After DNA damage, a number of proteins including BRCA1 are recruited to the DNA lesion where they are phosphorylated by protein kinases including ataxia telangiectasia mutated ATM protein. BRCA1 then forms a multi complex termed BRCA1-associated genomes surveillance complex [BASC] which influences the selection of the repair pathway. The presence of this BASC suggests that BRCA1 might have a key role as a coordinator of multiple processes required for the maintenance of genome integrity during the process of DNA replication. BRCA1 also participates in the regulation of the complex MRE-RAD50-NBS1 [MRN] in the initial steps of the foci formation. These foci are separate from BRCA-Rad51 foci, suggesting that BRCA1 might act in two distinct DNA double-strand breaks repair complexes. BRCA1 is also involved in other cellular functions including transcription, ubiquitination, and chromatin remodelling, among others. BRCA2 function is also important in DNA repair. Its role is to bind to the Rad51 protein, through its BRC motif and C-terminal domain, respectively. This complex has a central role in the locating and pairing of the resulting single-stranded ends with complementary sequences. The BCR motif controls the status of the Rad51 filament to govern DNA strand exchange. A 1, 127-amino-acid region of BRCA2 with all eight BRC repeats was found to promote recruitment of Rad51 to ssDNA while slowing the association with dsDNA, establishing a favourable state for DNA strand exchange. Structural analysis revealed that a second structural domain within α-helix context is also essential for BRC-Rad51 interaction. In addition to its interaction with the BRC repeats, Rad51 also interacts with an unrelated C-terminal region of BRCA2, referred to as the TR2 region. This interaction serves a regulatory role in the recombinational repair. During the S and early G2 phases of the cell cycle, or after DNA damage, the C-terminal TR2 region of BRCA2 is non-phosphorylated al serine 3291 and thus capable of interaction with Rad51, that leads to the stabilization of the filaments and as a consequence homologous recombination is active. As cells approach to mitosis, serine 3291 is phosphorylated, serving to downregulate homologous recombination, as TR2 cannot interact with and stabilize Rad51 filaments. BRCA2 has also been shown to participate in other cellular functions like cytokinesis and centrosome location. *BRCA1, BRCA2* and *Rad51* are co-expressed in development, and the three gene products also co-localize upon meiotic chromosomes during the formation of the synaptonemal complex, a structure connected with the tightly regulated events of meiotic recombination. Exposure of cells during S phase to agents that arrest DNA synthesis causes the rapid phosphorylation of BRCA1 and it coordinate relocalization, with BRCA2, Rad51 and BARD1, to sites of DNA synthesis. These observations point to a role for these protein complexes in homologous recombination during the S phase (Amir et al. 2010; Holloman 2011; Scully and Puget 2002; Tutt and Ashworth 2002; Scully et al. 2004; Thorslund and West 2007).

In cells with defective *BRCA1* and/or *BRCA2* repair of double-strand breaks occurs using alternative repair pathways, one of those routes is the base excision repair mechanism that usually repairs single-strand breaks. Inhibition of this pathway increases the number of unrepeated single-strand breaks, which leads to collapsed replication forks and produces double-strand breaks that can't be repaired. In this case, the impairment of both DNA repair pathways induces cell death. This approach follows the concept of synthetic lethality (Fig. 7) applied to anti-cancer drugs. This concept refers to a biological setting in when a mutation or lack of function in one of two pathways or genes has little effect on cell survival, but the combined inhibition leads to cell death. Pre-clinical studies have shown in *BRCA1*- and *BRCA2*-defective cells, PARP-1 inhibitors have shown to be therapeutically active. Ongoing studies are also in progress to assess the combination of PARP inhibitors and taxanes in patients with both *BRCA1/2*-positive and metastatic triple-negative breast cancer (ER, PR and HER2 negative) (Amir et al. 2010; Fong et al. 2009).

Emerging evidence indicates that in addition to the small number of germline *BRCA1* and *BRCA2* mutation-related tumors there is a wider group of sporadic cancers which manifest clinical features of *BRCA*-like functional loss either through epigenetic incativation of the *BRCA* genes or disruption of other non-redundant genes in the *BRCA*-associated homologous recombination repair pathway. These sporadic tumors display *BRCA*-like clinical properties reminiscent of hereditary cancers without *BRCA* gene loss, a phenomenon that has been described as BRCAness. Such sporadic tumors may therefore similarly benefit from the therapeutic approach of synthetic lethality with PARP inhibitors (Sandhu et al. 2010).

Although *BRCA1* and *BRCA2* are the most common lesions associated with genetic breast cancer, other germline mutations are also associated with familial breast cancer such as *TP53* (Malkin et al. 1990), *PTEN* (Liaw et al. 1997), *ATM* (Renwick et al. 2006), *CHEK2* (Meijers-Heijboer et al. 2002), *NBS1* (Steffen et al. 2004), *RAD50* (Heikkinen et al. 2006), *BRIP* (Seal et al. 2006), *PALB2* (Erkko et al. 2007), *RAD51C* (Somyajit et al. 2010), *RAD9* (Lieberman et al. 2011), and other are suspected (Walsh and King 2007).

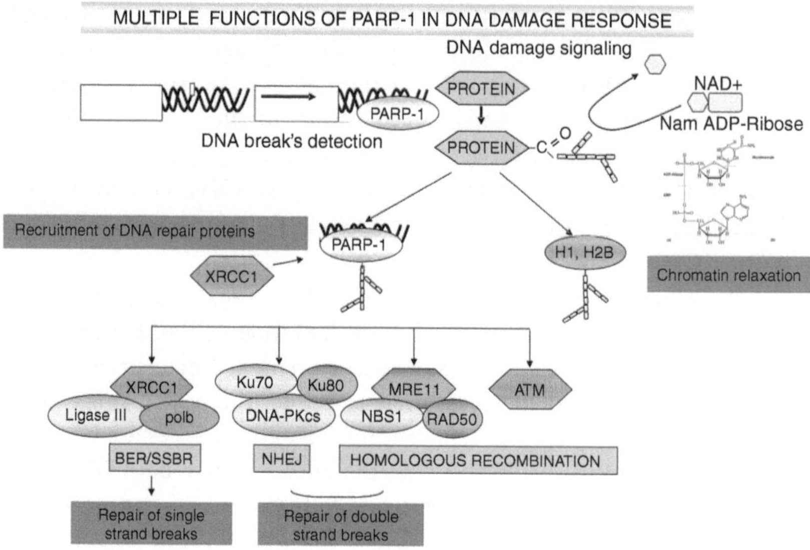

Fig. 7 PARP-1 function in DNA damage response.

PARP Inhibitors in the Treatment of DNA Repair Deficiency-related Cancers

Inhibition of PARP induces accumulation of large numbers of unrepaired SSBs, leading to the collapse of replication forks during S-phase and the consequent generation of DSBs. Single-strand breaks are usually repaired by the base excision repair pathway; therefore, inhibition of this pathway greatly increases the number of unrepaired single-strand breaks, which subsequently leads to DSBs at replication forks. Therefore, cells deficient in DNA DSB repair are highly sensitive to chemical inhibitors of PARP (31-33). In contrast, cells with intact DNA DSBs-response pathways repair damage with high fidelity and accordingly show very little sensitivity to PARP inhibitors.

Mutations in the breast cancer susceptibility genes BRCA1 and BRCA2 are responsible for the majority of hereditary breast cancers (34). Tumours in patients with heterozygous BRCA1 or BRCA2 germ-line mutations typically show somatic loss of heterozygosity at the BRCA1 or BRCA2 locus, respectively, resulting in loss of the wild-type allele (35, 36). BRCA1 and BRCA2 encode proteins that are required for efficient homologous recombination (HR), an error-free form of DSBs repair (37, 38). Because both BRCA1 and BRCA2 deficient cells have so elevated utilization of error-prone repair pathways (Turner et al. 2005), loss of BRCA function causes genome instability (Gudmundsdottir and Ashworth 2006), which renders viable cells susceptible to acquiring additional cancer initiating genetic lesions such as activation of oncogenes and inactivation of tumour suppressor genes. The tumours that arise are BRCA1 or BRCA2 deficient (BRCA1/2 -/-) whereas the remaining somatic cells still have functional BRCA proteins (BRCA1/2 +/-). BRCA1 and BRCA2-deficient cells are highly sensitive to agents that cause replication forks to stall. For example, cells with dysfunctional BRCA2 are characterized by an increased sensitivity to γ-irradiation or DNA-damaging agents (41-44) because the absences of error-free DSB repair mechanisms. PARP inhibitors are lethal in BRCA deficient cells as inhibition of PARP leads to the persistence of DNA lesions that would normally be repaired by BRCA-mediated homologous recombination. Repair of DSBs by alternative error-prone DSB repair mechanisms causes large numbers of chromatid breaks and aberrations, leading to cell cycle arrest and apoptosis. Cell lines lacking BRCA1 or BRCA2 are extremely sensitive to PARP inhibitors while the BRCA +/- and BRCA +/+ cells are relatively non-responsive to the treatment (31, 32). Indeed, *in vivo* experiments demonstrate highly efficient and selective deletion by PARP inhibition of nonneoplastic BRCA2-/- cells with no apparent deleterious effect on surrounding BRCA2-functional cells or whole animal physiology (45), and completely growth arrest of xenografted tumours derived from BRCA2-deficient cells with no effect on BRCA2-wild type tumours (31, 32). Despite these observations of selective cytotoxicity against nontumour cells with engineered BRCA2 mutations, relatively few data are available about the effect of PARP inhibitors on BRCA2-deficient mammary tumour cells. However, a recent report shows a potent growth arrest induced by PARP inhibition in BRCA2-deficient mammary cancer cell lines (46). If we exploit the fact that the PARP-1 pathway becomes essential only in homologous recombination-deficient cells, we could develop a treatment highly specific for the BRCA-deficient tumour cells (Bryant et al. 2005). The aim is to generate DNA damage that, in normal circumstances, is repaired with high fidelity but in a BRCA-dependent fashion (Turner et al. 2005). These observations suggest that PARP inhibitors might be an effective therapy for cancers arising in carriers of BRCA1 or BRCA2 mutations and have led to the assessment of these agents as monotherapy for cancer in BRCA mutation carriers (Ratnam and Low 2007). PARP inhibition may also be considered as a potential prophylactic treatment of non-neoplastic BRCA2-deficient cells in women heterozygous for BRCA2, as well as a primary follow-up treatment following surgery in patients in whom BRCA2 deficiency has led to tumorigenesis. It is one of the few examples where the defect causing the cancer is exploited for targeted therapy.

On the contrary, another report concluded that PARP inhibition is not a viable therapeutic approach based upon the insensitivity of one cell line treated with two weak small molecule inhibitors of PARP (Gallmeier and Kern 2005). However, the selective effect of PARP inhibition on BRCA deficiency is a property of only very potent PARP inhibitors (McCabe et al. 2005), though, clinically speaking, it is not the potency of a drug that is most important but the efficacy, such as some authors point (De Soto et al. 2006). Moreover, their studies indicate that the use of PARP inhibitors in mouse BRCA1 -/- mammary cancer cells seems to show only mild specificity in comparison with their controls in the xenografted nude mice. On the other hand, PARP inhibitors seem to uniformly inhibit growth of human breast cell lines regardless of the BRCA1 genotypes (De Soto et al. 2006). Further observations indicate that BRCA1-/-mammary cancer cells are indeed much more resistant to PARP inhibition than BRCA1-/- nontumour cells in the allograft model. These authors propose that the non-cancerous cells are generally naïve, whereas tumour cells represent cells that have had multiple genetic mutations and that in a tumour there may be numerous fundamentally different cancer cells when classified by the type and frequency of individual genomic changes (De Soto and Deng 2006). Another independent study has not demonstrated either inhibition of tumour growth with the chosen PARP inhibitor, although there was reduced but detectable polymer in this samples and the lack of response may be a result of incomplete PARP inhibition (50). These observations cast doubt about the uniformity of the specific cell killing of BRCA-deficient cells upon PARP inhibition. Maybe, some specific mutations during tumour progression due to genetic instability associated with BRCA1- and BRCA2-deficiency block the sensitivity of BRCA1/2-/- cancer cells to PARP inhibition. In fact, it has been suggested that mutation in BRCA2 and sensitivity to therapeutics in BRCA2 mutation carriers can be suppressed by intragenic deletion (51). In the absence of functional BRCA2, mechanisms such as single-strand annealing and non-homologous end joining compensate and repair DSBs (in absence of HR), by aligning short regions of homology flanking the DSB, deleting the intervening sequence (38, 52). As consequence of this same HR deficiency, new BRCA2 proteins with deletions are capable of restore HR, leading to resistance. In summary, BRCA associated tumorigenesis may be divided into two phases regarding sensitivity to PARP inhibition. In the early stages of tumorigenesis, BRCA mutant cells are sensitive to PARP inhibition.

However, in the second stage of tumorigenesis, BRCA deficient tumour cells may become insensitive to PARP inhibition due to additional mutations. Thus, the use of this strategy for therapeutic treatment for hereditary breast cancers is dependent on the continued susceptibility of BRCA mutant cells to PARP inhibitors (De Soto and Deng 2006).

PARP inhibitors have now entered clinical trials and initial results are promising, with frequent sustained responses in BRCA mutation carriers. Furthermore, PARP inhibitors are likely to produce few side effects, because PARP1 knockout mice are viable and healthy (53, 54) and inhibitors that are specific for PARP are relatively non-toxic and do not directly damage DNA; also, this approach is more specific than standard cytotoxic chemotherapy. Initial observations in patients with known BRCA-associated cancers, or those with a strong family history of disease suggestive of BRCA mutation, indicate that a novel, potent and orally bioavailable PARP inhibitor (Menear et al. 2008) shows low toxicity, and there are suggestions of significant antitumor activity, as assessed by radiography and by measurement of tumour biomarkers. This PARP inhibitor is being tested for efficacy as a monotherapy in BRCA-associated ovarian and breast cancer in Phase II clinical trials (56, 57). Many of the side effects are independent of the mechanism of action of these drugs and it appears that short term inhibition of PARP is well tolerated and that PARP inhibition can be achieved at reasonable doses. Perhaps more debatable is the suggestion that the low toxicity of potent PARP inhibitors may enable their prophylactic use in women heterozygous for BRCA mutations (Helleday et al. 2005). There is still no clinical experience regarding the tolerability of long term PARP inhibition. Prolonged inhibition of DNA repair could enhance the mutation rate and the potential for secondary malignancies. In fact, a recent study reveals an important role of PARP-1 in suppressing mammary tumorigenesis in mice and importantly, it suggests that long-term inhibition of PARP activity may bring increased risk for breast cancer formation (Tong et al. 2007). Nevertheless, short periods of prophylactic use may have utility but must obviously be weighed up against the actual risk a BRCA mutation carrier has of developing malignancy (Roukos and Briasoulis 2007).

Very recently an interesting report has settle the basis for the resistance to PARP inhibition acquired by certain mutation on BRCA2 (61). Mutations in BRCA2 and sensitivity to therapeutics BRCA2 mutations carriers can be suppressed by intragenic deletion. Multiple mechanisms have been described for the reversion of mutations in humans, but the size and nature of the deletions they observe are uncommon. Variant of BRCA2 lacking a significant fraction of the protein are competent in mediating PARP resistance and, surprisingly, in HR. The profound sensitivity of BRCA2 deficient cells to PARP inhibitors is determined by their defect in error free repair and deletions may also arise because of this same HR deficiency.

Despite the clinical promise of PARP inhibitors in the treatment of BRCA-related cancer, extending the utility of these agents to other cancers is challenging. Sensitivity to PARP inhibition depends on homologous recombination deficiency and not on inherited BRCA1 or BRCA2 deficiency *per se*. Therefore, this approach may be more widely applicable in the treatment of other impairments of the HR pathway. The identification of novel mediators of cellular response to PARP inhibitors may highlight additional patient populations that might benefit form this therapeutic approach. In this way, it has been used a high-throughput RNAi screen to identify new determinants of sensitivity to a PARP inhibitor (62).

Other Pathways Involved in HR

Loss of PARP-1 has been shown to cause an increase in Rad51 foci and sister chromatid exchanges (Schultz et al. 2003) as result of an increase in the number of lesions normally repaired by HR. The sensitivity of cells deficient in proteins involved in HR to PARP inhibition suggest that treatment with PARP inhibitors maybe a useful therapeutic strategy for tumors displaying defects in the HR pathway other than the BRCAness. The observation that ATM and CHK2 depletion resulted in sensitivity to PARP inhibitors further suggest that PARP inhibition would be beneficial for a wide variety of cancers with dysfunction on genes involved in DNA damage response (33). It has been examined the effects of deficiency of several proteins involved in HR on sensitivity to PARP inhibition and it has been shown that deficiency in RAD51, RAD54, DSS1, RPA1, NBS1, ATR, ATM, CHK1, CHK2, FANCD2, FANCA or FANCC induces such a sensitivity. These results suggest that PARP inhibition might be a useful therapeutic strategy not only for the treatment of BRCA mutations but also for the treatment of a wider range of tumors bearing a variety of deficiencies in the HR pathway.

Particular attention has received the interaction between PARP and ATM. These two proteins, together with DNA-PK, are involved in responding to DNA damage to activate pathways responsible for cellular survival. A defective activation of ATM has been described in PARP-1 deficient cells; on the other hand, ATM null cells display an extraordinary sensitivity to PARP inhibitors (Bryant and Helleday 2006), (65), suggesting that ATM and the collapse in the replication fork induced by the PARP inhibitor, function upstream of HR to repair certain types of DSBs (Bryant and Helleday 2006). PARP-1 appears to be important for HR levels; however, it is not directly involved in this process, and the increased in HR seen in PARP-1 deficient cells is probably due to an accumulation of recombinogenic substrates in these cells. It is important to note that this treatment differs from previous therapies in that no additional genotoxic drug is needed to cause cell death, and in this respect it would be expected the treatment to have few side effects.

PARP INHIBITORS IN BREAST AND OVARIAN CANCER

PARP catalyzes the cleavage of NAD^+ into ADP and ADP-ribose and attaches several molecules of the latter to the target protein in a process called poly(ADP-ribosyl)ation. The molecules that mimic NAD^+ block the binding of the NAD^+ to the enzyme, inhibiting PARP activity. The discovery of inhibitors of PARP was initially based on empirical, high-throughput screening, followed by optimization by chemical modifications based on structure-based design (see (Jagtap and Szabo 2005) for a review) (Figs. 8 and 9). Several studies have already shown the effectiveness of PARP inhibition in combination with either radiotherapy or chemotherapy in a range of human tumor mouse xenograft models (Miknyoczki et al. 2003; Calabrese et al. 2004). PARP Inhibitors are being used in different Clinical Trial in Breast and Ovarian Cancer and a summary is gathered in Table 1.

AG014699 (Pf01367338) Pfizer was the first agent studied clinically in 2003 (Table 1). A study evaluated 39 ovarian cancer cell lines with various molecular expressions either with single-agent AGO14699 or in combination with carboplatin, doxorubicin, gemcitabine, paclitaxel, or topotecan. The combination of AGO 146999 with carboplatin, topotecan, and doxorubicin appears to have greatest impact. The BRCA status of these cell lines was not reported. The study does suggest activity of AGO14699 in ovarian cancer and not confined to BRCA-deficient tumor or high-grade serous cancer. A phase II evaluation of single agent AG014699 in BRCA-deficient advanced breast and ovarian cancers is ongoing.

Olaparib (AZD2281, KU-0059436) AstraZeneca Olaparib is a selective and potent PARP inhibitor with a half maximal inhibitory concentration (IC_{50}) in the nanomolar range for both PARP1 and PARP2 enzymes (IC_{50} of 4.9 nM for PARP 1). Olaparib is a well-tolerated oral PARP inhibitor, which has shown promising monotherapy activity in patients with *BRCA1*

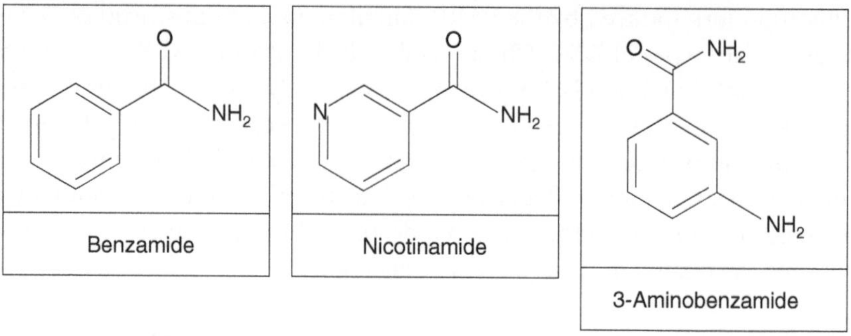

Fig. 8 Inhibitors of PARP. The classical inhibitors are nicotinamide, benzamide, and substituted benzamide, in particular 3-aminobenzamide. [12]

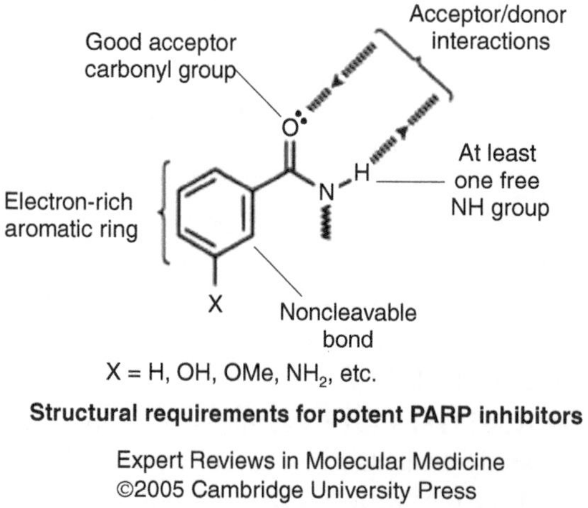

Fig. 9 Structural requirements for potent PARP inhibition. The benzamide pharmacophore is shown, illustrating the desired orientation of the carboxamide moiety for hydrogen bonding with critical residues in the NAD+–binding site of poly(ADP-ribose) polymerase 1 (PARP-1), and the desired unsaturation of the aromatic ring with a noncleavable bond at position 3. NAD+, oxidised nicotinamide adenine dinucleotide.

or *BRCA2* mutations who have breast and ovarian cancer (Tutt et al. 2010) PARP inhibitor to show activity in BRCA-related ovarian and breast cancers. It is been tested in combination with DNA damaging agents, such as topotecan, doxorubicin, carboplatin, carboplatin and paclitaxel, irinotecan, dacarbazine, and gemcitabine and cisplatin, as well as with antiangiogenesis agents (Weil and Chen 2011). Two parallel open-label, multicenter, phase 2 studies of olaparib in germline *BRCA1/2* mutation carriers with advanced breast (ICEBERG1) and ovarian (ICEBERG2)

Table 1 PARP Inhibitor. Clinical Trial in Breast and Ovarian Cancer.

Agent	*Company*	*Single/Combination therapy*	*Disease*	*Clinical status*	
AG014699	Pfizer [New York, NY]	Single agent/Combination with carboplatin, doxorubicin, gemcitabine, paclitaxel, or topotecan	Ovarian cancer Breast cancer	Phase 2	Ihnen 2010 Wiel 2010 (Weil and Chen) www.clinicaltrials.gov
Olaparip (KU59436)	AstraZeneca/Ku DOS [London, United Kingdom]	Single	Advanced solid tumors Women with advanced breast cancer	Phase 1 Phase 2	Fong et al. 2009 http://clinicaltrials.gov
(AZD2281)			measurable BRCA1- or BRCA2- advanced ovarian cancer TNBC	Phase 2 Phase 2	Tutt 2010 (Tutt et al.) Gelmorn 2010 Dent 2010
Veliparib (ABT 888)	Abbott	With cyclophosphamide	ovarian cancer breast cancer	Phase 2	Kummar 2010 Tan 2010
		With temozolomide	metastasic breast cancer	Phase 2	Isakoff 2010

Table 1 Contd. ...

Table 1 Contd. ...

			Ovarian cancer	Phase 1 complete; phase 2 ongoing	http:www.biparsciences.com
Iniparib (BSI 201)	BiPar (Brisbane, CA, USA)	Single Advanced	Breast	Phase 3	O'Shaughnessy 2010
			TNBC		
MK4827	Merck	Single agent	Ovarian cancer	Phase 1	Sandhu 2010 http://clinicaltrials.gov
INO-1001	Inotek/Genentech [Beverly, MA]	Combination with doxorubicin	p53-deficient breast cancer	Preclinical	Mason et al. 2008
BSI-201 and Temozolomide	Bipar sciences	Combination with temozolomide (TMZ)	Malignant glioma	Phase I and II	http://clinicaltrials.gov
ABT-888 and Temozolomide	Massachusetts General Hospital	Combination of ABT-888 and temozolomide is a safe and effective treatment.	Metastasic Breast cancer	Phase II	http://clinicaltrials.gov
ABT-888	Warren Grant Magnuson Clinical Center	Combined cyclophosphamide in treating patients who did not respond to previous therapy.	Solid tumors, leukemia, lymphoma	Phase I	http://clinicaltrials.gov
AZD2281 and Carboplatin	Warren Grant Magnuson Clinical Center	Combination with Carboplatin.	Breast cancer or ovarian epithelial cancer	Phase I	http://clinicaltrials.gov

cancers have provided further support for this antitumor synthetic lethal approach (Table 1)(Yap et al. 2011) .

Veliparib (ABT-888). Veliparib is an orally active PARP-1 and PARP-2 inhibitor (Donawho et al. 2007) (Table 1). A phase I study of veliparib in combination with metronomic cyclophosphamide in two patients with BRCA2 ovarian cancer achieved partial response. A phase II randomized trial evaluating the role of veliparib in combination with oral cyclophosphamide in ovarian cancer patients with BRCA mutation or high-grade serous ovarian cancer will be activated in the near future. Also veliparib in combination with TMZ was evaluated in metastatic breast cancer. Forty-one patients were treated (Weil and Chen 2011).

Iniparib (BSI 201, NSC-746045; IND-71677) sanofi-aventis Iniparib, also known as BSI 201, 4-iodo-3-nitrobenzamide, is an irreversible inhibitor of PARP1. It is a prodrug with a 4-minute half-life. Iniparib is given intravenously twice a week. It is the first PARP inhibitor to show survival advantage with TNBC and has entered phase III testing. The phase I study included 23 patients with solid tumors. Six of the 23 heavily pretreated patients had stable disease for at least 2 months (up to over 9 months in 1 patient). In another study patients with solid tumors were assigned to 1 of 4 combinations of iniparib with a cytotoxic agent, topotecan, gemcitabine, TMZ, or carboplatin with taxol.

MK4827 MK4827 is an oral PARP-1 and 2 inhibitor with IC50 of 3.8 nM for PARP 1. Preclinical data showed single-agent antitumor activity against BRCA mutant cell lines in culture and xenograft models. Also, MK4827 showed activity in combination with DNA damaging agents in cell cultures and xenograft models. It is currently being tested in phase 1 development as a single agent, in advanced ovarian tumors, and as combination therapy in patients with advanced solid tumors in combination with carboplatin, carboplatin with paclitaxel, and carboplatin with liposomal doxorubicin. MK4827 in Ovarian Cancer is in the phase I trial as a single agent enriched with BRCA1 or -2 mutation patients. An expansion cohort of high-grade serous ovarian cancer patients were added at the maximum toleranted dose (MTD). Responses were seen in all dose levels. The expansion cohort in high-grade serous ovarian cancer is ongoing. Once the biologically relevant dose is achieved, a phase Ib expansion in BRCA-mutant ovarian cancer patients is planned.

Currently there is a significative development in the PARP inhibitors; these agents are a promise in the cancer therapy, specifically the BRCA-associated tumorigenesis. However in the early stages of tumorigenesis, BRCA-mutant cells are sensitive to PARP inhibition. The problem is in the second stage of tumorigenesis when the cell acquired resistance to this inhibitor. Clinical disorders have been found to have the potencial to revert inherited mutation to a normal stated. A mutational event that converts BRCA2 mutation status to wild-type also converts the cells to being resistant to platinium agents and PARP inhibitor because there are a restoration of HR pathway (Weil and Chen 2011) BRCA deficient tumor cells may become insensitive to PARP inhibition owing to additional mutations. Thus, the use of this strategy for therapeutic treatment of hereditary breast cancers is dependent on the continued susceptibility of BRCA mutant cells to PARP inhibitors (De Soto and Deng 2006) .

The predictive Biomarker of Defective HR DNA Repair is very important in the therapeutic utility of PARP inhibitor. The identification of subgroup of patient with non-BRCA mutation-related HR repair-deficient tumor is possible with Array-Bassed Assay. The array-based strategies such gene expression profiling (Sorlie et al. 2003; Jazaeri et al. 2002), or comparative genomic hybridization (CGH) have been evaluated to define distinct BRCA1/2 – related profiled in ovarian and breast cancer (van Beers et al. 2005; Joosse et al. 2009; Yap et al. 2011).

PERSPECTIVES

Pharmacological inhibition of key proteins involved in the response to DNA damage has emerged as an effective tool for cancer treatment, as the resistance of cancer cells to DNA-damaging agents originates from the modulation of DNA repair pathways. PARP has important pro-survival and protective functions in terms of DNA repair. A multitude of novel pharmacological inhibitors of PARP has entered clinical testing either as adjunct antitumor therapeutics or as monotherapy in familiar breast and ovarian cancer. Besides the antineoplasic action of PARP inhibitors, the evidence summarized above strongly supports a crucial role of the ROS/NF-κB/PARP pathway in mediating multiple inflammatory diseases including type I diabetes and associated endothelial dysfunction, cardiovascular injury and heart failure, sepsis, rheumatoid arthritis and stroke. The information available to date supports the view that PARP over-activation during inflammatory damage is a pivotal feature of tissue damage in different inflammatory-based pathologies, and that the pharmacological inhibition of PARP may provide significant benefits in these conditions by salvaging affected tissues from necrosis, and by reducing, as well as by downregulating, the inflammatory responses. As with all new therapeutic areas, the usefulness of this target is unproven, but the potential effect of this class of agents is large. Whereas the clinical benefit of PARP inhibitors is being tested, additional new areas of research are also opening up in the preclinical front, which should eventually help in the consecution of new and more effective PARP inhibitors able to regulate the fine tuning of this pathway, including the specific inhibition of different PARP members, the association of PARP-1 in different complexes, and the activation/inactivation of poly(ADP-glycohydrolase).

ACKNOWLEDGEMENTS

This work has been supported by Grants PO7-CTS-0239 (Junta de Andalucía) to FJO; This work was supported by Ministerio de Ciencia e Innovación SAF2006-01094; SAF2009-13281-C02-01, Fundación La Caixa BM06-219-0; and Junta de Andalucía P07-CTS-0239 to and CTS-6602 FJO.

REFERENCES

Aguilar-Quesada, R., J.A. Munoz-Gamez, et al. 2007. Interaction between ATM and PARP-1 in response to DNA damage and sensitization of ATM deficient cells through PARP inhibition. BMC Mol Biol 8: 29.

Akiyama, T., S. Takasawa, et al. 2001. Activation of Reg gene, a gene for insulin-producing beta-cell regeneration: poly(ADP-ribose) polymerase binds Reg promoter and regulates the transcription by autopoly(ADP-ribosyl)ation. Proc Natl Acad Sci U S A 98: 48-53.

Alon, U. 2003. Biological networks: the tinkerer as an engineer. Science 301: 1866-1867.

Altmeyer, M. and M.O. Hottiger. 2009. Poly(ADP-ribose) polymerase 1 at the crossroad of metabolic stress and inflammation in aging. Aging (Albany NY) 1: 458-469.

Amir, E., B. Seruga, et al. 2010. Targeting DNA repair in breast cancer: a clinical and translational update. Cancer Treat Rev 36: 557-565.

Aylon, Y., B. Liefshitz, et al. 2003. Molecular dissection of mitotic recombination in the yeast *Saccharomyces cerevisiae*. Mol Cell Biol 23: 1403-1417.

Besson, V.C., N. Croci, et al. 2003. Deleterious poly(ADP-ribose) polymerase-1 pathway activation in traumatic brain injury in rat. Brain Res 989: 58-66.

Besson, V.C., Z. Zsengeller, et al. 2005. Beneficial effects of PJ34 and INO-1001, two novel water-soluble poly(ADP-ribose) polymerase inhibitors, on the consequences of traumatic brain injury in rat. Brain Res 1041: 149-156.

Blum, D., S. Torch, et al. 2001. Molecular pathways involved in the neurotoxicity of 6-OHDA, dopamine and MPTP: contribution to the apoptotic theory in Parkinson's disease. Prog Neurobiol 65: 135-172.

Bryant, H.E. and T. Helleday. 2006. Inhibition of poly(ADP-ribose) polymerase activates ATM which is required for subsequent homologous recombination repair. Nucleic Acids Res 34: 1685-1691.

Bryant, H.E., E. Petermann, et al. 2009. PARP is activated at stalled forks to mediate Mre11-dependent replication restart and recombination. Embo J 28: 2601-2615.

Bryant, H.E., N. Schultz, et al. 2005. Specific killing of BRCA2-deficient tumours with inhibitors of poly(ADP-ribose) polymerase. Nature 434: 913-917.

Burkle, A. 2001. PARP-1: a regulator of genomic stability linked with mammalian longevity. Chembiochem 2: 725-728.

Burkle, A. 2001. Physiology and pathophysiology of poly(ADP-ribosyl)ation. Bioessays 23: 795-806.

Calabrese, C.R., R. Almassy, et al. 2004. Anticancer chemosensitization and radiosensitization by the novel poly(ADP-ribose) polymerase-1 inhibitor AG14361. J Natl Cancer Inst 96: 56-67.

Canto, C. and J. Auwerx. 2011. Interference between PARPs and SIRT1: a novel approach to healthy ageing? Aging (Albany NY) 3: 543-547.

Clark, R.S., V.A. Vagni, et al. 2007. Local administration of the poly(ADP-ribose) polymerase inhibitor INO-1001 prevents NAD+ depletion and improves water maze performance after traumatic brain injury in mice. J Neurotrauma 24: 1399-1405.

Csiszar, A., M. Wang, et al. 2008. Inflammation and endothelial dysfunction during aging: role of NF-kappaB. J Appl Physiol 105: 1333-1341.

Chung, H.Y., B. Sung, et al. 2006. The molecular inflammatory process in aging. Antioxid Redox Signal 8: 572-581.

Davidson, J.M., K.L. Gorringe, et al. 2000. Molecular cytogenetic analysis of breast cancer cell lines. Br J Cancer 83: 1309-1317.

de Murcia, J.M., C. Niedergang, et al. 1997. Requirement of poly(ADP-ribose) polymerase in recovery from DNA damage in mice and in cells. Proc Natl Acad Sci U S A 94: 7303-7307.

De Soto, J.A. and C.X. Deng 2006. PARP-1 inhibitors: are they the long-sought genetically specific drugs for BRCA1/2-associated breast cancers? Int J Med Sci 3: 117-123.

De Soto, J.A., X. Wang, et al. 2006. The Inhibition and Treatment of Breast Cancer with Poly(ADP-ribose) Polymerase (PARP-1) Inhibitors. Int J Biol Sci 2: 179-185.

Donawho, C.K., Y. Luo, et al. 2007. ABT-888, an orally active poly(ADP-ribose) polymerase inhibitor that potentiates DNA-damaging agents in preclinical tumor models. Clin Cancer Res 13: 2728-2737.

Drake, J.W., B. Charlesworth, et al. 1998. Rates of spontaneous mutation. Genetics 148: 1667-1686.

Endres, M., G. Scott, et al. 1998. Role of peroxynitrite and neuronal nitric oxide synthase in the activation of poly(ADP-ribose) synthetase in a murine model of cerebral ischemia-reperfusion. Neurosci Lett 248: 41-44.

Erkko, H., B. Xia, et al. 2007. A recurrent mutation in PALB2 in Finnish cancer families. Nature 446: 316-319.

Faro, R., Y. Toyoda, et al. 2002. Myocardial protection by PJ34, a novel potent poly(ADP-ribose) synthetase inhibitor. Ann Thorac Surg 73: 575-581.

Fong, P.C., D.S. Boss, et al. 2009. Inhibition of poly(ADP-ribose) polymerase in tumors from BRCA mutation carriers. N Engl J Med 361: 123-134.

Forozan, F., E.H. Mahlamaki, et al. 2000. Comparative genomic hybridization analysis of 38 breast cancer cell lines: a basis for interpreting complementary DNA microarray data. Cancer Res 60: 4519-4525.

Gagne, J.P., M. Isabelle, et al. 2008. Proteome-wide identification of poly(ADP-ribose) binding proteins and poly(ADP-ribose)-associated protein complexes. Nucleic Acids Res 36: 6959-6976.

Gallmeier, E. and S.E. Kern. 2005. Absence of specific cell killing of the BRCA2-deficient human cancer cell line CAPAN1 by poly(ADP-ribose) polymerase inhibition. Cancer Biol Ther 4: 703-706.

Grube, K. and A. Burkle. 1992. Poly(ADP-ribose) polymerase activity in mononuclear leukocytes of 13 mammalian species correlates with species-specific life span. Proc Natl Acad Sci U S A 89: 11759-11763.

Gudmundsdottir, K. and A. Ashworth. 2006. The roles of BRCA1 and BRCA2 and associated proteins in the maintenance of genomic stability. Oncogene 25: 5864-5874.

Haaf, T., E.I. Golub, et al. 1995. Nuclear foci of mammalian Rad51 recombination protein in somatic cells after DNA damage and its localization in synaptonemal complexes. Proc Natl Acad Sci U S A 92: 2298-2302.

Haince, J.F., D. McDonald, et al. 2008. PARP1-dependent kinetics of recruitment of MRE11 and NBS1 proteins to multiple DNA damage sites. J Biol Chem 283: 1197-1208.

Hamby, A.M., S.W. Suh, et al. 2007. Use of a poly(ADP-ribose) polymerase inhibitor to suppress inflammation and neuronal death after cerebral ischemia-reperfusion. Stroke 38: 632-636.

Heikkinen, K., K. Rapakko, et al. 2006. RAD50 and NBS1 are breast cancer susceptibility genes associated with genomic instability. Carcinogenesis 27: 1593-1599.

Helleday, T., H.E. Bryant, et al. 2005. Poly(ADP-ribose) polymerase (PARP-1) in homologous recombination and as a target for cancer therapy. Cell Cycle 4: 1176-1178.

Holloman, W.K. 2011. Unraveling the mechanism of BRCA2 in homologous recombination. Nat Struct Mol Biol 18: 748-754.

Infante, J., P. Sanchez-Juan, et al. 2007. Poly(ADP-ribose) polymerase-1 (PARP-1) genetic variants are protective against Parkinson's disease. J Neurol Sci 256: 68-70.

Jagtap, P. and C. Szabo. 2005. Poly(ADP-ribose) polymerase and the therapeutic effects of its inhibitors. Nat Rev Drug Discov 4: 421-440.

Jazaeri, A.A., C.J. Yee, et al. 2002. Gene expression profiles of BRCA1-linked, BRCA2-linked, and sporadic ovarian cancers. J Natl Cancer Inst 94: 990-1000.

Joosse, S.A., E.H. van Beers, et al. 2009. Prediction of BRCA1-association in hereditary non-BRCA1/2 breast carcinomas with array-CGH. Breast Cancer Res Treat 116: 479-489.

Kameoka, M., S. Nukuzuma, et al. 2005. Poly(ADP-ribose) polymerase-1 is required for integration of the human immunodeficiency virus type 1 genome near centromeric alphoid DNA in human and murine cells. Biochem Biophys Res Commun 334: 412-417.

Khanna, K.K., M.F. Lavin, et al. 2001. ATM, a central controller of cellular responses to DNA damage. Cell Death Differ 8: 1052-1065.

Kim, M.Y., T. Zhang, et al. 2005. Poly(ADP-ribosyl)ation by PARP-1: 'PAR-laying' NAD+ into a nuclear signal. Genes Dev 19: 1951-1967.

Kolodner, R.D., C.D. Putnam, et al. 2002. Maintenance of genome stability in *Saccharomyces cerevisiae*. Science 297: 552-557.

Langelier, M.F., D.D. Ruhl, et al. 2011. The Zn3 domain of human poly(ADP-ribose) polymerase-1 (PARP-1) functions in both DNA-dependent poly(ADP-ribose) synthesis activity and chromatin compaction. J Biol Chem 285: 18877-18887.

LaPlaca, M.C., R. Raghupathi, et al. 1999. Temporal patterns of poly(ADP-ribose) polymerase activation in the cortex following experimental brain injury in the rat. J Neurochem 73: 205-213.

Li, X. and W.D. Heyer. 2008. Homologous recombination in DNA repair and DNA damage tolerance. Cell Res 18: 99-113.

Liaw, D., D.J. Marsh, et al. 1997. Germline mutations of the PTEN gene in Cowden disease, an inherited breast and thyroid cancer syndrome. Nat Genet 16: 64-67.

Lieberman, H.B., J.D. Bernstock, et al. 2011. The role of RAD9 in tumorigenesis. J Mol Cell Biol 3: 39-43.

Lindahl, T., M.S. Satoh, et al. 1995. Post-translational modification of poly(ADP-ribose) polymerase induced by DNA strand breaks. Trends Biochem Sci 20: 405-411.

Loveday, R.L., J. Greenman, et al. 2000. Genetic changes in breast cancer detected by comparative genomic hybridisation. Int J Cancer 86: 494-500.

Malkin, D., F.P. Li, et al. 1990. Germ line p53 mutations in a familial syndrome of breast cancer, sarcomas, and other neoplasms. Science 250: 1233-1238.

Martin-Oliva, D., R. Aguilar-Quesada, et al. 2006. Inhibition of poly(ADP-ribose) polymerase modulates tumor-related gene expression, including hypoxia-inducible factor-1 activation, during skin carcinogenesis. Cancer Res 66: 5744-5756.

Mason, K. A., D. Valdecanas, et al. 2008. INO-1001, a novel inhibitor of poly(ADP-ribose) polymerase, enhances tumor response to doxorubicin. Invest New Drugs 26: 1-5.

McCabe, N., C.J. Lord, et al. 2005. BRCA2-deficient CAPAN-1 cells are extremely sensitive to the inhibition of Poly(ADP-Ribose) polymerase: an issue of potency. Cancer Biol Ther 4: 934-936.

Meijers-Heijboer, H., A. van den Ouweland, et al. 2002. Low-penetrance susceptibility to breast cancer due to CHEK2(*)1100delC in noncarriers of BRCA1 or BRCA2 mutations. Nat Genet 31: 55-59.

Menear, K.A., C. Adcock, et al. 2008. 4-[3-(4-Cyclopropanecarbonylpiperazine-1-carbonyl)-4-fluorobenzyl]-2H-phth alazin-1-one: A Novel Bioavailable Inhibitor of Poly(ADP-ribose) Polymerase-1. J Med Chem.

Mesenge, C., C. Verrecchia, et al. 1996. Reduction of the neurological deficit in mice with traumatic brain injury by nitric oxide synthase inhibitors. J Neurotrauma 13: 11-16.

Miknyoczki, S.J., S. Jones-Bolin, et al. 2003. Chemopotentiation of temozolomide, irinotecan, and cisplatin activity by CEP-6800, a poly(ADP-ribose) polymerase inhibitor. Mol Cancer Ther 2: 371-382.

Molnar, A., A. Toth, et al. 2006. Activation of the poly(ADP-ribose) polymerase pathway in human heart failure. Mol Med 12: 143-152.

Morrison, C., E. Sonoda, et al. 2000. The controlling role of ATM in homologous recombinational repair of DNA damage. Embo J 19: 463-471.

Mortusewicz, O., J.C. Ame, et al. 2007. Feedback-regulated poly(ADP-ribosyl)ation by PARP-1 is required for rapid response to DNA damage in living cells. Nucleic Acids Res 35: 7665-7675.

Munoz-Gamez, J.A., J.M. Rodriguez-Vargas, et al. 2009. PARP-1 is involved in autophagy induced by DNA damage. Autophagy 5: 61-74.

Murphy, M.P. and L. Partridge. 2008. Toward a control theory analysis of aging. Annu Rev Biochem 77: 777-798.

Muthumani, K., A.Y. Choo, et al. 2006. The HIV-1 Vpr and glucocorticoid receptor complex is a gain-of-function interaction that prevents the nuclear localization of PARP-1. Nat Cell Biol 8: 170-179.

Oliver, F.J., J. Menissier-de Murcia, et al. 1999. Resistance to endotoxic shock as a consequence of defective NF-kappaB activation in poly(ADP-ribose) polymerase-1 deficient mice. Embo J 18: 4446-4454.

Outeiro, T. F., T.N. Grammatopoulos, et al. 2007. Pharmacological inhibition of PARP-1 reduces alpha-synuclein- and MPP+-induced cytotoxicity in Parkinson's disease in vitro models. Biochem Biophys Res Commun 357: 596-602.

Pacher, P., J.S. Beckman, et al. 2007. Nitric oxide and peroxynitrite in health and disease. Physiol Rev 87: 315-424.

Pacher, P., L. Liaudet, et al. 2002. The role of poly(ADP-ribose) polymerase activation in the development of myocardial and endothelial dysfunction in diabetes. Diabetes 51: 514-521.

Pacher, P. and C. Szabo. 2006. Role of peroxynitrite in the pathogenesis of cardiovascular complications of diabetes. Curr Opin Pharmacol 6: 136-141.

Pacher, P. and C. Szabo. 2008. Role of the peroxynitrite-poly(ADP-ribose) polymerase pathway in human disease. Am J Pathol 173: 2-13.

Pyriochou, A., G. Olah, et al. 2008. Inhibition of angiogenesis by the poly(ADP-ribose) polymerase inhibitor PJ-34. Int J Mol Med 22: 113-118.

Ratnam, K. and J.A. Low. 2007. Current development of clinical inhibitors of poly(ADP-ribose) polymerase in oncology. Clin Cancer Res 13: 1383-1388.

Renwick, A., D. Thompson, et al. 2006. ATM mutations that cause ataxia-telangiectasia are breast cancer susceptibility alleles. Nat Genet 38: 873-875.

Riballo, E., M. Kuhne, et al. 2004. A pathway of double-strand break rejoining dependent upon ATM, Artemis, and proteins locating to gamma-H2AX foci. Mol Cell 16: 715-724.

Rodriguez, M.I., A. Gonzalez-Flores, et al. 2011. Poly(ADP-ribose)-dependent regulation of Snail1 protein stability. Oncogene.

Romanowicz-Makowska, H., B. Smolarz, et al. 2011. Single nucleotide polymorphisms in the homologous recombination repair genes and breast cancer risk in polish women. Tohoku J Exp Med 224: 201-208.

Roukos, D.H. and E. Briasoulis. 2007. Individualized preventive and therapeutic management of hereditary breast ovarian cancer syndrome. Nat Clin Pract Oncol 4: 578-590.

Sandhu, S.K., T.A. Yap, et al. 2010. Poly(ADP-ribose) polymerase inhibitors in cancer treatment: a clinical perspective. Eur J Cancer 46: 9-20.

Satchell, M.A., X. Zhang, et al. 2003. A dual role for poly-ADP-ribosylation in spatial memory acquisition after traumatic brain injury in mice involving NAD+ depletion and ribosylation of 14-3-3gamma. J Neurochem 85: 697-708.

Scully, R. and N. Puget. 2002. BRCA1 and BRCA2 in hereditary breast cancer. Biochimie 84: 95-102.

Scully, R., A. Xie, et al. 2004. Molecular functions of BRCA1 in the DNA damage response. Cancer Biol Ther 3: 521-527.

Schultz, N., E. Lopez, et al. 2003. Poly(ADP-ribose) polymerase (PARP-1) has a controlling role in homologous recombination. Nucleic Acids Res 31: 4959-4964.

Seal, S., D. Thompson, et al. 2006. Truncating mutations in the Fanconi anemia J gene BRIP1 are low-penetrance breast cancer susceptibility alleles. Nat Genet 38: 1239-1241.

Simbulan-Rosenthal, C.M., D.S. Rosenthal, et al. 1999. Poly(ADP-ribose) polymerase upregulates E2F-1 promoter activity and DNA pol alpha expression during early S phase. Oncogene 18: 5015-5023.

Somyajit, K., S. Subramanya, et al. 2010. RAD51C: a novel cancer susceptibility gene is linked to Fanconi anemia and breast cancer. Carcinogenesis 31: 2031-2038.

Soriano, F.G., L. Liaudet, et al. 2002. Resistance to acute septic peritonitis in poly(ADP-ribose) polymerase-1-deficient mice. Shock 17: 286-292.

Sorlie, T., R. Tibshirani, et al. 2003. Repeated observation of breast tumor subtypes in independent gene expression data sets. Proc Natl Acad Sci U S A 100: 8418-8423.

Steffen, J., R. Varon, et al. 2004. Increased cancer risk of heterozygotes with NBS1 germline mutations in Poland. Int J Cancer 111: 67-71.

Stevens, M.J., F. Li, et al. 2007. Nicotinamide reverses neurological and neurovascular deficits in streptozotocin diabetic rats. J Pharmacol Exp Ther 320: 458-464.

Sung, Y.J. and R.T. Ambron. 2004. Poly(ADP-ribose) polymerase-1 (PARP-1) and the evolution of learning and memory. Bioessays 26: 1268-1271.

Szabo, C. and V.L. Dawson. 1998. Role of poly(ADP-ribose) synthetase in inflammation and ischaemia-reperfusion. Trends Pharmacol Sci 19: 287-298.

Tashiro, S., J. Walter, et al. 2000. Rad51 accumulation at sites of DNA damage and in postreplicative chromatin. J Cell Biol 150: 283-291.

Tentori, L., P.M. Lacal, et al. 2007. Poly(ADP-ribose) polymerase (PARP) inhibition or PARP-1 gene deletion reduces angiogenesis. Eur J Cancer 43: 2124-2133.

Thorslund, T. and S.C. West. 2007. BRCA2: a universal recombinase regulator. Oncogene 26: 7720-7730.

Tong, W.M., Y.G. Yang, et al. 2007. Poly(ADP-ribose) polymerase-1 plays a role in suppressing mammary tumourigenesis in mice. Oncogene 26: 3857-3867.

Trenz, K., E. Smith, et al. 2006. ATM and ATR promote Mre11 dependent restart of collapsed replication forks and prevent accumulation of DNA breaks. Embo J 25: 1764-1774.

Turner, N., A. Tutt, et al. 2005. Targeting the DNA repair defect of BRCA tumours. Curr Opin Pharmacol 5: 388-393.

Tutt, A. and A. Ashworth. 2002. The relationship between the roles of BRCA genes in DNA repair and cancer predisposition. Trends Mol Med 8: 571-576.

Tutt, A., M. Robson, et al. 2010. Oral poly(ADP-ribose) polymerase inhibitor olaparib in patients with BRCA1 or BRCA2 mutations and advanced breast cancer: a proof-of-concept trial. Lancet 376: 235-244.

van Beers, E.H., T. van Welsem, et al. 2005. Comparative genomic hybridization profiles in human BRCA1 and BRCA2 breast tumors highlight differential sets of genomic aberrations. Cancer Res 65: 822-827.

Virag, L. and C. Szabo. 2002. The therapeutic potential of poly(ADP-ribose) polymerase inhibitors. Pharmacol Rev 54: 375-429.

Waldman, A.S. and B.C. Waldman. 1991. Stimulation of intrachromosomal homologous recombination in mammalian cells by an inhibitor of poly(ADP-ribosylation). Nucleic Acids Res 19: 5943-5947.

Walsh, T. and M.C. King. 2007. Ten genes for inherited breast cancer. Cancer Cell 11: 103-105.

Wang, Z.Q., L. Stingl, et al. 1997. PARP is important for genomic stability but dispensable in apoptosis. Genes Dev 11: 2347-2358.

Weil, M.K. and A.P. Chen. 2011. PARP inhibitor treatment in ovarian and breast cancer. Curr Probl Cancer 35: 7-50.

Whatcott, C.J., M.L. Meyer-Ficca, et al. 2009. A specific isoform of poly(ADP-ribose) glycohydrolase is targeted to the mitochondrial matrix by a N-terminal mitochondrial targeting sequence. Exp Cell Res 315: 3477-3485.

Yap, T.A., S.K. Sandhu, et al. 2011. Poly(ADP-ribose) polymerase (PARP) inhibitors: Exploiting a synthetic lethal strategy in the clinic. CA Cancer J Clin 61: 31-49.

CHAPTER 10

Statins in Prostate Cancer: Prevention and Treatment

Khaled Aziz and Phuoc T. Tran*

Department of Radiation Oncology and Molecular Radiation Sciences, Sidney Kimmel Comprehensive Cancer Center, Johns Hopkins School of Medicine, 1550 Orleans Street, CRB2, Rm 406, Baltimore, MD, 21231, USA

ABSTRACT

Prostate cancer is the most common noncutaneous cancer diagnosed in men and the second leading cause of cancer related deaths in the US. The progressive steps of prostate cancer development from a pre-neoplastic stage to invasive adenocarcinoma has been intensively studied and offers opportunities for earlier preventive interventions as well as may divulge more ideal targets for the treatment of malignant disease. Some efforts using hormonal manipulations and dietary supplements have been tested directly for efficacy in prostate cancer prevention trials with mixed results. HMG-CoA reductase inhibitors (statins) have been observed to confer benefits for prostate cancer prevention and possess therapeutic potential in retrospective studies. The mechanism of action of this promising class of drugs has been the subject of several preclinical studies over the past decade. In this chapter we highlight the current status of knowledge in prostate cancer prevention, detail the anti-cancer properties of statins including the known mechanisms of action, and discuss the role of MYC as a predictive biomarker to identify patients who may be expected to respond to statin therapy.

INTRODUCTION

Prostate cancer is the second leading cause for cancer related deaths for men in the United States with 241,740 new cases diagnosed in 2011

**Corresponding author*: E-mail: tranp@jhmi.edu

(Siegel et al. 2012). Prostate cancer specific mortality is generally very low for patients with localized low-risk disease following local therapies such as radical prostatectomy or radiation. The risk is significantly greater for intermediate-risk and high-risk, localized prostate cancer patients and essentially universal for metastatic patients. Given the frequency and latency of prostate cancer development an alternative approach to treatment are preventative strategies.

PROSTATE CANCER DEVELOPMENT

Prostate cancer development and progression of the disease has been intensively studied. Understanding this multistep process of prostate tumorigenesis may aid in identifying critical time points or targets for prevention. Prostatic intraepithelial neoplasia (PIN) is a histologic entity viewed as an early precursor to prostatic malignancy (Park et al. 2001; Bishara et al. 2004; Akhavan et al. 2007). PIN is observed within the lining of prostatic ducts and acini as progressive abnormalities of prostatic cellular differentiation and proliferation characterized by neoplastic cells with cytologic atypia, nuclear irregularity and prominent nucleoli (Bostwick and Cheng 2012). The appearance of aneuploidy and angiogenesis increases with progressing grade of PIN and supports PIN as a precancerous lesion (Alcaraz et al. 2001; Gonzalez-Moreno et al. 2010). Apoptotic and proliferative indices in PIN also resemble prostate cancer in comparison to benign hyperplastic prostate tissue (Zeng et al. 2004). The normal orientation of epithelial proliferation from the basal cell compartment to the luminal surface is inverted in PIN. In addition, disruption of the basal cell layer and increased chromatin density correlates with increasing aggressiveness of PIN (Montironi et al. 2007).

PIN does not cause an elevation in prostate-specific antigen (PSA) but is detectable by biopsy. Diagnosis of PIN may warrant repeated biopsy to exclude under-sampled cancer depending on the clinical context. On average there is a 9% prevalence of PIN at needle biopsies performed at screening in patients without prostate cancer (Steiner 2001; Bostwick and Cheng 2012). The incidence varies within a population, increasing in prevalence and volume with age (Sakr et al. 1994). Racial origin and geographic location also influence the incidence of PIN (Tan et al. 2006).

Treatment of patients diagnosed with PIN using finasteride, a 5α-reductase inhibitor, does not alter the morphological characteristics of PIN (Yang et al. 1999; Alberts et al. 2006). However, leuprolide (GnRH analog) and flutamide (anti-androgen) treatment results in 50% reduction in detection of PIN at biopsy (Steiner and Raghow 2003).

PROSTATE CANCER PREVENTION

Prevention strategies are an attractive option in prostate cancer disease management since surgical resection and radiation therapy can have adverse effects on quality of life. Prevention measures that serve to slow disease development and progression can also be useful since active surveillance, restricted typically to older men with low grade and low volume disease, is becoming an increasingly accepted approach for these prostate cancer patients (Cooperberg et al. 2007).

Hormonal Interventions

Two large randomized prospective clinical trials have studied the utility of 5α-reductase inhibitors (5-ARI) as a cancer prevention strategy. 5α-reductase inhibitors have been indicated for male patterned baldness and benign prostatic hyperplasia. They work by suppressing the conversion of testosterone to dihydrotestosterone, a potent androgen with key roles in the development and growth of prostate cells.

The Prostate Cancer Prevention Trial (PCPT) studied the 5-ARI finasteride (Proscar) and a similar study, the Reduction by Dutasteride of Prostate Cancer Events (REDUCE) studied dutasteride (Avodart) for prostate cancer prevention (Thompson et al. 2003; Andriole et al. 2010). Taking these studies together, the results show an approximately 25% reduction in prostate cancer diagnosis. However, this is mainly due to the decreased incidence of low grade prostate cancer (Gleason score =<6). Surprisingly there was an absolute increase in the detection of high-grade prostate cancer in the prevention group.

It was originally postulated that 5α-reductase inhibitors reduce serum levels of PSA and prostate volume leading to increased biased detection of high-grade disease. However, the fact that 56% of prostate cancer in the PCPT trial and 90% in the REDUCE trial were diagnosed by scheduled biopsies argued against the bias promoted by PSA levels. Also, it was suggested that increased sampling density may have caused core needle biopsies to uncover more cancer in smaller prostates. Re-analysis of the data by adjusting for the reduced prostate volume did not reveal that the increased sampling density was accountable for the increased detection of high-grade prostate cancer. Thus the reduction in prostate cancer risk for both drugs was limited to low grade disease with Gleason scores of 6 or lower. These findings ultimately lead to the failure of either drug to get a second indication from the Oncologic

Drugs Advisory Committee of the US Food and Drug Administration as chemopreventative agents for prostate cancer.

In the REDUCE trial 80% of the tumors fell under the category of "extremely low grade disease" therefore questioning the clinical utility of 5α-reductase inhibitors. The Oncologic Drugs Advisory Committee and the FDA agreed with the assessment that finasteride and dutasteride does not confer a significant risk-benefit profile for the proposed use as a chemopreventative agent for prostate cancer in healthy men. Three to four low grade disease can be prevented for the acceptance of one high-grade disease patient (Theoret et al. 2011). There has been debate on the issue with researchers pointing out that patients diagnosed with low grade disease still choose for intervention that can potentially lead to further complications (Kim et al. 2011).

The association between use of 5α-reductase inhibitors and low-risk prostate cancer has been studied in retrospective analyses. Finelli et al. reported a decrease in progression and delay of time to progression for patients taking 5α-reductase inhibitors (Finelli et al. 2011). In the contrary, a recent study by Ross et al. has reported no significant decrease in grade re-classification of biopsies in men with very low-risk prostate cancer. However, it does suggest that men with very low-risk disease are unlikely to be at increased risk for the development of high-grade disease due to androgen suppression (Ross et al. 2012).

Dietary Interventions

Dietary supplements are particularly attractive in prostate cancer prevention considering the slow pace of tumor development and progression of disease as well as the prevalence of the disease in western countries with unique diets and lifestyles. In addition, these non-pharmacological means of prostate cancer prevention are associated with less cost and perceived toxicity. Several agents such as vitamin E, selenium and soy have been tested for efficacy in large scale, population based randomized trials.

The utilization of micronutrients with antioxidant properties have long been implicated in cancer prevention. The Alpha tocopherol Beta Carotene trial (ATBC), a lung cancer prevention study in smokers had previously reported that vitamin E had no effect on lung cancer risk while beta carotene increased the risk (Albanes et al. 1995). Surprisingly, there was a reduced risk of prostate cancer in the vitamin E group which led to further studies on the subject. Interestingly, in vitro studies suggest Vitamin E as a potential chemotherapeutic agent that can cause oxidative DNA damage and induce cell death in a caspase dependent or independent manner in prostate cancer cells. The cellular microenvironment under consideration and structure of the isoform of vitamin E used in this study by Constantinou et al. determined the effect (Constantinou et al. 2011).

Selenium was identified as a potential chemopreventative agent for prostate cancer when a phase III trial for nonmelanoma skin cancer patients reported reduced incidence of prostate cancer (Clark et al. 1996). It has been shown that combination treatment of selenium and vitamin E potentiates cell cycle arrest in prostate cancer cells irrespective of androgen receptor status in a synergistic fashion (Venkateswaran et al. 2004).

Utilization of soy for prostate cancer prevention is supported by observation of low incidence of prostate cancer in men in Asia where soy is widely consumed (Perkins et al. 1995; Magi-Galluzzi et al. 2011). It has also been shown that soy-based isoflavones cause cell cycle arrest, decrease angiogenesis and cause gene promoter demethylation in prostate cancer cells (Adjakly et al. 2011).

Fleshner et al. recently reported a phase III trial testing the daily intake of three agents: selenium (200 μg), vitamin E (800 U) and soy (40g) in high-grade PIN patients. The randomized trial showed no significant benefit in prevention of invasive prostate cancer with utilization of the three agents in combination (Fleshner et al. 2011). Results in this trial are congruous with reports from the selenium and vitamin E cancer prevention trial (SELECT). The four arm trial in healthy patients compared selenium alone, Vitamin E alone, the combination of both and a placebo arm with a median length of 5.5 years of supplementation. There was no significant benefit reported from use of the agents in combination or separately with regard to prostate cancer prevention. A non-significant increase in prostate cancer incidence was reported for vitamin E supplementation group which did not hold for the combined treatment arm. A statistically non-significant increase in risk of type 2 diabetes mellitus was reported for the selenium group (Lippman et al. 2009). Additional follow up later confirmed that use of vitamin E did indeed significantly increase the incidence of prostate cancer. At the same time combined use of vitamin E and selenium did not show a statistically significant increase in prostate cancer risk (Klein et al. 2011).

Use of green tea catechins has shown significantly decreased risk of prostate cancer in men with high-grade PIN. Intake of 600mg/day of green tea catechins capsules for one year resulted in the diagnosis of a single prostate cancer case in the experimental arm as opposed to 9 in the control placebo group. There were 30 patients in each cohort (Bettuzzi et al. 2006). Further follow-up at 2 years reported maintenance of significant prostate cancer inhibition for subjects in the experimental group (Brausi et al. 2008). A phase II randomized trial is currently underway to confirm the findings [NCT00596011]. Interestingly, tissue bioavailability of green tea polyphenols was poor after 3 to 6 weeks of treatment in pre-radical prostatectomy patients. This suggests that long term treatment may be necessary for beneficial responses or that polyphenon E does not act directly on prostate tissue (Nguyen et al. 2012).

Epigenetic reactivation by HDAC inhibition upon treatment with green tea polyphenols has been suggested to cause cell cycle arrest and induce apoptosis in prostate cancer cells. Expression of p21/waf1, a cyclin-dependent kinase inhibitor, and Bax, a proapoptotic gene is increased, promoting cell cycle arrest and apoptosis, respectively (Thakur et al. 2012).

STATINS AND CANCER PREVENTION

Statins competitively inhibit the 3-hydroxy-3-methylglutaryl coenzyme-A (HMG-CoA) reductase enzyme in the mevalonate pathway which is responsible for cholesterol synthesis. Reduction of cholesterol has been observed to reduce the incidence of cardiovascular and cerebrovascular events resulting in these agents to be one of the most highly prescribed drugs currently in the US after first entering the US market in 1987. Statins are prescribed very commonly for hyperlipidemic conditions, including patients who present with hypertension, obesity, diabetes and other cardiovascular risk factors (Alberti et al. 2005).

The hydrophobic family of statins include atorvastatin, lovastatin, simvastatin and fluvastatin. The hydrophilic family includes rosuvastatin and pravastatin. Hydrophobic statins, in comparison to hydrophilic statins, have greater intracellular access and have been used more often for oncologic studies.

The inhibition of mevalonate formation by statins causes generation of downstream isoprenoid molecules to be inhibited. Isoprenoids are long, hydrophobic molecules such as geranylgeranyl groups and farnesyl groups that are covalently bound to a variety of proteins to help them adhere to the cell membrane resulting often times in the activation of downstream targets (Roy et al. 2011). Biological pathways influencing carcinogenesis and metastasis have been reported to be regulated by statin drugs. The implications of isoprenoid signaling inhibition and cholesterol inhibition by statins are discussed in detail later, with a focus on prostate cancer.

Initially, large randomized controlled trials of non-statin cholesterol lowering agents in cardiovascular disease (CVD) prevention showed an increase in non-CVD disease related death, such as cancer, in some of these cohorts (Oliver 1991). Conversely, many studies specific for statins have reported reductions in the risk of cancers in statin users. Some large population based studies have reported the benefits conferred by statin use in either total cancer prevention and for specific types of cancer including colorectal cancer, breast cancer, melanoma and prostate cancer. The results have varied with respect to individual cancer types but overall cancer prevention has been reported for statin users in several studies (Blais et al. 2000; Graaf et al. 2004; Friis et al. 2005). Selected studies are discussed in detail later. On the other hand, a number of studies have reported an absence of a statistically significant positive correlation between statin use and reduced total cancer risk (Strandberg et al. 2004; Sacks et al. 1996; Downs et al. 1998; LIPID Research Group 1998; Jacobs et al. 2011).

A nested case-control study using the Quebec health database compared patients prescribed on statins to patients using bile acid-binding resins, which also reduce cholesterol levels. It was reported that HMG-CoA reductase inhibitor use reduced the risk of malignant neoplasm diagnosis by 28% compared to use of bile acid-binding resins. Except for lymphoma most cancer types including colorectal, breast and skin showed reduced risk of occurrence, albeit non-significant, in statin users (Blais et al. 2000).

Another observational study using the Dutch PHARMO database compared cancer incidence between patients treated with statins and controls. It was reported that there was a 20% reduced risk of diagnosis in subjects taking statins for at least 6 months. The risk decreased with increasing period of statin use although the p-value was not significant ($P = 0.08$). Interestingly, patients who were past users of statins and stopped the use for 6 months had the same cancer risk as non-users. Significant risk reduction for cumulative dose was achieved for users who received more than 1,350 defined daily doses. With regard to individual cancer types, a statistically significant risk reduction was reported for renal cancer. Reduced risk was also reported for colorectal, skin and prostate cancer (Graaf et al. 2004).

A Danish population-based study used the Prescription Database of North Jutland County and Danish Cancer Registry for the period of 1989-2002 to compare cancer incidence between statin users and users of other lipid-lowering drugs. Although the individual types of cancers: liver, colorectal, lung, breast, prostate, lymphatic and hematopoietic tissues, did not show a statistically significant decrease in cancer risk; an overall significant reduction in cancer risk was reported. Consistent trends in risk estimates with regard to number of prescriptions and duration of follow-up were not found in this study (Friis et al. 2005).

STATINS AND PROSTATE CANCER PREVENTION

The effect of cholesterol-lowering drugs on prostate cancer risk has been studied in several large epidemiologic studies. Studies supporting the positive correlation of statin use with decreased advanced prostate cancer incidence (Bjerre and LeLorier 2001; Carroll et al. 2005; Bonovas et al. 2006; Dale et al. 2006; Kuoppala et al. 2008; Taylor et al. 2008) and studies reporting a negative correlation or no correlation exist as well (Flick et al. 2007; Jacobs et al. 2007). We have restricted our discussion to a few selected studies due to space limitations.

Participants of The Health Professionals Follow-up Study program were used as subjects to study the effect of cholesterol-lowering drugs in prostate carcinogenesis (n = 34,989 men). Although there was no significant reduction in overall prostate cancer risk, there was a significantly decreased risk of advanced prostate cancer [Relative Risk (RR) = 0.51] in current statin users. The association was even stronger between current statin users and reduced risk of fatal or metastatic prostate cancer. Risk of advanced disease was reduced by increasing duration of statin use (RR = 0.25). However, use of statin was not correlated with histological grade of disease. Analysis of the observations required accounting for factors such as PSA testing differences between statin users and non-users since PSA testing (and frequency of PSA testing) has an inverse correlation with diagnosis of advanced prostate cancer. In addition, users of statins are more likely to undergo PSA testing than the control group. The association between statin use and observed effects on total prostate cancer and advanced prostate cancer remained after this analysis (Platz et al. 2006).

A 2,399 patient retrospective study from Johns Hopkins reported an inverse association between statin use (16.1% of patients) and nonorgan confined disease. An inverse relationship between statin use and high grade disease was only significant for patients who had a preoperative PSA concentration of 10ng/ml or more. Statins did not show a significant inverse association with recurrence or progression to metastasis or death. However, longer term use of statin after prostatectomy suppressing recurrence was considered a possibility (Mondul et al. 2011).

A population-based case-control study using the Finnish Cancer Registry and Population Register Center, performed a matched pair analysis for 24,723 prostate cancer cases and matched case controls. 2,622 (10.6%) of the subjects were statin users. A marginally elevated total prostate cancer risk [odds ratio (OR) 1.07] was reported in statin users. However, the risk of advanced prostate cancer was decreased for atorvastatin, lovastatin and simvastatin users (OR 0.61, 95% CI 0.37-0.98; OR 0.61, 95% CI 0.43-0.85; and OR 0.78, 95% CI 0.61-1.01 respectively). The authors acknowledged absence of PSA testing records as a limitation of their study (Murtola et al. 2007).

A study in the California Men's Health Study group involving 888 prostate cancer cases reported no significant association between prostate cancer incidence and less than 5 years of statin use. Interestingly, there was a 28% reduction in prostate cancer risk for statin use exceeding 5 years. There was no distinct difference in response towards prevention of advanced prostate cancer from total prostate cancer cases. A small cohort of advanced disease patients was cited as a possible reason for the lack of statistical significance. For patients taking statins for at least 5 years, the advanced prostate cancer RR was 0.57, while total prostate cancer cases the RR was 0.72. This study also reported the observed benefits of statin to be restricted to the patients who were regularly taking non steroidal anti-inflammatory agents (Flick et al. 2007).

A large population based study in the Cancer Prevention Study II Nutrtion Cohort reported an absence of a significant risk reduction for prostate cancer for patients taking statins for duration of at least 5 years. The study examined the risk for 10 types of cancer, reported reduced RR for advanced prostate cancer compared to total prostate cancer cases [0.81 compared to 0.98 for total prostate cancer (<5y statin use); 0.86 compared to 1.02 for total prostate cancer cases (>5y statin use)]. Former statin users had a RR of 1.05 for advanced prostate cancer, while the RR was 0.95 for total prostate cancer cases. However, the difference was not statistically significant (Jacobs et al. 2011). A previous study focusing particularly on advanced prostate cancer from the same population had reported a marginally statistically significant reduction (not significant at the $p<0.05$ level) in advanced prostate cancer (RR = 0.6, 95% CI, 0.36-1) for long term statin use of at least 5 years (Jacobs et al. 2007).

A Veterans Affairs New England Healthcare System study compared statin users (n = 41,078) with patients taking antihypertensive medication (n =14,797) for prostate cancer incidence. They reported a 31% reduced risk of total prostate cancer in statin users. There was a 14% (not statistically significant) reduction in risk of low grade prostate cancer and a 60% reduction in risk for high grade prostate cancer. Higher level of serum cholesterol was directly correlated to increased risk of high grade disease and total prostate cancer incidence. Furthermore, the dose of statin achieved at 1 year was inversely correlated with incidence of high grade prostate cancer (Farwell et al. 2011).

A Veteran Affairs Medical Center study based in Oregon compared 100 prostate cancer cases with 202 clinical controls to see an association of statin use with total prostate cancer and aggressive prostate cancer. A significant benefit was found to be conferred by statins in prostate cancer prevention, both for total cases and advanced prostate cancer cases. Stratification by Gleason score showed that the relationship between prostate cancer prevention and statin use only held for cancers having a Gleason score of 7 or more. In fact there was a 76% reduced risk of aggressive disease for Gleason score of 7 or more. The cumulative dose of statin and duration of dose had a significant effect on prostate cancer risk. There was a less pronounced effect of average daily dose (Shannon et al. 2005).

Similarly, a retrospective study for men who underwent prostate biopsy between 2000 and 2007 (n = 4,204) at the Cleveland Clinic reported a significant reduction (61.4% vs 72.4%) in Gleason score of 7 or greater prostate cancer for statin users (n = 1,022). The probability of high volume prostate cancer was also reduced from 31.4% to 27.2%. Greater than 5 years of statin use predicted a lower rate of high grade prostate cancer (Tan et al. 2011).

In support of the inverse association between statin use with advanced prostate cancer incidence, a matched case-control study reported a 50% reduction in prostate cancer deaths for statin users. 387 cases of fatal prostate cancer were matched to 442 control subjects from New Jersey. The analysis was then adjusted for confounders such as obesity, educational level and use of antihypertensive medication and the inverse association between statin use and advanced prostate cancer incidence remained. When taking into account the use of hypertensive medications the apparent protective effect of statins actually increased. Distinction between use of lipophilic and hydrophilic statins suggested negligible differences. Potency of statins correlated significantly with risk reduction (Marcella et al. 2011).

In conclusion, these retrospective studies suggest statins may have preventative effects for prostate cancer development and may limit progression that is reflected by decreased incidence of prostate cancer cases, particularly advanced prostate cancer.

MECHANISMS OF STATIN ANTI-CANCER EFFECTS

HMG-CoA reductase is a membrane bound enzyme which catalyzes the rate limiting step in the mevalonate pathway involving biosynthesis of sterol and isoprenoid molecules. Even though the synthesis of cholesterol involves 20 enzymes, HMG-CoA is an ideal target because it catalyzes an irreversible rate-limiting reaction at the beginning of the pathway. A detailed overview of the regulation of HMG-CoA can be found in the recent review by Burg and Espenshade (Burg and Espenshade 2011). Due to the relevance of hypercholesterolemia to human diseases, the sterol biosynthesis pathway has been the subject of intensive studies and statins have emerged as the most commonly prescribed drug class for lowering serum cholesterol.

Isoprenoid Signaling

Isoprenoid molecules are lipid moieties that are post-translationally attached to various proteins. These isoprenoid intermediates like farnesyl pyrophosphate (FPP) and geranyl geranyl pyrophosphate (GGPP) are covalently bound to various G proteins and G protein subunits by farnesyl and geranyl geranyl protein transferases, respectively. Post-translational modification of GTP binding proteins in the Ras, Rho and Rab families by prenylation at the carboxyl terminal enables them to anchor to cell membranes. This has been described to be a crucial step in the regulation of apoptosis, angiogenesis and inflammation of cancer cells by statins. A detailed overview of Ras prenylation is covered in an early review by Khosravi-Far et al. (Khosravi-Far et al. 1992).

A feedback loop exists where isoprenoids can suppress HMG-CoA reductase by post translational downregulation. HMG-CoA reductase is subject to complex feedback regulation at the transcriptional, translational and post-translational levels as well as degradation (Goldstein and Brown 1990; Goldstein et al. 2006). Both sterol and non-sterol products of the mevalonate pathway have specific feedback regulatory mechanisms in normal cells. Tumor cells are resistant to sterol-mediated feedback and are more sensitive to isoprenoid mediated suppression than normal cells (Edwards and Ericsson 1999; Wong et al. 2002; Mo and Elson 2004; Demierre et al. 2005).

A handful of studies support an autophagic mechanism of cell death in prostate cancer cells by statin treatment (Kim et al. 2009; Parikh et al. 2010). Atorvastatin treatment showed a dose dependent increase in autophagy marker expression and caused a G1 arrest in PC3 cells. When intermediate products of the mevalonate pathway like squalene, ubiquinone and farnesol was added back, there was no change in the observed effect. However, addition of geranylgeranol reversed autophagic cell death. Interestingly androgen receptor (AR) positive cancer cell lines like LNCaP and normal prostate epithelial lines like RWPE1 do not show the same effect (Parikh et al. 2010). Taking the clinical studies that suggest an effect of statins to be restricted to reduction of advanced prostate cancer into account, the described observations suggest an important role for geranylgeranylation in the progression of prostate cancer to advanced disease.

Cholesterol Deprivation

Cholesterol is a necessary intermediate in androgen biosynthesis. Therefore suppressing cholesterol synthesis by statin treatment may reduce the levels of circulating androgen and also causes a drop in the levels of dihydrotestosterone (DHT) intracellularly in the prostate. DHT is a strong ligand for AR, a transcription factor, which localizes to the nucleus after binding to androgenic hormones in the cytoplasm. PSA production is regulated by AR, which may explain the observed decrease in PSA levels in patients upon statin treatment in several clinical studies described in previous sections.

Cholesterol inhibition has been studied in the context of PI3K/AKT signaling in prostate cancer cells. The PI3K pathway is instrumental in prostate cancer tumorigenesis and prostate cancer tumor maintenance. Cholesterol-rich membranes (lipid rafts) enable tyrosine kinase epidermal growth factor induced constitutive signaling through the PI3K/AKT signal transduction pathway (Zhuang et al. 2002). Increasing cholesterol levels in mice has been correlated with higher levels of phosphorylated AKT and reduced apoptosis in xenograft tumors (Zhuang et al. 2005). AR has been shown to interact directly with AKT1 in a PI3K independent fashion in lipid raft subcellular compartments of prostate cancer cells. Androgen stimulation significantly strengthens this association (Cinar et al. 2007).

A recent study in mice suggests a direct correlation between circulating cholesterol levels and intra-tumoral androgen levels. Based on reports that prostate cancer cells are capable of de novo or intracrine steroid synthesis, the findings suggest that inhibition of cholesterol along with androgen deprivation strategies may be an attractive option to inhibit growth of aggressive prostate cancer in patients (Mostaghel et al. 2012).

Mondul et al. has evaluated the association between plasma cholesterol and prostate cancer in the CLUE II cohort of patients that included 6,816 males. There was a reduced likelihood for men with low cholesterol levels to develop high grade disease. The significance was strengthened when the analysis was restricted to organ confined disease. The results support a role for cholesterol in influencing the development of high grade prostate cancer and further supports that decreased cholesterol is a mechanism by which statins may prevent prostate cancer development (Mondul et al. 2010).

Myc Inhibition

The oncoprotein c-Myc (also referred to as Myc) is a transcription factor overexpressed in a variety of cancers. It has been demonstrated that a region encompassing the *c-MYC* locus (8q24) is amplified in advanced/aggressive prostate tumors (Nesbit et al. 1999; Sato et al. 1999). Recent data also suggests that *MYC* overexpression may be associated with inferior outcomes following surgery for localized prostate cancer (Hawksworth et al. 2010). Myc is involved early in prostate tumorigenesis, as *MYC* mRNA is overexpressed in a majority of human prostate tumors and high grade prostatic intraepithelial neoplasia (PIN) lesions (Fleming et al. 1986; Gurel et al. 2008). In transgenic mouse models, *MYC* overexpression in the prostate causes PIN that progresses to adenocarcinoma (Zhang et al. 2000; Ellwood-Yen et al. 2003).

In addition to the evidence suggesting the role of Myc in prostate cancer tumorigenesis and progression, studies have also shown that Myc is a prognostic marker for poor outcomes and distant metastasis in many other cancers (Bourhis et al. 1990; Gamberi et al. 1998; Herms et al. 2000; Wolfer et al. 2010; Wang et al. 2011). Several different transgenic mouse models have shown that inactivation of Myc can induce sustained tumor regression in different cancer types (D'Cruz et al. 2001; Jain et al. 2002; Giuriato et al. 2004; Shachaf et al. 2004). Genomic and proteomic analysis have revealed that a threshold level of Myc is required for tumor maintenance (Shachaf et al. 2008). The preponderance of this evidence has driven intense interest in finding agents able to target Myc for cancer therapy. However, Myc has been difficult to target directly and thus efforts focused at targeting upstream mediators of Myc such as Ras are intriguing. Ras is active in a number of progrowth regulatory signal transduction pathways some of which converge on Myc (Bachireddy et al. 2005). Taking into account that statins inhibit prenylation of Ras family proteins as previously described, statins could theoretically be a successful pharmaceutical agent for Myc downregulation.

Targeting Myc for therapeutic purposes may also be an effective means of targeting the microenvironment to kill tumor cells. Thrombospondin-1 (TSP-1) is a potent antiangiogenic protein that has been shown to play an important role in modulating the vascular network of the tumor microenvironment. Some studies have suggested progressive downregulation of TSP-1 as prostate epithelium goes from PIN to low grade ($\leq$ Gleason 6) to higher grade (>Gleason 6) carcinoma (Kwak et al. 2002; Vallbo et al. 2004). Myc has been shown to inhibit the expression of TSP-1 and this presumably allows tumor cells to develop and sustain a vascular network via angiogenesis (Baudino et al. 2002; Watnick et al. 2003). On the other hand, p53 has been shown to suppress tumor angiogenesis partly via repression of TSP-1 (Browder et al. 2002). Interestingly, it has recently been shown that Myc inhibition results in regression of experimental lymphomas at least in part through a p53-dependent induction of TSP-1 expression (Giuriato et al. 2006).

Statins have been shown to have activity in preclinical models of Myc-induced lymphomas (Shachaf et al. 2007). The survival advantage seen in these Myc models with statins was limited to Myc-induced lymphoma and failed to reverse or prevent tumorigenesis in the presence of a similar activated K-Ras^{G12D} transgenic lymphoma mouse model. These data suggest that statin anti-tumors effects may be specific to Myc overexpressing cancers. More preclinical data confirming that statins target Myc activation in mouse tumors comes from *in vivo* monitoring of Myc phosphorylation using a split firefly-luciferase complementation strategy to detect and quantify Myc activation (Fan-Minogue et al. 2010).

Due to the heterogenous nature of prostate cancer, ranging from indolent low-risk disease unlikely to cause mortality to aggressive higher-risk disease that quickly develops systemic metastasis, it is important to pin-point the sub-population of prostate cancer patients most likely to benefit from statin therapy. These observations make Myc an attractive candidate as a predictive biomarker for statin directed therapies in prostate cancer. We are testing the hypothesis that Myc is a predictive biomarker and modifiable target for statin-induced cytotoxicity and radiation response in prostate cancer with our upcoming phase 0 clinical trial using the pre-prostatectomy model (ClinicalTrials.gov ID: NCT01478828).

STATINS AND PROSTATE CANCER THERAPY

Statins have been suggested to have therapeutic potential in addition to its prevention properties for prostate cancer. In recent years, it has been proposed to improve outcomes following local therapies for prostate cancer.

Retrospective clinical studies have associated improved clinical outcomes in men undergoing radiation therapy for prostate cancer with use of statins. In a cohort of 691 prostate cancer patients who underwent curative intent radiation therapy, 27% had been using statins (n = 189). Men taking statins had a 93% 4-year freedom from biochemical failure compared to 80% for non-users. Relapse-free survival for statin users was 84% compared to 69% at 4-years for non-users. Statins did not show an effect in distant metastasis prevention between the two groups. Patients taking statins before radiation therapy (n = 150) were compared to patients who started taking statins after radiation therapy (n = 39). There was no significant difference in freedom from biochemical failure when stratifying by the timing of statins with respect to radiation therapy. A direct association was reported between lower cholesterol, low density lipoprotein (LDL) and improved outcome in statin users (Gutt et al. 2010).

In another study of 1711 men who received radiation therapy for prostate cancer, statin use (23% of patients, n = 382) was correlated to increased PSA relapse free survival (PRFS). The 5-year PRFS was 89% for statin users compared to 80% for controls. The 8-year PRFS was 83% for statin users compared to 74% for non-users. Interestingly, only the high-risk patients showed a significant beneficial response to statins. There was no association of statin use with overall survival. Similar to the study described above, distant metastasis free survival was not observed to be associated with use of statins (Kollmeier et al. 2011).

In contrast, a study conducted on 968 prostate cancer patients undergoing radiation therapy, 23% of whom were taking statins, reported no significant benefit conferred by statin use on progression free survival (PFS). Although, the 5-year PFS was 70% versus 59% in favor of statin users, adjusting for treatment year and risk group nullified the benefit. A short follow-up time of 3.1 years for statin users is possibly a limiting factor for this study (Soto et al. 2009).

Using the Shared Equal Access Regional Cancer Hospital (SEARCH) database, 236 men taking statins at time of radical prostatectomy (RP) were found and compared to 1083 men non-taking statins at the time of RP. After adjusting for multiple clinical and pathological factors, statin use was associated with a 30% lower risk of PSA recurrence ($p = .03$), which was dose dependent. These results suggest that statin use has anti-cancer effects in the perioperative period during RP (Hamilton et al. 2010).

The Cancer of the Prostate Strategic Urologic Research Endeavor (CaPSURE) registry was used to study the association of statin use and nonsteroidal anti-inflammatory drugs (NSAIDs) among patients undergoing RP or radiation therapy (RT). Among 7,042 men, 4,611 men underwent RP while 2,431 men underwent RT. In a 4-year median follow-up analysis, statin use was associated with a reduced risk of all cause mortality for both RP (HR = 0.35, 95% CI = 0.21-0.58) and RT (HR = 0.59, 95% CI = 0.37-0.94) patients. Use of NSAID was also associated with reduced risk of both subsets of patients (Katz et al. 2010). Taken together, these studies suggest a therapeutic role for statins in reducing biochemical failure for prostate cancer patients being treated with radiation therapy or undergoing radical prostatectomy.

CONCLUSION

Prostate cancer is a heterogenous disease that ranges from indolent low-risk disease to advanced high-risk disease. Although a large number of retrospective studies have reported the beneficial effects conferred by statin use with regard to prevention of prostate cancer progression, especially risk of advanced prostate cancer, the studies have admittedly had several limitations. Therapeutic benefits of statin use for patients undergoing radical prostatectomy or radiation therapy has also been supported by retrospective studies. Prospective randomized trials are warranted for confirming these promising results and for validation of predictive biomarkers that are needed to direct patients expected to respond to statins.

REFERENCES

Adjakly, M., R. Bosviel, et al. 2011. DNA methylation and soy phytoestrogens: quantatative study in DU-145 and PC-3 human prostate cancer cell lines. Epigenomics 3(6): 795-803.

Akhavan, A., J.D. Keith, et al. 2007. The proportion of cores with high-grade prostatic intraepithelial neoplasia on extended-pattern needle biopsy is significantly associated with prostate cancer on site-directed repeat biopsy. BJU International 99(4): 765-769.

Albanes, D., O. Heinonen, et al. 1995. Effects of alpha-tocopherol and beta-carotene supplements on cancer incidence in the Alpha-Tocopherol Beta-Carotene Cancer Prevention Study. The American Journal of Clinical Nutrition 62(6): 1427S-1430S.

Alberti, K.G.M.M., P. Zimmet, et al. 2005. The metabolic syndrome—a new worldwide definition. The Lancet 366(9491): 1059-1062.

Alberts, S.R., P.J. Novotny, et al. 2006. Flutamide in Men with Prostatic Intraepithelial Neoplasia: A Randomized, Placebo-Controlled Chemoprevention Trial. American Journal of Therapeutics 13(4): 291-297.

Alcaraz, A., M.A. Barranco, et al. 2001. High-grade prostate intraepithelial neoplasia shares cytogenetic alterations with invasive prostate cancer. Prostate 47(1): 29-35.

Andriole, G.L., D.G. Bostwick, et al. 2010. Effect of Dutasteride on the Risk of Prostate Cancer. New England Journal of Medicine 362(13): 1192-1202.

Bachireddy, P., P.K. Bendapudi, et al. 2005. Getting at MYC through RAS. Clinical Cancer Research 11(12): 4278-4281.

Baudino, T.A., C. McKay, et al. 2002. c-Myc is essential for vasculogenesis and angiogenesis during development and tumor progression. Genes & Development 16(19): 2530-2543.

Bettuzzi, S., M. Brausi, et al. 2006. Chemoprevention of Human Prostate Cancer by Oral Administration of Green Tea Catechins in Volunteers with High-Grade Prostate Intraepithelial Neoplasia: A Preliminary Report from a One-Year Proof-of-Principle Study. Cancer Research 66(2): 1234-1240.

Bishara, T., D.M. Ramnani, et al. 2004. High-grade Prostatic Intraepithelial Neoplasia on Needle Biopsy: Risk of Cancer on Repeat Biopsy Related to Number of Involved Cores and Morphologic Pattern. The American Journal of Surgical Pathology 28(5): 629-633.

Bjerre, L.M. and J. LeLorier. 2001. Do statins cause cancer? A meta-analysis of large randomized clinical trials. The American Journal of Medicine 110(9): 716-723.

Blais, L., A. Desgagne, et al. 2000. 3-Hydroxy-3-Methylglutaryl Coenzyme A Reductase Inhibitors and the Risk of Cancer: A Nested Case-Control Study. Arch Intern Med 160(15): 2363-2368.

Bonovas, S., K. Filioussi, et al. 2006. Statins and Cancer Risk: A Literature-Based Meta-Analysis and Meta-Regression Analysis of 35 Randomized Controlled Trials. Journal of Clinical Oncology 24(30): 4808-4817.

Bostwick, D.G. and L. Cheng. 2012. Precursors of prostate cancer. Histopathology 60(1): 4-27.

Bourhis, J., M.G. Le, et al. 1990. Prognostic value of c-myc proto-oncogene overexpression in early invasive carcinoma of the cervix. Journal of Clinical Oncology 8(11): 1789-1796.

Brausi, M., F. Rizzi, et al. 2008. Chemoprevention of Human Prostate Cancer by Green Tea Catechins: Two Years Later. A Follow-up Update. European Urology 54(2): 472-473.

Browder, T., J. Folkman, et al. 2002. Antiangiogenic therapy and p53. Science 297(471): 5581.

Burg, J.S. and P.J. Espenshade. 2011. Regulation of HMG-CoA reductase in mammals and yeast. Progress in Lipid Research 50(4): 403-410.

Carroll, M.D., D.A. Lacher, et al. 2005. Trends in Serum Lipids and Lipoproteins of Adults, 1960-2002. JAMA: The Journal of the American Medical Association 294(14): 1773-1781.

Cinar, B., N.K. Mukhopadhyay, et al. 2007. Phosphoinositide 3-Kinase-independent Non-genomic Signals Transit from the Androgen Receptor to Akt1 in Membrane Raft Microdomains. Journal of Biological Chemistry 282(40): 29584-29593.

Clark, L.C., G.F. Combs, et al. 1996. Effects of Selenium Supplementation for Cancer Prevention in Patients with Carcinoma of the Skin. JAMA: The Journal of the American Medical Association 276(24): 1957-1963.

Constantinou, C., C.M. Neophytou, et al. 2011. Induction of DNA Damage and Caspase-Independent Programmed Cell Death by Vitamin E. Nutrition and Cancer 64(1): 136-152.

Cooperberg, M.R., J.M. Broering, et al. 2007. Contemporary Trends in Low Risk Prostate Cancer: Risk Assessment and Treatment. The Journal of Urology 178(3, Supplement): S14-S19.

D'Cruz, C.M., E.J. Gunther, et al. 2001. c-MYC induces mammary tumorigenesis by means of a preferred pathway involving spontaneous Kras2 mutations. Nat Med 7(2): 235-239.

Dale, K.M., C.I. Coleman, et al. 2006. Statins and Cancer Risk. JAMA: The Journal of the American Medical Association 295(1): 74-80.

Demierre, M.-F., P.D.R. Higgins, et al. 2005. Statins and cancer prevention. Nat Rev Cancer 5(12): 930-942.

Downs, J.R., M. Clearfield, et al. 1998. Primary Prevention of Acute Coronary Events with Lovastatin in Men and Women with Average Cholesterol Levels. JAMA: The Journal of the American Medical Association 279(20): 1615-1622.

Edwards, P.A. and J. Ericsson. 1999. STEROLS AND ISOPRENOIDS: Signaling Molecules Derived from the Cholesterol Biosynthetic Pathway. Annual Review of Biochemistry 68(1): 157-185.

Ellwood-Yen, K., T.G. Graeber, et al. 2003. Myc-driven murine prostate cancer shares molecular features with human prostate tumors. Cancer Cell 4(3): 223-238.

Fan-Minogue, H., Z. Cao, et al. 2010. Noninvasive molecular imaging of c-Myc activation in living mice. Proceedings of the National Academy of Sciences 107(36): 15892-15897.

Farwell, W.R., L.W. D'Avolio, et al. 2011. Statins and Prostate Cancer Diagnosis and Grade in a Veterans Population. Journal of the National Cancer Institute 103(11): 885-892.

Finelli, A., G. Trottier, et al. 2011. Impact of 5α-Reductase Inhibitors on Men Followed by Active Surveillance for Prostate Cancer. European Urology 59(4): 509-514.

Fleming, W.H., A. Hamel, et al. 1986. Expression of the c-myc Protooncogene in Human Prostatic Carcinoma and Benign Prostatic Hyperplasia. Cancer Research 46(3): 1535-1538.

Fleshner, N.E., L. Kapusta, et al. 2011. Progression From High-Grade Prostatic Intraepithelial Neoplasia to Cancer: A Randomized Trial of Combination Vitamin-E, Soy, and Selenium. Journal of Clinical Oncology 29(17): 2386-2390.

Flick, E.D., L.A. Habel, et al. 2007. Statin Use and Risk of Prostate Cancer in the California Men's Health Study Cohort. Cancer Epidemiology Biomarkers & Prevention 16(11): 2218-2225.

Friis, S., A.H. Poulsen, et al. 2005. Cancer risk among statin users: A population-based cohort study. International Journal of Cancer 114(4): 643-647.

Gamberi, G., M.S. Benassi, et al. 1998. C-myc and c-fos in human osteosarcoma: prognostic value of mRNA and protein expression. Oncology 55(6): 556-563.

Giuriato, S., K. Rabin, et al. 2004. Conditional animal models: a strategy to define when oncogenes will be effective targets to treat cancer. Seminars in Cancer Biology 14(1): 3-11.

Giuriato, S., S. Ryeom, et al. 2006. Sustained regression of tumors upon MYC inactivation requires p53 or thrombospondin-1 to reverse the angiogenic switch. Proceedings of the National Academy of Sciences 103(44): 16266-16271.

Goldstein, J.L. and M.S. Brown. 1990. Regulation of the mevalonate pathway. Nature 343(6257): 425-430.

Goldstein, J.L., R.A. DeBose-Boyd, et al. 2006. Protein Sensors for Membrane Sterols. Cell 124(1): 35-46.

Gonzalez-Moreno, O., J. Lecanda, et al. 2010. VEGF elicits epithelial-mesenchymal transition (EMT) in prostate intraepithelial neoplasia (PIN)-like cells via an autocrine loop. Experimental Cell Research 316(4): 554-567.

Graaf, M.R., A.B. Beiderbeck, et al. 2004. The Risk of Cancer in Users of Statins. Journal of Clinical Oncology 22(12): 2388-2394.

Gurel, B., T. Iwata, et al. 2008. Nuclear MYC protein overexpression is an early alteration in human prostate carcinogenesis. Mod Pathol 21(9): 1156-1167.

Gutt, R., N. Tonlaar, et al. 2010. Statin Use and Risk of Prostate Cancer Recurrence in Men Treated with Radiation Therapy. Journal of Clinical Oncology 28(16): 2653-2659.

Hamilton, R.J., L.L. Banez, et al. 2010. Statin medication use and the risk of biochemical recurrence after radical prostatectomy. Cancer 116(14): 3389-3398.

Hawksworth, D., L. Ravindranath, et al. 2010. Overexpression of C-MYC oncogene in prostate cancer predicts biochemical recurrence. Prostate Cancer Prostatic Dis 13(4): 311-315.

Herms, J., I. Neidt, et al. 2000. C-MYC expression in medulloblastoma and its prognostic value. International Journal of Cancer 89(5): 395-402.

Jacobs, E.J., C.C. Newton, et al. 2011. Long-term Use of Cholesterol-Lowering Drugs and Cancer Incidence in a Large United States Cohort. Cancer Research 71(5): 1763-1771.

Jacobs, E.J., C. Rodriguez, et al. 2007. Cholesterol-Lowering Drugs and Advanced Prostate Cancer Incidence in a Large U.S. Cohort. Cancer Epidemiology Biomarkers & Prevention 16(11): 2213-2217.

Jain, M., C. Arvanitis, et al. 2002. Sustained Loss of a Neoplastic Phenotype by Brief Inactivation of MYC. Science 297(5578): 102-104.

Katz, M.S., P.R. Carroll, et al. 2010. Association of statin and nonsteroidal anti-inflammatory drug use with prostate cancer outcomes: results from CaPSURE. BJU International 106(5): 627-632.

Khosravi-Far, R., A. Cox, et al. 1992. Protein prenylation: key to ras function and cancer intervention? Cell Growth Differ 3(7): 461-469.

Kim, J., C.I. Amos, et al. 2011. 5α-Reductase Inhibitors for Prostate-Cancer Prevention. New England Journal of Medicine 365(24): 2340-2340.

Kim, R.H., J.M. Coates, et al. 2009. Arginine Deiminase as a Novel Therapy for Prostate Cancer Induces Autophagy and Caspase-Independent Apoptosis. Cancer Research 69(2): 700-708.

Klein, E.A., I.M. Thompson, et al. 2011. Vitamin E and the Risk of Prostate Cancer. JAMA: The Journal of the American Medical Association 306(14): 1549-1556.

Kollmeier, M.A., M.S. Katz, et al. 2011. Improved Biochemical Outcomes with Statin Use in Patients with High-Risk Localized Prostate Cancer Treated with Radiotherapy. International Journal of Radiation Oncology, Biology, Physics 79(3): 713-718.

Kuoppala, J., A. Lamminpää, et al. 2008. Statins and cancer: A systematic review and meta-analysis. European Journal of Cancer 44(15): 2122-2132.

Kwak, C., R.J. Jin, et al. 2002. Thrombospondin-1, vascular endothelial growth factor expression and their relationship with p53 status in prostate cancer and benign prostatic hyperplasia. BJU International 89(3): 303-309.

LIPID Research Group. 1998. Prevention of Cardiovascular Events and Death with Pravastatin in Patients with Coronary Heart Disease and a Broad Range of Initial Cholesterol Levels. New England Journal of Medicine 339(19): 1349-1357.

Lippman, S.M., E.A. Klein, et al. 2009. Effect of Selenium and Vitamin E on Risk of Prostate Cancer and Other Cancers. JAMA: The Journal of the American Medical Association 301(1): 39-51.

Magi-Galluzzi, C., T. Tsusuki, et al. 2011. TMPRSS2–ERG gene fusion prevalence and class are significantly different in prostate cancer of caucasian, african-american and japanese patients. The Prostate 71(5): 489-497.

Marcella, S.W., A. David, et al. 2011. Statin use and fatal prostate cancer. Cancer: n/a-n/a.

Mo, H. and C.E. Elson. 2004. Studies of the Isoprenoid-Mediated Inhibition of Mevalonate Synthesis Applied to Cancer Chemotherapy and Chemoprevention. Experimental Biology and Medicine 229(7): 567-585.

Mondul, A., S. Clipp, et al. 2010. Association between plasma total cholesterol concentration and incident prostate cancer in the CLUE II cohort. Cancer Causes and Control 21(1): 61-68.

Mondul, A.M., M. Han, et al. 2011. Association of Statin Use with Pathological Tumor Characteristics and Prostate Cancer Recurrence after Surgery. The Journal of Urology 185(4): 1268-1273.

Montironi, R., R. Mazzucchelli, et al. 2007. Mechanisms of Disease: high-grade prostatic intraepithelial neoplasia and other proposed preneoplastic lesions in the prostate. Nat Clin Pract Urol 4(6): 321-332.

Mostaghel, E.A., K.R. Solomon, et al. 2012. Impact of Circulating Cholesterol Levels on Growth and Intratumoral Androgen Concentration of Prostate Tumors. PLoS ONE 7(1): e30062.

Murtola, T.J., T.L.J. Tammela, et al. 2007. Cholesterol-Lowering Drugs and Prostate Cancer Risk: A Population-based Case-Control Study. Cancer Epidemiology Biomarkers & Prevention 16(11): 2226-2232.

Nesbit, C., J. Tersak, et al. 1999. MYC oncogenes and human neoplastic disease. Oncogene 18(19): 3004-3016.

Nguyen, M.M., F.R. Ahmann, et al. 2012. Randomized, Double-Blind, Placebo-Controlled Trial of Polyphenon E in Prostate Cancer Patients before Prostatectomy: Evaluation of Potential Chemopreventive Activities. Cancer Prevention Research 5(2): 290-298.

Oliver, M.F. 1991. Might treatment of hypercholesterolaemia increase non-cardiac mortality? The Lancet 337(8756): 1529-1531.

Parikh, A., C. Childress, et al. 2010. Statin-induced autophagy by inhibition of geranylgeranyl biosynthesis in prostate cancer PC3 cells. The Prostate 70(9): 971-981.

Park, S., K. Shinohara, et al. 2001. Prostate Cancer Detection in Men with Prior High Grade Prostatic Intraepithelial Neoplasia or Atypical Prostate Biopsy. The Journal of Urology 165(5): 1409-1414.

Perkins, C., C. Morris, et al. 1995. Cancer incidence and mortality in California by detailed race/ethnicity, 1988-1992. Sacramento, CA: California Department of Health Services, Cancer, Surveillance Section.

Platz, E.A., M.F. Leitzmann, et al. 2006. Statin Drugs and Risk of Advanced Prostate Cancer. Journal of the National Cancer Institute 98(24): 1819-1825.

Ross, A.E., Z. Feng, et al. 2012. Effect of treatment with 5-α reductase inhibitors on progression in monitored men with favourable-risk prostate cancer. BJU International: no-no.

Roy, M., H.-J. Kung, et al. 2011. Statins and prostate cancer: role of cholesterol inhibition vs. prevention of small GTP-binding proteins. American Journal of Cancer Research 1(4): 542-561.

Sacks, F.M., M.A. Pfeffer, et al. 1996. The Effect of Pravastatin on Coronary Events after Myocardial Infarction in Patients with Average Cholesterol Levels. New England Journal of Medicine 335(14): 1001-1009.

Sakr, W., D. Grignon, et al. 1994. High grade prostatic intraepithelial neoplasia (HGPIN) and prostatic adenocarcinoma between the ages of 20-69: an autopsy study of 249 cases. *In Vivo* 8: 439-443.

Sato, K., J. Qian, et al. 1999. Clinical Significance of Alterations of Chromosome 8 in High-Grade, Advanced, Nonmetastatic Prostate Carcinoma. Journal of the National Cancer Institute 91(18): 1574-1580.

Shachaf, C.M., A.J. Gentles, et al. 2008. Genomic and Proteomic Analysis Reveals a Threshold Level of MYC Required for Tumor Maintenance. Cancer Research 68(13): 5132-5142.

Shachaf, C.M., A.M. Kopelman, et al. 2004. MYC inactivation uncovers pluripotent differentiation and tumour dormancy in hepatocellular cancer. Nature 431(7012): 1112-1117.

Shachaf, C.M., O.D. Perez, et al. 2007. Inhibition of HMGcoA reductase by atorvastatin prevents and reverses MYC-induced lymphomagenesis. Blood 110(7): 2674-2684.

Shannon, J., S. Tewoderos, et al. 2005. Statins and Prostate Cancer Risk: A Case-Control Study. American Journal of Epidemiology 162(4): 318-325.

Siegel, R., D. Naishadham, et al. 2012. Cancer statistics, 2012. CA: A Cancer Journal for Clinicians 62(1): 10-29.

Soto, D.E., S. Daignault, et al. 2009. No Effect of Statins on Biochemical Outcomes After Radiotherapy for Localized Prostate Cancer. Urology 73(1): 158-162.

Steiner, M. 2001. High grade prostatic intraepithelial neoplasia is a disease. Current Urology Reports 2: 195-198.

Steiner, M.S. and S. Raghow. 2003. Antiestrogens and selective estrogen receptor modulators reduce prostate cancer risk. World Journal of Urology 21(1): 31-36.

Strandberg, T.E., K. Pyörälä, et al. 2004. Mortality and incidence of cancer during 10-year follow-up of the Scandinavian Simvastatin Survival Study (4S). The Lancet 364(9436): 771-777.

Tan, N., E.A. Klein, et al. 2011. Statin Use and Risk of Prostate Cancer in a Population of Men Who Underwent Biopsy. The Journal of Urology 186(1): 86-90.

Tan, P.H., H.W. Tan, et al. 2006. Is high-grade prostatic intraepithelial neoplasia on needle biopsy different in an Asian population: A clinicopathologic study performed in Singapore. Urology 68(4): 800-803.

Taylor, M.L., B.J. Wells, et al. 2008. Statins and cancer: a meta-analysis of case-control studies. European Journal of Cancer Prevention 17(3): 259-268 10.1097/CEJ.0b013e3282b721fe.

Thakur, V.S., K. Gupta, et al. 2012. Green tea polyphenols causes cell cycle arrest and apoptosis in prostate cancer cells by suppressing class I histone deacetylases. Carcinogenesis 33(2): 377-384.

Theoret, M.R., Y.-M. Ning, et al. 2011. The Risks and Benefits of 5α-Reductase Inhibitors for Prostate-Cancer Prevention. New England Journal of Medicine 365(2): 97-99.

Thompson, I.M., P.J. Goodman, et al. 2003. The Influence of Finasteride on the Development of Prostate Cancer. New England Journal of Medicine 349(3): 215-224.

Vallbo, C., W. Wang, et al. 2004. The expression of thrombospondin-1 in benign prostatic hyperplasia and prostatic intraepithelial neoplasia is decreased in prostate cancer. BJU International 93(9): 1339-1343.

Venkateswaran, V., N.E. Fleshner, et al. 2004. Synergistic effect of vitamin E and selenium in human prostate cancer cell lines. Prostate Cancer Prostatic Dis 7(1): 54-56.

Wang, W., L. Xue, et al. 2011. Prognostic value of β-catenin, c-myc, and cyclin D1 expressions in patients with esophageal squamous cell carcinoma. Medical Oncology 28(1): 163-169.

Watnick, R.S., Y.-N. Cheng, et al. 2003. Ras modulates Myc activity to repress thrombospondin-1 expression and increase tumor angiogenesis. Cancer Cell 3(3): 219-231.

Wolfer, A., B.S. Wittner, et al. 2010. MYC regulation of a "poor-prognosis" metastatic cancer cell state. Proceedings of the National Academy of Sciences 107(8): 3698-3703.

Wong, W., J. Dimitroulakos, et al. 2002. HMG-CoA reductase inhibitors and the malignant cell: the statin family of drugs as triggers of tumor-specific apoptosis. Leukemia 16: 508-519.

Yang, X.J., K. Lecksell, et al. 1999. Does long-term finasteride therapy affect the histologic features of benign prostatic tissue and prostate cancer on needle biopsy? Urology 53(4): 696-700.

Zeng, L., R.G. Rowland, et al. 2004. Apoptosis incidence and protein expression of p53, TGF-β receptor II, p27Kip1, and Smad4 in benign, premalignant, and malignant human prostate. Human Pathology 35(3): 290-297.

Zhang, X., C. Lee, et al. 2000. Prostatic neoplasia in transgenic mice with prostate-directed overexpression of the c-myc oncoprotein. The Prostate 43(4): 278-285.

Zhuang, L., J. Kim, et al. 2005. Cholesterol targeting alters lipid raft composition and cell survival in prostate cancer cells and xenografts. The Journal of Clinical Investigation 115(4): 959-968.

Zhuang, L., J. Lin, et al. 2002. Cholesterol-rich Lipid Rafts Mediate Akt-regulated Survival in Prostate Cancer Cells. Cancer Research 62(8): 2227-2231.

CHAPTER 11

Mechanistic Aspects of Oxidative Stress Involvement in Human Environmental and Occupational Carcinogenesis

Dominique Ziech[1], Aglaia Pappa[2] and Mihalis I. Panayiotidis[3,*]

[1]Faculty of Medicine, School of Medicine, Universidad Diego Portales, Santiago, Chile
[2]Department of Molecular Biology & Genetics, Democritus University of Thrace, University Campus, 68100, Alexandroupolis, Greece
[3]Department of Pathology, Medical School, University of Ioannina, University Campus, 45110, Ioannina, Greece

ABSTRACT

Cancer formation is a rather complex process that requires triggering of multiple phases including "initiation", "promotion" and "progression". In addition, there are a number of cellular processes that are required during each of these phases for ultimately resulting in malignant transformation. On the other hand, an integral process during carcinogenesis is the generation of oxidative stress which in turn induces alterations in gene expression profiles of major cellular pathways including those of DNA repair, cell death, epigenetics, etc. In this chapter, we aim to focus on the mechanistic aspects of the involvement of oxidative stress generation during man-made environmental and occupational agents inducing environmental and occupational carcinogenesis respectively.

INTRODUCTION

When cells are removed from their natural environment they detach from their basement membranes and the surrounding extracellular matrix and thus result in loss of normal cell-matrix interactions that are necessary in

*Corresponding author: E-mail: mpanagiot@cc.uoi.gr

providing signals for growth and survival. This process is termed *anoikis* (Greek for homelessness) and it is a form of programmed cell death or apoptosis. On the other hand, cancer cells are known to escape *anoikis* and instead they survive and divide uncontrollably through their ability to invade other tissues (a process called *metastasis*). Thus, in essence, cancer cells have the capacity to proliferate and invade both of which are critical characteristics for malignant transformation (Gottlieb 2009).

Overall, cancer is identified as a disease of epidemic proportions. Today, cancer is present in all human populations across the globe and cancer rates are expected to continue to rise, within the next few decades, due to the ever expanding amount of toxic and cancer-causing agents present in our everyday life. To this end, the relationship between cancer-causing agents (referred to as carcinogens), their ability to cause cellular damage (via various mechanisms including induction of oxidative stress), and the subsequent development of cancer (carcinogenesis) are the three main topics of discussion in this chapter. More specifically, our focus will be directed towards understanding the mechanistic aspects linking various man-made carcinogens to oxidative stress-induced cellular injury and their relationship to human carcinogenesis (Franco and Panayiotidis 2010).

AN OVERVIEW OF OXIDATIVE STRESS

The mechanisms underlying transformation of a normal cell to a malignant one are rather intricate and complex and thus, to a large degree, poorly understood. However, extensive research over the past few decades has allowed for the identification and elucidation of a common mechanistic aspect involving the generation of reactive oxygen species (ROS) and the induction of oxidative stress. Overall, ROS are byproducts of cellular metabolism and represent an integral component in maintaining tissue oxygen homeostasis which, in turn, is maintained through a series of reactions called "redox" (reduction-oxidation). These reactions encompass the transfer of electrons between two chemical species, one that loses electrons (oxidized) and consequently one that gains electrons (reduced). When oxygen homeostasis is not maintained, the cellular environment becomes oxidatively stressed. To this end, it is estimated that approximately 1-3% of the oxygen we consume is generating various ROS molecules including superoxide anion (O_2^-), hydroxyl radical ($OH\cdot$), hydrogen peroxide (H_2O_2) and singlet oxygen ($^1O_2$) (Tudek et al. 2010; Seifreid et al. 2007; Seifreid et al., 2003; Pappa et al. 2007).

Under normal and physiologic conditions, an equilibrium exists between ROS production and metabolism (by our antioxidant defense systems) where ROS act as signaling molecules that mediate cell growth, migration and differentiation. However, when this delicate balance is interrupted and ROS concentrations become excessive (generation of oxidative stress) they can contribute to the pathophysiology of a number of human diseases (including a variety of human cancers like lung, breast, pancreatic, esophageal, etc.) primarily by means of causing cell death-induced damage to cellular macromolecules like DNA, proteins, lipids, carbohydrates, etc. (Ziech et al. 2010a; Ziech et al. 2010b; Tudek et al. 2010). Finally, sources known to generate such excessive concentrations of ROS are both of endogenous (e.g. mitochondrial oxidative phosphorylation, P450 metabolism, peroxisomes, activation of inflammatory cells) as well as of exogenous origins (e.g. ozone, industrial chemicals, xenobiotics and radiation) (Ziech et al. 2010b; Franco et al. 2008; Panayiotidis 2008; Franco and Panayiotidis 2009; Franco et al. 2009; Ziech et al. 2011).

AN OVERVIEW OF HUMAN ENVIRONMENTAL AND OCCUPATIONAL CARCINOGENESIS

Normal cells are characterized by controlled cellular growth and division whereby new cells are constantly produced in order to replace old and/or damaged ones. Cancer arises when this orderly process becomes deregulated due to the presence of a number of mutations. For example, DNA can oftentimes become modified (damaged) in ways that produce mutations which in turn compromise normal cellular growth and division. It is therefore the presence of these mutations that, if not effectively repaired, can contribute to human disease including carcinogenesis. To this end, cancer is defined as a disease whereby mutated cells divide uncontrollably and also invade other tissues via the lymph system thus ultimately leading to malignancy. Carcinogenesis is a multistage process that involves all of the "initiation", "promotion" and "progression" phases. Both the "initiation" and "promotion" phases compose the two "early" carcinogenic states. More specifically, the "initiation" phase triggers carcinogenic induction of DNA mutations whereas the "promotion" phase induces an expansion in mutated cell numbers (either through increased, uncontrolled cell division and/or decreased apoptotic cell death) (Panayiotidis 2008). Normal cells with acquired DNA mutations can then undergo cellular transformation and at this point are referred to as "initiated cells". Finally, the "progression" state is characterized by the transformation of initiated cells into malignant ones and it is the result of continued acquisition of mutations by exposure to various chemical and physical environmental insults. In general, the underlying cause of such cellular transformation lies within the ability of various carcinogens to induce direct DNA damage and thus mutations. Such carcinogens are commonly classified as genotoxic agents (e.g. various metals and chemicals, UV light, ionizing radiation, etc.) and result in populations of initiated cells. When such mutations are not repaired but are allowed a round of DNA synthesis they become "fixed" and as a result can cause generation of irreversible DNA damage (Klaunig et al. 2010).

A vast number of carcinogens have been identified to be involved in a wide range of environmental and occupational settings including: alcohol consumption, diet, tobacco use, air pollution and various industrial chemical compounds (e.g. metals, formaldehyde, asbestos, plastic polymers, etc.). Interestingly, tobacco and diet together account for almost two-thirds of all cancer-related deaths. Today, the media rarely misses an opportunity to report on newly discovered carcinogenic substances and it often seems that "as if everything causes cancer". However, the truth is that (among all suspected carcinogenic agents present today) only a few have been confirmed to possess carcinogenic activities. According to the International Agency for Research on Cancer (IRAC), 107 agents have been classified as confirmed human carcinogens, with an additional 59 agents categorized as probable and another 267 as possible human carcinogens. Furthermore, it has also been reported that 10% of some 80,000 chemicals utilized today are potentially carcinogenic to the human species (Ziech et al. 2010a).

One of the very first instances in introducing the concept of exposure to environmental compounds and their ability to act as cancer-causing agents was a report issued in 1775 by Sir Percivall Pott (1714-1788). In his report, Pott suggested that a strong link exists between cancer occurrence and exposures to residues produced by chimney sweeps. Today, researchers have been able to identify and associate benzo(a)pyrene (a chemical found in soot) as a potent carcinogen and certainly a risk factor in human carcinogenesis. Consequently, these results have led that chimneys being deemed unsafe and that their use should be prohibited (Miller and Ramos 2001). Another prime example of a well-established environmental agent associated with carcinogenesis is ionizing radiation. It was first suspected to be a carcinogenic source around the turn of the 20th century when the use of X-rays and radium in medicine was linked to a high incidence of skin cancer development. Today, it is known that ionization radiation causes deleterious effects (mutations in DNA, apoptotic induction, etc.) at each of the "initiation", "promotion" and "progression" phases of carcinogenesis not only in skin but rather in a wide range of human tissues (Klaunig et al. 2010). Other very well-established environmental carcinogens include cigarette smoke and tobacco. Currently, cigarette smoking is defined as the single most preventable cause of illness and premature death and, as stated by the United States Surgeon General, it is estimated to cause 30% of all cancer-related deaths in the US. To this end, tobacco smoke (a well-known source of carcinogens and ROS) has been estimated to increase the rates of oxidative DNA damage by about 35-50% (Valko et al. 2007). More specifically, in 1915, a group of Japanese scientists documented that tar (a main tobacco component) could cause tumor development in rabbits. In general, tar is the particulate matter derived from burning organic compounds and is the leading cancer-causing chemical in tobacco smoke. To date, more than 4,000 chemicals have been identified in tobacco smoke, with some 60 of which have been confirmed as carcinogens. Specifically, it is estimated that each cigarette delivers approximately 10mg of soot, tar, ash, phenols, benzopyrene, hydrogen cyanide, formaldehyde, etc. into the lungs. Moreover, researchers have also identified 4-(N-nitrosomethylamino)-1-(3-pyridyl)-1-butanone (NNK) as a potent carcinogen formed during the production, curing, and aging of tobacco. On the other hand, indirect inhalation of cigarette smoke (passive smoking or second-hand smoke) has also been strongly implicated in the carcinogenic process. In fact, in 1997, scientists showed evidence linking increased risk for lung cancer development with passive, second-hand smoke inhalation. This association was based upon findings showing the presence of tobacco metabolites in the urine of non-smokers who had been exposed to second-hand smoke. Interestingly, activities associated with chewing tobacco and snuff have also been shown to contain large quantities of cancer-causing chemicals called N-nitrosamines. In fact, cancers of the lip, mouth, tongue, and throat have been strongly linked to the use of smokeless tobacco products, however, it has been estimated that a lifetime smoker has 20 to 30 times greater a risk of developing cancer when compared to a non-smoker (Oppeltz and Jatoi 2011; Jatoi et al. 2009). Taken together, total tobacco-attributed deaths are projected to rise from 5.4 million in 2005 to 6.4 million in 2015 and to 8.3 million in 2030, with an estimated 600,000 deaths attributed to second-hand smoke (Oppeltz and Jatoi 2011; Mathers and Loncar 2006; Oberg et al. 2011). The combination of tobacco smoke with alcohol consumption has been shown to increase the risk for cancer significantly. This is not surprising, as one may assume that the effects of two carcinogens may act synergistically and thus enhance their carcinogenic potential. In fact, in a study performed by the International Head and Neck Cancer Epidemiology (INHANCE) consortium in 2011, it was confirmed that when alcohol and tobacco were administered alone they act as independent carcinogenic agents but when co-administered they lead to an even greater cancer risk (Pelucchi et al. 2011). Alcohol is a strong carcinogen and a well-known factor involved in human carcinogenesis with its consumption being established as a second leading risk factor associated with neck and head cancers (Ziech et al. 2010b). In 2002, about 390,000 new cancer developments and 230,000 cancer-related deaths were attributed to alcohol drinking worldwide. Additionally, a large number of epidemiological studies have associated heavy alcohol consumption (more or equal to 4 drinks per day) with an increased risk of pharyngeal and esophageal squamous cell carcinoma (5-fold), laryngeal (2.5-fold), colorectal, breast (50%) and pancreatic cancers (30%). Alternatively, low doses of alcohol consumption (less than or equal to one drink per day) were found to be associated with an increased risk of pharyngeal (20%) and esophageal (30%) squamous cell carcinomas (Pelucchi et al. 2011).

On another note, the percentage of all cancers that can be attributed to work-related factors has been estimated to vary between 4-15%. One of the first occupational carcinogens to be identified was, in 1930, by Ernest Kennaway who effectively documented the carcinogenic ability of a class of industrial compounds known as hydrocarbons. Since then, IRAC has categorized various occupational chemicals into a database which includes many agents from industries like shoe, tire, furniture-making, diesel fuel, nickel production and dry-cleaning processes. More specifically, these agents include arsenic, asbestos, benzene, chromium, 2-naphthylene, vinylchloride, cadmium, formaldehyde, etc. all of which have been confirmed as occupational carcinogens (Ziech et al. 2010a). For example, asbestos (a small silicate fiber) has been in use since the 1800s namely in construction sites (housing, industrial, etc.), and it is now a well-known carcinogenic agent that upon inhalation and deposition in the lung tissue can lead to cancer development. Furthermore, occupational exposure to asbestos increases the risk of asbestosis, the second most important cause of lung cancer (Valko et al. 2007) although today its exposure is tightly regulated by the Occupational Safety and Health Administration (OSHA) agency. Formaldehyde is another common chemical substance utilized in various construction sites, the manufacturing of formaldehyde-based resins, wood-composites, furniture, fiberglass and plastics, within histology, pathology and biological laboratories as well as in agricultural and many environmental settings including indoor (both residential and occupational) and outdoor exposures (industrial and automobile emissions, home cooking and heating, cigarette smoke, etc.), treated and bottled drinking water, groundwater, etc. Because of its ubiquitous nature, in nearly all work places, its use is regulated by various agencies including the Department of Homeland Security, CPSC, DOT, EPA, FDA, HUD, the Mine Safety and Health Administration, OSHA, ACGIH, and NIOSH (National Toxicology Program 2010). Another class of occupational carcinogens is polychlorinated biphenyls (PCBs), which are man-made chemicals that encompass approximately 209 toxic compounds. These were utilized as coolants and lubricants, however, their production was halted, in 1977, due to scientific evidence suggesting a link between their cancer-causing properties and their environmental accumulation. Another group of cancer-causing contaminants include the dioxins which are byproducts of various chemical reactions such as during the production of herbicides. In particular, one dioxin called TCDD (2, 3, 7, 8-tetrachlorodibenzo-para-dioxin) has been identified (by IARC) as a confirmed human carcinogen (widely used as a solvent and reactant) with the ability to cause lung cancer. Moreover, benzene is a fundamental unit of dioxins used widely as a solvent and a reactant. Research compiled over the last few decades has associated benzene with various blood cancers, including leukemia (National Toxicology Program 2010).

MECHANISMS OF OXIDATIVE STRESS-INDUCED HUMAN CARCINOGENESIS

Mounting evidence suggests that one of the common denominators linking various environmental and/or occupational agents to human carcinogenesis is the generation of oxidative stress and its involvement in disease pathophysiology. Under physiological conditions, our antioxidant defenses counterbalance the generation of oxidative stress by being able to inhibit ROS formation. One of the most commonly studied antioxidant defense mechanisms is that of phase II detoxification system, which is composed of several enzymes such as, catalase (CAT), superoxide dismutases (SODs), glutathione peroxidases (GPx) and glutathione-S-transferases (GSTs). The expression of these enzymes protects cells from oxidative damage and can prevent mutagenesis and carcinogenesis. However, when cells are under chronic conditions of oxidative stress, these defense mechanisms are impaired and are therefore unable to effectively protect against oxidative stress-induced cellular damage (Acharya et al. 2010, Pappa et al. 2007).

While oxidative injury to major macromolecules plays a crucial role in cellular viability and disease, oxidative DNA damage in particular has been the subject of immerse research. In fact, elevated levels of oxidative DNA lesions have been observed in various human tumors suggesting their direct involvement in the carcinogenic process (Tudek et al. 2010). The estimated frequency of oxidative DNA damage in human cells is 10^4 lesions per cell per day (Klaunig et al. 2010) and to date more than 100 oxidized DNA products have been identified including: depurinations, depyrimidinations, single- and double-stranded DNA breaks, base and sugar modifications and DNA-protein crosslinks (Valko et al. 2007). Currently, approximately 20 purine- and pyrimidine-derived oxidative lesions have been implicated in the etiology of cancer. However, two of them namely 7, 8-dihydro-8-oxoguanine (8-oxo-G) and thymineglycol (TG) have been characterized as the most frequent DNA mutations induced by oxidative stress. TG is known to block transcription and cause cell death in both cancer and non-cancer cells whereas the 8-oxo-G mutation has been shown to block DNA polymerase activity thus resulting in the mismatched pairing of adenine causing G→T and C→A genomic substitutions. The 8-oxo-G lesion has been observed within oncogenes and tumor suppressor genes both of which have key roles in carcinogenesis (Seifried et al. 2007). Additionally, other studies have found a positive correlation between the size of the tumor and the amount of 8-oxo-G present (Valko et al. 2007). In addition, under normal conditions, oxidative DNA damage would be efficiently repaired by our cellular repair machinery however, under chronic conditions of oxidative stress generation these repair mechanisms are hindered and thus DNA mutations are propagated in the genome. In fact, various studies have demonstrated decreased efficacy in the repair of oxidatively damaged DNA lesions in cancer patients where, for example, the presence of 8-oxo-G can lead to base mispairing and if not efficiently repaired (before replication commences) then will lead to mutation(s) (Tudek et al. 2010; Seifried et al. 2007; Acharya et al. 2010).

Oxidative stress has been shown to also mediate gene expression by causing changes in the regulation of various genes key in the multistage carcinogenic process such as in the case of oncogenes and tumor suppressor genes (Seifried et al. 2007). For example, *p53* (a known tumor suppressor gene with a role in protecting against tumorigenesis) is known to be adversely affected by oxidative stress. Under normal conditions, the *p53* gene halts the cell cycle and/or initiates apoptosis under conditions of cellular stress whereas under conditions of excessive ROS production, it becomes inactivated (i.e. down-regulated) resulting in loss of its tumor suppressor activity thus allowing for uncontrolled cellular growth. The 8-oxo-G mutation has been commonly detected in the *p53* gene. To date, many *p53* mutations have been identified in more than half of all human cancers including breast, liver and stomach to name a few (Seifried et al. 2007; Valko et al. 2007). On the other hand, over-expression of the *Ras* oncogene has also been implicated in a variety of human cancers. Specifically, it has been associated with the generation of high levels of O_2^- leading to subsequent increase in H_2O_2 levels which, in turn, have been associated with tumor invasiveness by allowing for increased degradation of the cellular matrix, a key step in metastasis (Seifried et al. 2007). Finally, oxidative stress has also been shown to cause point mutations in a number of antioxidant genes such as the manganese superoxide dismutase (*MnSOD*) and *CAT*. In case of the *MnSOD* gene, such point mutations have been associated with an elevated risk for breast and lung cancers as well as non-Hodgkin's lymphoma. Finally, point mutations to the *CAT* gene are hypothesized to contribute to the etiology of ROS-induced human carcinogenesis (Klaunig et al. 2010). Finally, the 8-oxo-G mutation is primarily repaired by the base excision repair mechanism (BER) which is composed of a variety of repair enzymes including the 8-oxo-guanine DNA glycosylase (OGG1) and apurinic/apyrimidinic (AP) endonuclease 1 (APE1), to name a few. Oxidative stress-induced point mutations in these two enzymes results in the impairment of their functional repair ability, and as a result, are strongly implicated to contribute to carcinogenesis. In fact, various epidemiologic studies have associated *OGG1* and *APE1* point mutations with significant increase in the risk of a number of human cancers, including lung, esophageal, prostate and gastric cancers (Klaunig et al. 2010).

Thus far, we have seen how oxidative stress alters gene expression (i.e. the silencing of tumor suppressor genes, the activation of oncogenes and defects in DNA repair mechanisms). However, recent data strongly suggest that oxidative stress also has adverse effects through modulation of epigenetic mechanisms. In fact, it is currently hypothesized that oxidative stress-induced alterations in the epigenetic pathway(s) play a crucial role in cancer development. For example, mounting evidence has linked ROS-induced mediated DNA damage with alterations in DNA methylation patterns and consequently, modulations in transcriptional activity (Klaunig et al. 2010). In general, epigenetics are defined as the study of heritable changes in gene expression without affecting the underlying DNA sequence itself and hence the name "epi- (Greek for "over", "above", etc.) genetics". Examples of such alterations include changes in DNA methylation content (e.g. hyper- *vs* hypomethylation) and various histone modifications (e.g. deacetylations, methylations, etc.), both of which can affect gene expression (Ziech et al. 2010a). To this end, many genes in tumor cells have been shown to contain alterations in their DNA methylation patterns in a way where global hypomethylation is followed by regional hypermethylation in otherwise unmethylated gene promoter regions (e.g. CpG islands). These events can result in the transcriptional silencing of tumor suppressor genes and/or the activation of oncogenes respectively both of which have been proposed to be a common mechanism during human carcinogenesis. For example, hypermethylation-induced inactivation of both *p16* and *p53* has been shown to allow cancer to proliferate uncontrollably and also has been identified in a wide variety of human tumors. Alternatively, genome-wide hypomethylation has been shown to increase mutation rates and lead to genomic instability. Oxidative stress has been implicated in global hypomethylation through its ability to directly impair the ability of DNA to function as a substrate for methyl acquisition. Under normal conditions, DNA is methylated by DNA methyltransferases (DNMTs) which transfer methyl groups to specific CpG dinucleotide sequences called CpG islands. Overall, DNMT enzymes are key regulators involved in both hyper- and hypomethylation patterns and their alteration has been strongly suggested to play a crucial role in the acquisition of malignant transformation (Franco et al. 2008). For example, exposure to carcinogenic agents (e.g. X- and γ-rays, ultraviolet, etc.) has been shown to hinder the ability of DNA to function as a substrate for DNMTs and thus resulting in global hypomethylation. Also, the presence of the 8-oxo-G mutation (a common ROS-mediated lesion) has been shown to strongly inhibit the methylation of specific nucleic acids (e.g. cytosine residues) thus interfering with the ability of DNA to serve as a substrate for the DNMT complex. The hypomethlylation of genes (e.g. *Ras*) and/or along repetitive DNA sequences (e.g. CpG islands) have been documented to cause the induction and reactivation of genes that are otherwise silenced in non-malignant cells. For example, recent reports have correlated hypomethylation-induced activation of various single-copy genes (e.g. *c-myc* and *H-Ras*) with the pathophysiology of several human cancers. In addition, hypomethlation of the *ps2* gene has been correlated with breast adenocarcinomas, while hypomethylation of the *CAGE* genes have been associated with breast, gastric, hepatocellular and lung carcinomas. In addition, hypomethylation has been detected in the early stages of cancer development as well as in late-stage metastasis (Ziech et al. 2010a). Taken together, mounting evidence has linked changes in global DNA methylation patterns as highly prevalent across many cancer types, and highly specific with regard to tumor type. For example, colon adenomas (e.g. benign precursors of colon carcinomas) display substantial global DNA hypomethylation in addition to frequent gene specific hypermethylation. While the precise mechanistic details linking hyper- and hypomethylation have yet to be fully elucidated, there is strong evidence correlating these epigenetics events with the development of environmental and occupational carcinogenesis (Ziech et al. 2010a). Thus, it is evident that alterations to DNA methylation patterns (either via hyper- or hypomethylation events) serve as an underlying source of aberrant gene expression and so are hypothesized to act as key mediators involved during malignant transformation.

In addition, changes in the delicate organization of the histone protein complex have also been correlated with carcinogenesis via mechanisms involving oxidative stress-induced injury. In general, chromatin is a highly organized complex consisting of DNA and chromosomal proteins. On the other hand, histones are basic chromosomal proteins that have been extremely conserved throughout evolution and are classified into: H1, H2A, H2B, H3 and H4. Histone proteins can undergo numerous post-translation modifications (e.g. methylation, acetylation, phosphorylation, ubiquitination, etc.), all of which play a significant role in chromatin structure and function. Recent evidence suggest that various environmental factors (e.g. exposures to heavy metals, UV radiation, etc.) can cause changes in chromatin structure, gene expression, and protein pattern (Pawlaki and Deckert, 2007). In essence, it is suggested that oxidative stress-induced alterations of histone proteins cause chromatin condensation and transcriptional inactivation both of which can cause the impairment of gene expression by either inducing or repressing oncogenes and tumor suppressor genes respectively (Klaunig et al. 2010). For example, there are very well established histone modifications implicated in gene silencing and include: (1) Histone H3 methylation (at lysines 9 and 27), (2) Histone H3 deacetylation (at lysine 9), (3) Histone H4 methylation (at lysine 20) and (4) Histone H4 deacetylation (at lysine 16). Finally, to date, one of the most well-established histone modifications is the methylation of histone H3 (at lysine 9) and as a result has been though to serve as a reliable biomarker for gene silencing (Ziech et al. 2010a; Franco et al. 2008; Ziech et al. 2011).

CONCLUSIONS

Transformation of a normal cell to a malignant one is a multiphase process that requires alterations in major cellular pathways including, but not limited, to DNA repair, cell death, etc. On the other hand, we are often exposed to various environmental and occupational chemicals known to be carcinogenic.

Over the years, much of the scientific effort has been targeted in identifying the underlying mechanism(s) for man-made chemicals capable of inducing the carcinogenic process. To this end, we now know that an integral mechanism during malignant transformation is the generation of oxidative stress which in turn is known to induce alterations in major cellular cascades and associated gene products. Although we attempt to elucidate some of these mechanistic aspects of oxidative stress involvement during carcinogenesis, there is still a lot more work that needs to be done from scientists across the globe. For instance, although we have known, for some time, the role of oxidative stress in inducing genetic alterations we still lack to understand how such alterations affect gene expression patterns characteristic of cancer cells. To this end, only recently it became apparent that one way may be through the deregulation of the epigenetic pathway by oxidative stress in ways that include changes in DNA methylation patterns and/or histone modifications both of which are absolutely important in regulating gene expression. Therefore, it is of outmost importance to continue to investigate as to how oxidative stress constitutes an integral part of human environmental and occupational carcinogenesis in an attempt to elucidate the mechanistic aspects of its involvement in order to design better treatment options.

REFERENCES

Acharya, A., I. Das, D. Chandhok, and T. Saha. 2010. Redox regulation in cancer: a double-edged sword with therapeutic potential. Oxid Med Cel Longev. 3: 23-34.

Franco, R., and M.I. Panayiotidis. 2010. Cell death or survival: The double-edged sword of environmental and occupational toxicity. Chem Biol Interact. 188: 265-266.

Franco, R., and M.I. Panayiotidis. 2009. Environmental toxicity, oxidative stress, human disease and the "black box" of their synergism: How much have we revealed? Mutat Res. 674: 1-2.

Franco, R., R. Sanchez-Olea, E.M. Reyes-Reyes, and M.I. Panayiotidis. 2009. Environmental toxicity, oxidative stress and apoptosis: menage a trois. Mutat Res. 674: 3-22.

Franco, R., O. Schoneveld, A.G. Georgakilas, and M.I. Panayiotidis. 2008. Oxidative stress, DNA methylation and carcinogenesis. Cancer Lett. 266: 6-11.

Gottlieb, E. 2009. Cancer: The fat and the furious. Nature. 461: 109-113.

Jatoi, I., K.M. Cummings, and E. Cazap. 2009. Global tobacco problem getting worse, no better. J. Oncol Pract. 5: 21-23.

Klaunig, J.E., L.M. Kamendulis, and B.A. Hocevar. 2010. Oxidative stress and oxidative damage in carcinogenesis. Toxicol Pathol. 38: 96-109.

Mathers, C.D., and D. Loncar. 2006. Projections of global mortality and burden of disease from 2002 to 2030. PLoS Med. 3: e442.

Miller, K.P., and K.S. Ramos. 2001. Impact of cellular metabolism on the biological effects of benzo[a]pyrene and related hydrocarbons. Drug Metab Rev. 33: 1-35.

National Toxicology Program. 2010. Final Report on Carcinogens Background Document for Formaldehyde. Rep Carcinog Backgr Doc. 10-5981: i-512.

Oberg, M., M.S. Jaakkola, A. Woodward, A. Peruga, and A. Pruss-Ustun. 2011. Worldwide burden of disease from exposure to second-hand smoke: a retrospective analysis of data from 192 countries. Lancet. 377: 139-146.

Oppeltz, R.F., and I. Jatoi. 2011. Tobacco and the escalating global cancer burden. J Oncol. 2011: 408104.

Panayiotidis, M.I. 2008. Reactive oxygen species (ROS) in multistage carcinogenesis. Cancer Lett. 266: 3-5.

Pappa, A., R. Franco, O. Schoneveld, A. Galanis, R. Sandaltzopoulos, and M.I. Panayiotidis. 2007. Sulfur-containing compounds in protecting against oxidant-mediated lung diseases. Curr Med Chem. 14: 2590-2596.

Pawlaki, S., and J. Deckert. 2007. Histone modifications under environmental stress. Biol Lett. 44: 65-73.

Pelucchi, C., I. Tramacere, P. Boffetta, E. Negri, and C. La Vecchia. 2011. Alcohol consumption and cancer risk. Nutr Cancer. 63: 983-990.

Seifried, H.E., D.E. Anderson, E.I. Fisher, and J.A. Milner. 2007. A review of the interaction among dietary antioxidants and reactive oxygen species. J Nutr Biochem. 18: 567-579.

Seifried, H.E., S.S. McDonald, D.E. Anderson, P. Greenwald, and J.A. Milner. 2003. The antioxidant conundrum in cancer. Cancer Res. 63: 4295-4298.

Tudek, B., A. Winczura, J. Janik, A. Siomek, M. Foksinski, and R. Olinski. 2010. Involvement of oxidatively damaged DNA and repair in cancer development and aging. Am J Transl Res. 2: 254-284.

Valko, M., D. Leibfritz, J. Moncol, M.T. Cronin, M. Mazur, and J. Telser. 2007. Free radicals and antioxidants in normal physiological functions and human disease. Int J Biochem Cell Biol. 39: 44-84.

Ziech, D., R. Franco, A. Pappa, and M.I. Panayiotidis. 2011. Reactive oxygen species (ROS)-induced genetic and epigenetic alterations in human carcinogenesis. Mutat Res. 711: 167-173.

Ziech, D., R. Franco, A. Pappa, V. Malamou-Mitsi, S. Georgakila, A.G. Georgakilas, and M.I. Panayiotidis. 2010a. The role of epigenetics in environmental and occupational carcinogenesis. Chem Biol Interact. 188: 344-349.

Ziech, D., R. Franco, A.G. Georgakilas, S. Georgakila, V. Malamou-Mitsi, O. Schoneveld, A. Pappa, and M.I. Panayiotidis. 2010b. The role of reactive oxygen species and oxidative stress in environmental carcinogenesis and biomarker development. Chem Biol Interact. 188: 334-339.

Index

Color Plate Section

Chapter 2

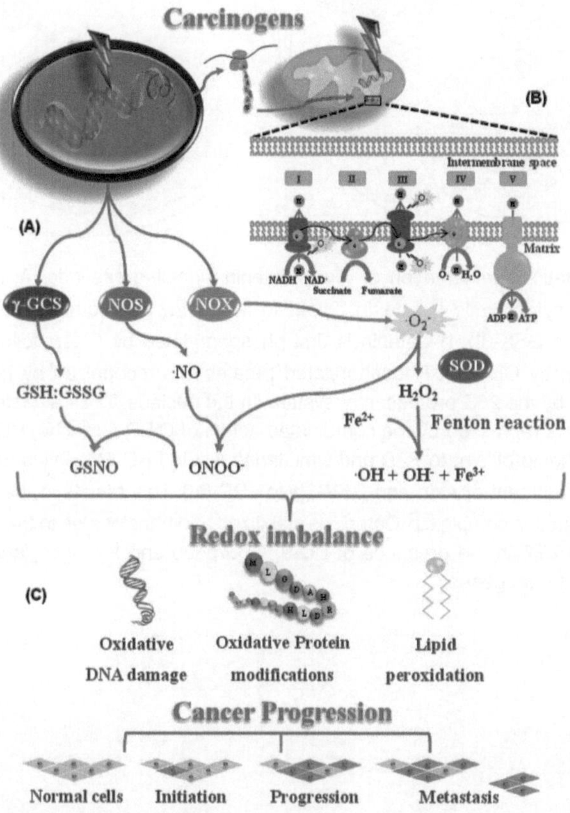

Fig. 1 Cancer cells exhibit increased generation of ROS and RNS compared with normal cells, which provides favorable conditions for cancer initiation, progression and metastasis. Increased generation of RS is given particularly by alterations in the protein levels of RS generating enzymes such as NOX and NOS (A). In the mitochondria, $O_2^{\bullet-}$ is formed by monoelectronic reduction of O_2 by Complexes I and III of the respiratory chain. Mutations in nuclear or mitochondrial genes encoding components of the ETC can also lead to an increase in ROS generation by inhibition of electron transfer (B). In addition, alterations in the levels of antioxidant enzymes such as SODs and γ-GCS (involved in GSH synthesis) also disrupt the redox balance in cancer cells. Although oxidative DNA-damage is a central element in the transformation process towards malignancy, it is also evident that RS can promote/predispose to cancer development by oxidative damage/modification to lipids and proteins/enzymes (C).

Chapter 3

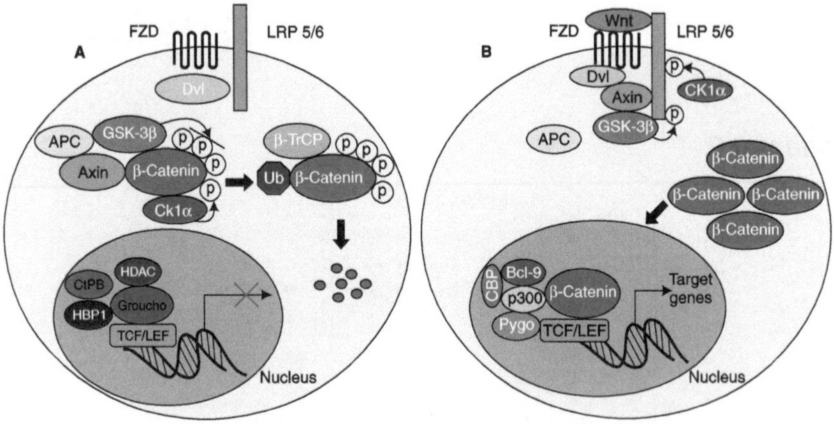

Fig. 1 Schematic representation of Wnt/β-catenin signaling cascade. A. In the absence of Wnt ligand, cytoplasmic β-catenin bound to its destruction complex consists of APC, Axin, CK1α and GSK-3β. β-Catenin is first phosphorylated by CK1α followed by further phosphorylation by GSK-3β. Phosphorylated β-catenin is recognized by β-TrCP followed by degradation by the 26S proteosomal system. In the nucleus, TCF-regulated transcription of target genes is repressed by the coordinated action of CtBP, Groucho, HBP1 and HDAC. B. Following binding of Wnt to FZD and simultaneously to LRP 5/6, Dvl is recruited to FZD followed by recruitment of Axin and GSK-3β to LRP 5/6. This results in the dissociation of β-catenin destruction complex. β-Catenin is stabilized and translocates to the nucleus, where it binds to TCF/LEF in the presence of Bcl-9, CBP/p300 and Pygogus, and activates the transcription of target genes.

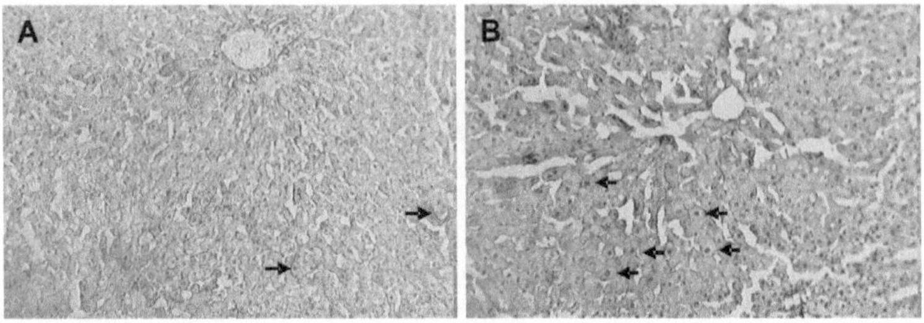

Fig. 2 Induction of hepatic β-catenin expression during DENA hepatocarcinogenesis in male Sprague-Dawley rats. Hepatocarcinogenesis was initiated with DENA (200 mg/kg) followed by promotion with PB in drinking water (0.05%, w/v) as per our published protocol (Bishayee et al. 2011a). Rats were sacrificed 18 weeks following DENA exposure, liver sections were prepared, and β-catenin expression was detected by immunohistochemistry. A. Normal hepatocyte show membrane-bound weak or moderate immunopositivity of β-catenin (black arrows). B. Accumulation of β-catenin in cytoplasm (green arrows) and nucleus (black arrows) during early phase of DENA-initiated liver cancer. Magnification: 100 × (unpublished laboratory observation).

Chapter 6

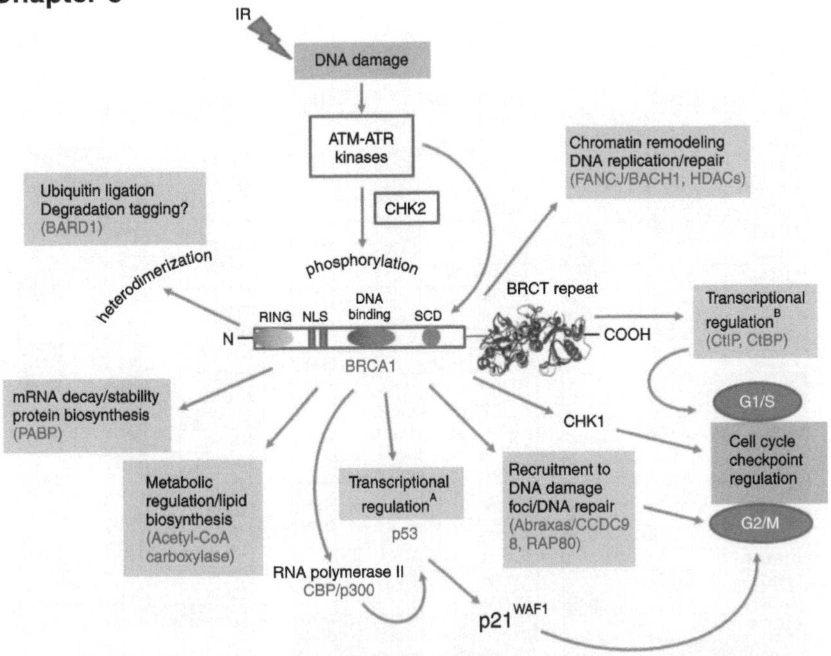

Fig. 1 Protein interactions mediated by the tandem BRCT repeat of BRCA1 tumor suppressor: A schematic outline of the multiple BRCT-mediated interactions that modulate DNA damage sensing and repair and ionizing radiation (IR)-dependent cell cycle checkpoint regulation.

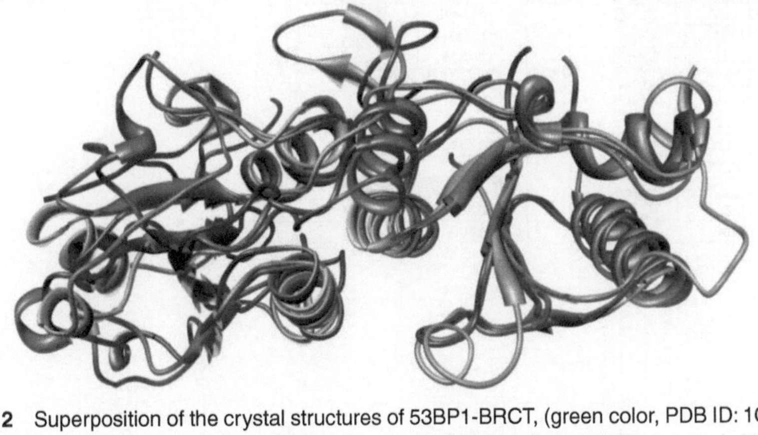

Fig. 2 Superposition of the crystal structures of 53BP1-BRCT, (green color, PDB ID: 1GZH, (Derbyshire et al. 2002)), BRCA1-BRCT (blue color, PDB ID: 1JNX, (Williams et al. 2001)), and BARD1-BRCT (purple color, PDB ID: 2NTE, (Birrane et al. 2007)). The different structural organization of the inter-BRCT linker of 53BP1 is shown in the middle-top part of the figure with the characteristic β-ribbon structure. The helical insertion -also an exclusive characteristic of the 53BP1-BRCT structure- can be observed at the bottom-right part of the figure (by permission from Thanassoulas et al. 2010).

Table 2 Missense variants of the BRCT domain of BRCA1.

	[1]	[2]		
N1647K				
S1651F				
M1652T				
M1652I			[3]	[13]
V1653M				
S1655F				
G1656D				
F1662S				
M1663K				[13]
M1663L				[13]
L1664P				[13]
V1665M				[13]
A1669S			[4]	[13]
E1682K				
E1682V				
T1685I				
T1685A				
M1689R				
M1689T				
T1691K				
T1691I				
D1692Y				[13]
D1692N				[13]
D1692H				
F1695L				
V1696L				[14]
C1697R				
R1699W				[15]
R1699Q			[5]	[13]
R1699L				[13]
T1700A				
G1706E				
G1706A			[6]	[13]
A1708E				[13
A1708V				
V1713A				
V1714G				
S1715R				[13]
S1715N				[13]
S1715C				[13]
W1718C				
W1718S				
T1720A				
S1722F				
R1726G				
N1730S				
D1733G			[3]	
F1734S				
V1736A				[13]
V1736G				
G1738R				
G1738E				
D1739Y				
D1739V				
D1739G				
D1739E				
V1741G				
H1746N				
P1749R				
R1751P				
R1751Q				[13]
A1752P				
A1752V				
R1753T				
F1761S				
G1763V				
L1764P				[13]
I1766S				[13]
P1771R				
P1771L				
T1773S				[13]
T1773I				
M1775R			[7]	[15]
M1775K			[8]	[14]
D1778G				
D1778Y				
D1778N				[13]
L1780P				[13]
M1783T				[14]
M1783I				
M1783L				
Q1785H				
C1787S				[13]
G1788V				[13]
G1788D				[13]
A1789S				
E1794D				
G1803A				
V1804D				
H1805P				
P1806A				[13]
V1808A				[13]
V1809A				
V1809F				[14]
V1810G				
Q1811R				
P1812A			[9]	[14]
D1818G			[10]	
N1819S				
A1823T			[11]	
Q1826H				
A1830T				
V1833M				[13]
R1835P				
E1836K				
W1837G				[13]
W1837R				[13]
W1837C				
V1838E			[12]	
S1841N				[13]
S1841R				
A1843P				[13]
L1844R				
D1851E				
Y1853C				[13]
L1854P				
P1856S				
P1859R				

Chapter 8

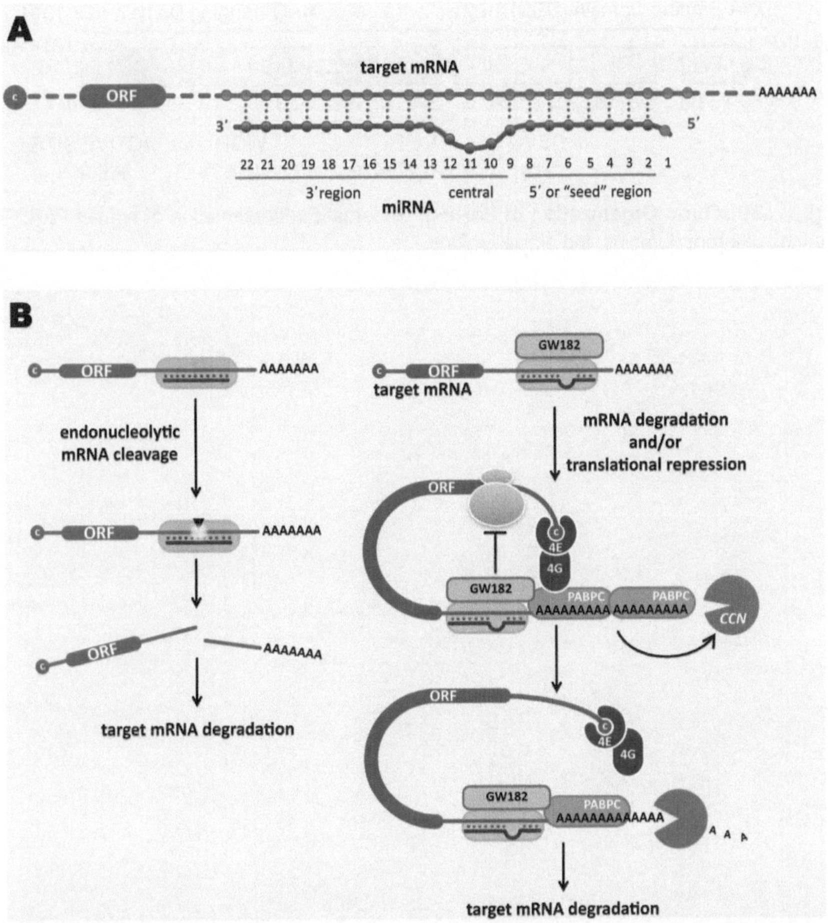

Fig. 2 Target recognition and pathways of miRNA-mediated silencing. A. Animal miRNAs (lower strand) recognize their target mRNAs (upper strand) through complementary regions, generally located in their 3′UTRs, by Watson-Crick base pairing. The 5′ or 'seed', and the 3′ regions flank the central region. Full complementarity to seed sequence (nt 2-7/8) is an almost absolute determinant for silencing. The central region (nt 9/10-12) generally bulges out preventing endonucleolytic cleavage by AGOs. In case of endonucleolytic cleavage, it determines the cleavage position of the target at nt 10 from the 5′ end of miRNA. The 5′ terminal nucleotide (nt 1, light grey sphere) is not available for pairing as it is bound by AGOs. Spheres indicate mRNA or miRNA nucleotides (nt). Numbers 1-22 indicate positions of miRNA nucleotides. Sphere with "c" indicates the 5′-cap. ORF, Open reading frame. B. The main pathways of miRNA-mediated silencing. In case of full complementarity between miRNA and target mRNA, the latter is degraded via direct endonucleolytic cleavage by AGO2 (left). Most often, in case of partial complementarity silencing is mediated by mRNA degradation by deadenylation and translational repression (right).

Chapter 9

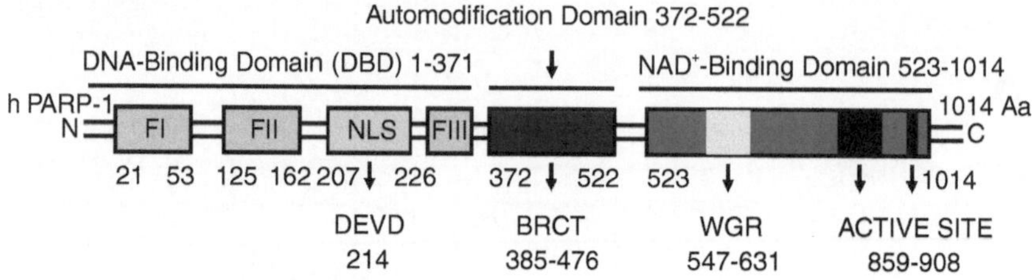

Fig. 1 Structural Organization of PARP-1: Schematic representation of human PARP-1 with the different domains and active regions.

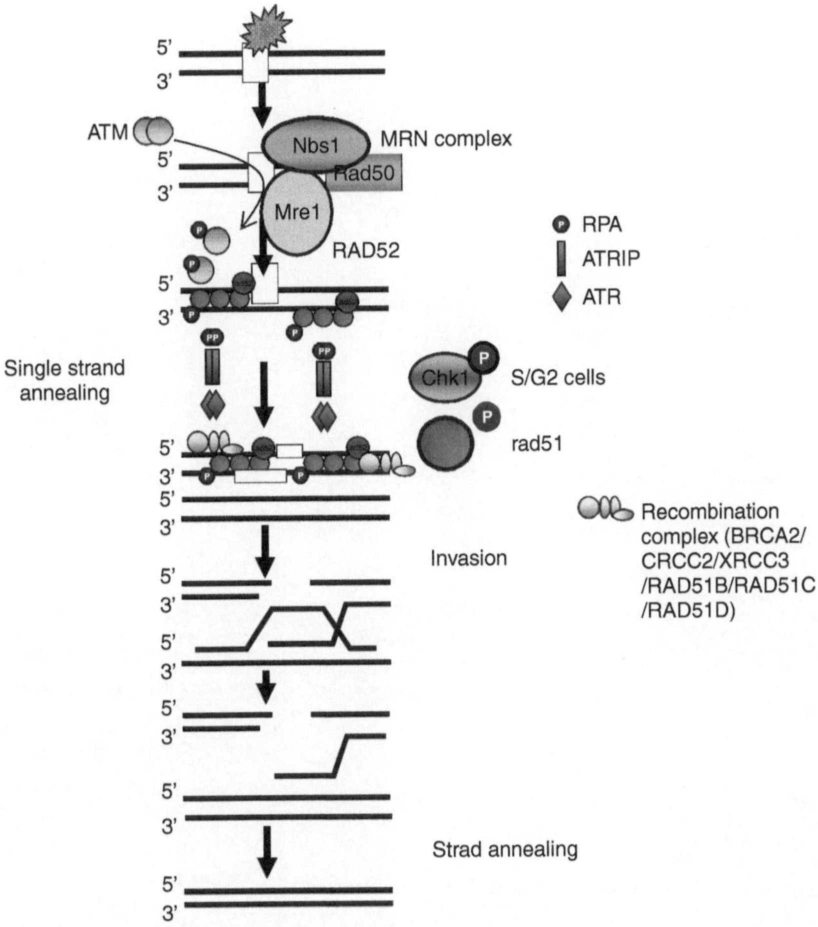

Fig. 4 Homologous recombination pathway. DSBs or those occurring in heterochromatic (packed) DNA activate the ATM kinase that inturn activates the full DNA damage response, allowing recruitment of endprocessing factors at the DNA ends to facilitate end joining (C). (D) Cells in the S, G2 or M phases have a high cyclin-dependent kinase activity, which trigger end resection. The exposed single stranded DNA (ssDNA) is coated with the ssDNA-binding protein RPA, which in turn triggers an ATR response that activates HR repair.